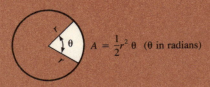

3.1 Area of a Sector

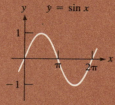

$$A = \frac{1}{2}r^2\theta \quad (\theta \text{ in radians})$$

3.2 Linear and Angular Velocity

Linear velocity $= v = r\omega$ (ω in units of radians per unit time.)

3.4, 3.5 Graphs of the Trigonometric Functions

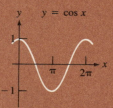

$y = \sin x$

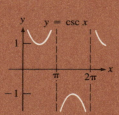

$y = \csc x$

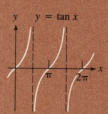

$y = \cos x$

$y = \sec x$

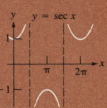

$y = \tan x$

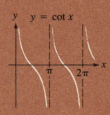

$y = \cot x$

3.4, 3.5 Period and Amplitude

The period of $y = a \sin bx$ and $y = a \cos bx$ is $\left|\frac{2\pi}{b}\right|$.
The amplitude of $y = a \sin bx$ and $y = a \cos bx$ is $|a|$.
The period of $y = a \tan bx$ and $y = a \cot bx$ is $\left|\frac{\pi}{b}\right|$.
The period of $y = a \csc bx$ and $y = a \sec bx$ is $\left|\frac{2\pi}{b}\right|$.

3.6 Vertical S

The graph of $y = $
of $y = a \sin bx$, exce

$$\left.\begin{matrix} \text{up} \\ \text{down} \end{matrix}\right\} \text{if } k \text{ is} \left\{\begin{matrix} \text{positive} \\ \text{negative} \end{matrix}\right\}$$

A similar statement is true for the other trigonometric functions.

The graph of $y = a \sin b(x + c)$ is identical to the graph of $y = a \sin bx$, except that it is shifted $|c|$ units to the

$$\left.\begin{matrix} \text{left} \\ \text{right} \end{matrix}\right\} \text{if } c \text{ is} \left\{\begin{matrix} \text{positive} \\ \text{negative} \end{matrix}\right\}$$

A similar statement is true for the other trigonometric functions.

4.2 Functions of Two Angles

$$\cos(A + B) = \cos A \cos B - \sin A \sin B$$
$$\cos(A - B) = \cos A \cos B + \sin A \sin B$$
$$\sin(A + B) = \sin A \cos B + \cos A \sin B$$
$$\sin(A - B) = \sin A \cos B - \cos A \sin B$$
$$\tan(A + B) = \frac{\tan A + \tan B}{1 - \tan A \tan B}$$
$$\tan(A - B) = \frac{\tan A - \tan B}{1 + \tan A \tan B}$$

4.3 The Double-Angle Identities

$$\sin 2A = 2 \sin A \cos A$$
$$\cos 2A = \cos^2 A - \sin^2 A$$
$$= 2 \cos^2 A - 1$$
$$= 1 - 2 \sin^2 A$$
$$\tan 2A = \frac{2 \tan A}{1 - \tan^2 A}$$

4.4 The Half-Angle Identities

$$\sin \frac{A}{2} = \pm \sqrt{\frac{1 - \cos A}{2}}$$
$$\cos \frac{A}{2} = \pm \sqrt{\frac{1 + \cos A}{2}}$$
$$\tan \frac{A}{2} = \frac{1 - \cos A}{\sin A}$$
$$= \frac{\sin A}{1 + \cos A}$$

PLANE TRIGONOMETRY
SECOND EDITION

Books in the Gustafson and Frisk Series

BEGINNING ALGEBRA
INTERMEDIATE ALGEBRA
ALGEBRA FOR COLLEGE STUDENTS
COLLEGE ALGEBRA, SECOND EDITION
PLANE TRIGONOMETRY, SECOND EDITION
COLLEGE ALGEBRA AND TRIGONOMETRY

PLANE TRIGONOMETRY
SECOND EDITION

R. DAVID GUSTAFSON
Rock Valley College

PETER D. FRISK
Rock Valley College

BROOKS/COLE PUBLISHING COMPANY
MONTEREY, CALIFORNIA

To the teachers who most inspired our interest in mathematics
Theodosia Keeler
Alice Iverson

Contemporary Undergraduate Mathematics Series, Robert J. Wisner, Consulting Editor

Brooks/Cole Publishing Company
A Division of Wadsworth, Inc.

Printed in the United States of America

10   9   8   7   6   5   4   3   2

Library of Congress Cataloging In Publication Data

Gustafson, R. David (Roy David), [date]
   Plane trigonometry.

   Includes index.
   1. Trigonometry, Plane.   I. Frisk, Peter D.,
[date].   II. Title.
QA533.G85   1984   516.2′4   84-9446

ISBN 0-534-03606-6

Sponsoring Editor: *Craig Barth*
Production Editor: *David Hoyt*
Manuscript Editor: *Marilu Uland*
Permissions Editor: *Carline Haga*
Interior and Cover Design: *Vernon T. Boes*
Cover Illustration: *David Aguero*
Art Coordinator: *Rebecca Tait*
Interior Illustration: *Lori Heckelman*
Typesetting: *Graphic Typesetting Service, Los Angeles, California*
Printing and Binding: *R. R. Donnelley & Sons Co., Crawfordsville, Indiana*

PREFACE

TO THE INSTRUCTOR

The principal reason for revising any textbook is, of course, to improve it. Based on the suggestions of reviewers and other users representing every area of the country, on student evaluations, and on our own classroom experience, we have rewritten and expanded many sections of the original edition of *Plane Trigonometry*. However, the underlying philosophy of the book remains the same. The introduction to the trigonometric functions continues to be the popular angle-in-standard-position approach with heavy emphasis on right-triangle applications. We continue to consider angles in their decimal forms; for those who wish to discuss minutes and seconds, this material is included in an appendix. Because a modern course in trigonometry must emphasize the trigonometric functions as functions with real number domains, we discuss the concept of radian measure early in the course, but not before the students need it. To convince students that the trigonometric functions can have real number domains, we include a special section discussing the unit circle.

The major changes in this new edition are as follows.

1. The material on right-triangle applications has been expanded and separated into two sections. The triangles of Section 2.5 are limited to one plane. Section 2.6 discusses problems involving triangles located in different planes. This section also includes many problems that require answers to be expressed in terms of other variables.
2. The discussion on graphing has been expanded. The treatment of vertical and horizontal translations has been reorganized into a new section.
3. The chapter on inverse trigonometric relations and functions has been extensively rewritten to make it clearer. A discussion of the inverse cotangent, secant, and cosecant functions has been included as an optional section.
4. More formal work with vectors is provided in a new section in Chapter 6.
5. The material on complex numbers has been expanded to include division of complex numbers written in trigonometric form. A table of general polar curves has been included in the section on the polar-coordinate system.

We think you will like this book for the following reasons.

Many Examples The book includes over 285 worked examples that are keyed to the exercise sets. By studying these examples, students can learn a great deal on their own, thereby easing the burden on the classroom teacher.

Many Exercises The book includes over 2500 exercises, with the odd-numbered answers provided. For the review exercises, all answers are provided.

Use of Second Color A second color is used not only to highlight important definitions and theorems, but also to "point" to items that you would point to in a classroom discussion.

Use of Calculators The use of calculators is encouraged throughout the book. Exercises in which calculators should not be used are clearly marked.

Writing Style The writing is informal and nontechnical. This makes it possible for the student to learn directly from the book, rather than depending entirely on the teacher for explanations.

Teacher Support Auxiliary materials include a test booklet containing chapter tests, a student solutions manual with complete solutions to the even-numbered exercises, and a student study guide.

Summary of Information For quick reference, key formulas and ideas are listed inside the front and back covers of the book.

Accuracy Every effort has been made to ensure that this book is as free from error as possible. Both authors have independently worked each exercise. In addition, Brooks/Cole employed a problem checker who checked each answer a third time.

TO THE STUDENT

We have tried to write a book that you can use and understand. We have provided an extensive number of worked examples and have tried to present them in a way that will make sense to you. If you do not read the explanations carefully, however, much of the value of the book will be lost.

What you learn here will be of great value both in other course work and in your chosen occupation. Therefore, we suggest that you consider keeping your book after completing this course. It is the one piece of reference material that will keep at your fingertips the material that you have learned here.

We wish you well.

ACKNOWLEDGMENTS

We wish to thank those who have reviewed *Plane Trigonometry* at various stages of its development:

Wilson Banks
Illinois State University

John S. Cross
University of Northern Iowa

Grace De Velbiss
Sinclair Community College

Emily Dickinson
University of Arkansas

Russ Diprizio
Oakton Community College

Ray Edwards
Chabot College

Jerry Gustafson
Beloit College

Jerome Hahn
Bradley University

Douglas Hall
Michigan State University

David Hansen
Monterey Pennisula College

William Hinrichs
Rock Valley College

Arthur M. Hobbs
Texas A & M

Jack Hofer
California Polytechnic State University

Warren Jaech
Tacoma Community College

Marcus McWaters
University of Southern Florida

Eldon Miller
University of Mississippi

Stuart Mills
Louisiana State University

Gilbert W. Nelson
North Dakota State University

Anthony Peressini
University of Illinois

William D. Popejoy
University of Northern Colorado

Donald Sherbert
University of Illinois

L. Thomas Shiflett
Southwestern Missouri State University

Richard Slinkman
Bemidji State University

Jack Snyder
Sinclair Community College

Warren Strickland
Del Mar College

Lee Topham
North Harris County College

Robert J. Wisner
New Mexico State University

Robert Zink
Purdue University

We are also grateful to the following professors for their helpful suggestions on the revision: John H. Biggs, Frostburg State College; Jerry Frang, Rock Valley College; Gerald Skidmore, Alvin Community College; John Spellmann, Southwest Texas State University; Arnold Wendt, Western Illinois University; Edward T. White, Frostburg State College; and Jim Yarwood, Rock Valley College.

For invaluable help in the production process, we express our appreciation for the staff of Brooks/Cole, especially Craig Barth, Vernon Boes, David Hoyt, and Rebecca Tait.

R. David Gustafson
Peter D. Frisk

A WORD TO STUDENTS ABOUT CALCULATORS

Although it is not essential that you have a calculator to complete this course, it would make the work much easier. If you do not already own a calculator, this is probably the time to consider buying one. There are many types of calculators available, and you should choose one that you will not outgrow.

HP-11C   Photo courtesy of Hewlett-Packard Company.

TI-55II   Photo courtesy of Texas Instruments.

EL-5100S   Photo courtesy of Sharp Electronics Corporation.

Avoid inexpensive calculators that only add, subtract, multiply and divide; they are not adequate for mathematics. Instead, look for a scientific calculator that is capable of evaluating trigonometric, inverse trigonometric, logarithmic, and exponential functions. It should be able to find reciprocals, roots, powers, and factorials; display numbers in scientific notation; and save intermediate results in memory.

Some calculators are really small programmable computers. These machines are more powerful than simple calculators and are also more expensive. You will not need the extra power of a programmable calculator in this course, but you might find one useful in later work.

Calculators differ in the ways they handle numbers and even in the way that numbers are entered. Each model has its own special features. Whatever calculator you purchase, be sure to read its instruction manual carefully.

Remember that calculators are intended to relieve the drudgery of doing mathematical calculations. Use your calculator freely when performing calculations, but know when *not* to use it. Know also when an answer is unreasonable, thus indicating that an error has occurred. An answer is not correct just because a calculator produced it. No calculator can be a substitute for thinking.

CONTENTS

PLANE TRIGONOMETRY
SECOND EDITION

CHAPTER ONE
PRELIMINARY CONCEPTS

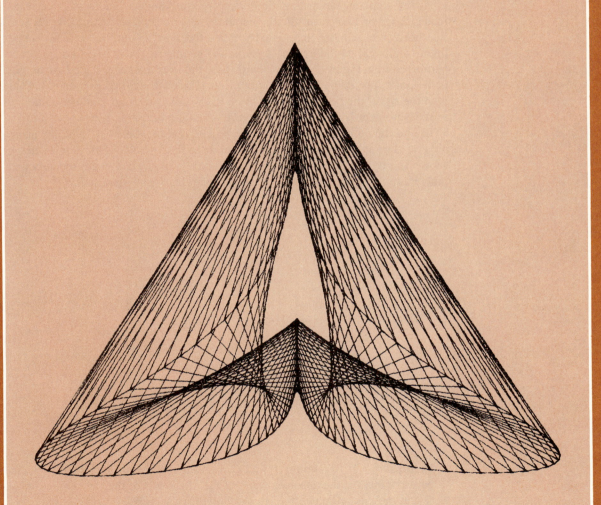

Many topics from plane geometry and algebra are applied in the study of trig-onometry. The most basic of these ideas are discussed in Appendix I, while other topics, ones with which you might not be familiar, are discussed in this first chapter.

1.1   THE RECTANGULAR COORDINATE SYSTEM AND THE DISTANCE FORMULA

It was the work of René Descartes (1596–1650) that merged algebra and geom-etry into a single unified subject. The genius of Descartes lay in his idea of a coordinate system. In such a system every point (a geometric concept) is assigned a pair of numbers (an arithmetic concept) as its unique "address." Descartes' idea is based on two perpendicular number lines, one horizontal and one vertical. The horizontal number line (usually called the *x*-**axis**) and the vertical number line (usually called the *y*-**axis**) separate the geometric plane into four quadrants numbered as in Figure 1-1. The point where the axes intersect, called the **origin,** is the point labeled 0 on each number line. The positive direction on the *x*-axis is to the right, the positive direction on the *y*-axis is upward, and the same unit of measurement is usually used on each axis (although this is not necessary).

To find the point associated with the pair of real numbers $(-3, 2)$, for exam-ple, we start at the origin and count 3 units to the left, and then 2 units up. As shown in Figure 1-2, point *P* with coordinates $(-3, 2)$ lies in the second quad-rant. Point *Q* with coordinates $(4, -3)$ lies in the fourth quadrant.

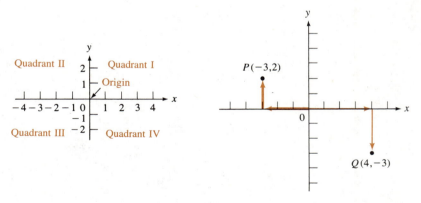

Figure 1-1   **Figure 1-2**

(1.1)   **Definition.**   If two points *P* and *Q* lie on a horizontal line, the distance between them is the absolute value of the difference of the *x* coordinate of *Q* and the *x* coordinate of *P*. In symbols, $d(PQ) = |x_Q - x_P|$.

Example 1   Find the distance between $P(2, -3)$ and $Q(-4, -3)$.

Solution   Plot the points as in Figure 1-3 and note that PQ is a horizontal segment. By definition, the distance between P and Q, represented by $d(PQ)$, is the absolute value of the difference of the x coordinate of Q and the x coordinate of P.

$$d(PQ) = |-4 - 2| = |-6| = 6$$

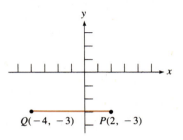

Figure 1-3

(1.2)   **Definition.**   If two points P and Q lie on a vertical line, the distance between them is the absolute value of the difference of the y coordinate of Q and the y coordinate of P. In symbols, $d(PQ) = |y_Q - y_P|$.

Example 2   Find the distance between $P(4, -3)$ and $Q(4, 7)$.

Solution   Plot the points as in Figure 1-4 and note that PQ is a vertical segment. By definition, the distance between P and Q, represented by $d(PQ)$, is the absolute value of the difference of the y coordinate of Q and the y coordinate of P.

$$d(PQ) = |7 - (-3)| = |10| = 10$$

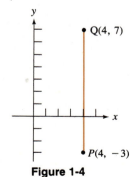

Figure 1-4

We now consider the case when we remove the restriction that the line segment be either horizontal or vertical.

(1.3)

> **The Distance Formula.**   The distance between two points $P(x_1, y_1)$ and $Q(x_2, y_2)$ is given by the formula
>
> $$d(PQ) = \sqrt{(x_2 - x_1)^2 + (y_2 - y_1)^2}$$

Proof   We plot points P and Q and construct right triangle PRQ as in Figure 1-5. Note that $d(PR) = |x_2 - x_1|$ and $d(RQ) = |y_2 - y_1|$. We use the Pythagorean Theorem to find the distance $d(PQ)$.

$$d(PQ)^2 = |x_2 - x_1|^2 + |y_2 - y_1|^2$$

Because $|k|^2 = k^2$ for any number k, this result can be written as

$$d(PQ)^2 = (x_2 - x_1)^2 + (y_2 - y_1)^2$$

or, taking the square root of both sides, as

$$d(PQ) = \sqrt{(x_2 - x_1)^2 + (y_2 - y_1)^2}$$

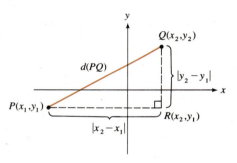

Figure 1-5

This formula for finding the distance between two points is called the **distance formula.** It is an important formula and should be memorized.

Example 3   Find the distance between points $P(-2, 3)$ and $Q(3, -5)$.

Solution   Use the distance formula.

$$\begin{aligned}
d(PQ) &= \sqrt{(x_2 - x_1)^2 + (y_2 - y_1)^2} \\
&= \sqrt{[3 - (-2)]^2 + (-5 - 3)^2} \\
&= \sqrt{5^2 + (-8)^2} \\
&= \sqrt{89}
\end{aligned}$$

The distance between points P and Q is $\sqrt{89}$ units.

Exercise 1.1

In Exercises 1–18, find the distance between P and Q.

1. $P(2, 5)$ and $Q(7, 5)$
2. $P(2, -2)$ and $Q(10, -2)$
3. $P(2, 5)$ and $Q(2, 13)$
4. $P(-4, -2)$ and $Q(-4, 10)$
5. $P(-2, -3)$ and $Q(-14, -3)$
6. $P(-2, 8)$ and $Q(-2, -8)$
7. $Q(-5, -2)$ and $P(-5, 10)$
8. $Q(2, -4)$ and $P(-8, -4)$
9. $P(2, 4)$ and $Q(5, 8)$
10. $P(5, 9)$ and $Q(8, 13)$
11. $P(-2, -8)$ and $Q(3, 4)$
12. $P(-5, -2)$ and $Q(7, 3)$
13. $P(6, 8)$ and $Q(12, 16)$
14. $P(10, 4)$ and $Q(2, -2)$
15. $Q(-3, 5)$ and $P(-5, -5)$
16. $Q(2, -3)$ and $P(4, -8)$
17. $Q(0, 0)$ and $P(3, -4)$
18. $Q(0, 0)$ and $P(12, -35)$

19. Show that a triangle with vertices at $A(-2, 4)$, $B(2, 8)$, and $C(6, 4)$ is isosceles.

20. Show that a triangle with vertices at $A(-2, 13)$, $B(-8, 9)$, and $C(-2, 5)$ is isosceles.

21. Every point on the line CD in Illustration 1 is equidistant from points A and B. Use the distance formula to find the equation of line CD.

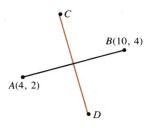

Illustration 1

22. Use the distance formula to find the equation of a circle with center at $(2, 4)$ and passing through $(6, 9)$.

1.2 RELATIONS AND FUNCTIONS

Some mathematical concepts have limited theoretical value, while others are basic to the whole structure of mathematics. One of these important basic concepts is the idea of a function. We must begin, however, by discussing a more general concept—that of a relation.

(1.4)

> **Definition.** A **relation** in R, the set of real numbers, is any nonempty set of ordered pairs of real numbers.

Example 1   Suppose that $A = \{1, 2, 3\}$ and $B = \{5, 6\}$. Any set of ordered pairs with first components taken from set A and second components taken from set B is a relation in the set of real numbers. Write three relations with first components taken from set A and second components taken from set B.

Solution   There are many such relations. Here are three.

    a. $\{(1, 5), (1, 6)\}$
    b. $\{(1, 5), (2, 5), (3, 6)\}$
    c. $\{(2, 6), (3, 5), (3, 6), (1, 6)\}$ ■

(1.5)

> **Definition.**   The **domain** of a relation is the set of all first components of the ordered pairs.

(1.6)

> **Definition.**   The **range** of a relation is the set of all second components of the ordered pairs.

In Example 1, the domain of relation **a** is $\{1\}$ and its range is $\{5, 6\}$. The domain of relation **b** is $\{1, 2, 3\}$ and its range is $\{5, 6\}$. The domain of relation **c** is $\{1, 2, 3\}$ and its range is $\{5, 6\}$.

Example 2   Does the equation $y = x^2$ define a relation?

Solution   The equation $y = x^2$ does define a relation because it determines a set of ordered pairs (x, y):

$$\{(0, 0), (1, 1), (-1, 1), (2, 4), (-2, 4), (\pi, \pi^2), (-\pi, \pi^2), \ldots\}$$

The replacement set for x, the set of real numbers, is the domain of this relation. The set of corresponding values of y, the nonnegative real numbers, is the range of this relation.   ■

The equation given in Example 2 determines a special kind of relation called a **function.** In the set of ordered pairs given in Example 2, note that to each number x there corresponds exactly one value of y.

(1.7)

> **Definition.**   A **function** is a relation in which to each value of a first component there corresponds exactly one value of a second component.

The relation $\{(1, 5), (2, 6), (3, 7)\}$ is a function because to each value of a first component there corresponds exactly one value of a second component.

    To the number 1 corresponds the value 5.
    To the number 2 corresponds the value 6.
    To the number 3 corresponds the value 7.

The relation $\{(\mathbf{1}, 5), (2, 5), (3, 5), (\mathbf{1}, 6)\}$ is not a function because the first component—**1**—pairs with two different second components—5 and 6.

Example 3   Is the relation determined by pairs (x, y) such that $x^2 = y + 2$ a function? Find its domain and range.

Solution   Solve the equation for y. Because $y = x^2 - 2$, to each number x there corresponds a single value of y. Because no first component (a value of x) pairs with more than one second component (a value of y), this relation is a function. The domain of this function is the set of real numbers, and the range is the set of all real numbers greater than or equal to -2. ◼

Example 4   Find the domain and range of the relation determined by pairs (x, y) such that $3(y - 6x) - 2 = x$, and tell if this relation is a function.

Solution   Perform the algebraic operations necessary to solve the equation for y.

$$3(y - 6x) - 2 = x$$
$$3y - 18x = x + 2 \qquad \text{Remove parentheses and add 2 to both sides.}$$
$$3y = x + 18x + 2 \qquad \text{Add } 18x \text{ to both sides.}$$
$$y = \frac{19}{3}x + \frac{2}{3} \qquad \text{Combine terms and divide both sides by 3.}$$

From this equation, you can see that x can be any real number. Thus, the domain of the relation is the set of real numbers. Because y also can be any real number, the range of this relation is the set of real numbers as well. From the equation it follows that, since to any number x there corresponds only one value of y, this relation is a function. ◼

Example 5   Find the domain and range of the relation determined by pairs (x, y) such that $y^2 = x + 2$, and tell if this relation is a function.

Solution   Solve the equation for y.

$$y^2 = x + 2$$
$$y = \pm \sqrt{x + 2} \qquad \text{Take the square root of both sides.}$$

Because $\pm \sqrt{x + 2}$ must be a real number, $x + 2 \geq 0$ and $x \geq -2$. Thus, the domain of the relation is all x such that $x \geq -2$. The range of this relation is the set of all real numbers because y can be any real number. Note that to some numbers x there correspond two values of y. For example, if $x = 7$, then y can be either 3 or -3. Therefore, this relation is not a function. ◼

Example 6   Suppose that y is the function of x defined by

$$y = f(x) = \frac{x + 1}{x}$$

Find the values of   **a.** $f(2)$,   **b.** $f(h)$, and   **c.** $f(0)$.

Solution   **a.** $f(2)$ is the value of y obtained when 2 is substituted for x.

$$f(2) = \frac{2 + 1}{2} = \frac{3}{2}$$

b. $f(h)$ is the value of y obtained when h is substituted for x.

$$f(h) = \frac{h + 1}{h}$$

c. $f(0)$ is meaningless, because 0 is not an acceptable replacement for x. The number 0 is not an element of the domain because division by 0 is not allowed.

Example 7   Let $y = f(x) = x^2 - 2x - 3$. Find   **a.** $f(a)$,   **b.** $f(h)$, and   **c.** $f(a + h)$.

Solution   **a.** Because $f(a)$ is the value of y obtained when a is substituted for x,

$$f(a) = a^2 - 2a - 3$$

And similarly:

b. $f(h) = h^2 - 2h - 3$

c. $f(a + h) = (a + h)^2 - 2(a + h) - 3$
$$= a^2 + 2ah + h^2 - 2a - 2h - 3$$
Note that $f(a + h) \neq f(a) + f(h)$.

If both the domain and the range of a relation are subsets of the set of real numbers, it is possible to graph the ordered pairs of that relation on a rectangular coordinate system. This graph consists of all points in the coordinate plane with coordinates $(x, y) = (x, f(x))$. The next two examples show how to graph a relation that is determined by a given equation. In Example 8, the given relation is a function; in Example 9, the given relation is not a function.

Example 8   Graph the function that is defined by the equation $y = x^2 - 10x + 24$.

Solution   Make a table of values of x and y and plot the points. For example, if $x = 2$, then $f(x) = 8$. Remember that $f(x)$ is a symbol for the value of y when a number is substituted for x. If $x = 7$, then $f(x) = 3$. Other pairs of coordinates are shown in Figure 1-6. After plotting these points, join them with a smooth curve to obtain the parabola in Figure 1-6. Note that, for each different value of x, the function yields a single y value.

$y = x^2 - 10x + 24$

x	y
2	8
3	3
5	-1
7	3
8	8

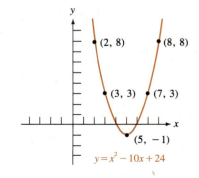

Figure 1-6

Example 9 Graph the relation that is defined by the equation $x^2 + y^2 = 13$.

Solution Make a table of values of x and y and plot the points. For example, if $x = 2$, then y is either 3 or -3. If $x = -2$, then y is 3 or -3. Other pairs of coordinates are shown also in Figure 1-7. After plotting these points, join them with a smooth curve to obtain the circle in Figure 1-7. Note that the equation $x^2 + y^2 = 13$ does not determine a function. This is because, in the interval $-\sqrt{13} < x < \sqrt{13}$, there are two values of y associated with each value of x.

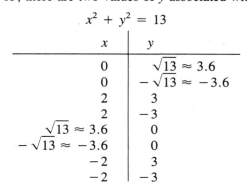

$$x^2 + y^2 = 13$$

x	y
0	$\sqrt{13} \approx 3.6$
0	$-\sqrt{13} \approx -3.6$
2	3
2	-3
$\sqrt{13} \approx 3.6$	0
$-\sqrt{13} \approx -3.6$	0
-2	3
-2	-3

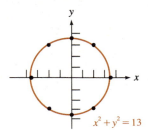

Figure 1-7

Note that the symbol $\approx$ means "is approximately equal to." ◼

Perhaps you already know this easier way to graph the equation of Example 9. Any equation of the form $x^2 + y^2 = r^2$ has a circle for its graph. The center of the circle is the origin, and its radius is $|r|$. Hence, the graph of $x^2 + y^2 = 13$ is a circle with center at the origin, and with a radius of $\sqrt{13}$. This is the circle shown in Figure 1-7.

Vertical line test A graph indicates pictorially whether a correspondence given by a set of ordered pairs (x, y) is a function. If any vertical line intersects the graph more than once, the correspondence is *not* a function. This is because to one number x there corresponds more than one value of y. See Figure 1-8. This test to determine whether a graph represents a function is called the **vertical line test.**

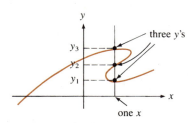

Figure 1-8

Note that every vertical line drawn in Figure 1-6 would intersect the graph exactly once. Thus, the equation of that graph determines a function. However, some vertical lines drawn in Figure 1-7 would intersect the graph more than once. Thus, the equation of that graph cannot represent a function.

Exercise 1.2

In Exercises 1–12, give the domain and the range of the given relation. Assume that x and y represent real numbers in each ordered pair (x, y).

1. $\{(1, 2), (2, 3)\}$

2. $\{(1, 1), (1, 2), (1, 3)\}$

3. $\{(3, 4), (5, 6), (8, 9), (10, 12)\}$

4. $\{(5, 1), (4, 1), (3, 1), (2, 1), (1, 1)\}$

5. $y = 3x + 1$

6. $y = 7x - 2$

7. $y = 3x^2 + 1$

8. $y = 2x^2 - 2$

9. $x^2 = y$

10. $3x^2 = 2y$

11. $y = \sqrt{x - 3}$

12. $x = \sqrt{y + 2}$

In Exercises 13–24, tell whether the given relation is a function. Assume that x and y represent real numbers in each ordered pair (x, y).

13. $\{(1, 2), (2, 3)\}$

14. $\{(1, 1), (1, 2), (1, 3)\}$

15. $\{(3, 4), (5, 6), (3, 9), (10, 12)\}$

16. $\{(5, 1), (4, 1), (3, 1), (2, 1), (1, 1)\}$

17. $y = 3x + 1$

18. $y = 7x - 2$

19. $y = 3x^2 + 1$

20. $y = 2x^2 - 2$

21. $x = y^2$

22. $3x = 2y^2$

23. $y = \sqrt{x - 3}$

24. $x = \sqrt{y + 2}$

In Exercises 25–32, find the required value, if possible, when $f(x) = \dfrac{2x - 1}{x}$.

25. $f(2)$

26. $f(5)$

27. $f(-3)$

28. $f(-8)$

29. $f(0)$

30. $f\left(\frac{1}{2}\right)$

31. $f(h)$

32. $f(a + h)$

In Exercises 33–38, find the required value, if possible, when $f(x) = \dfrac{\sqrt{x - 1}}{2}$. Assume that f(x) must be a real number.

33. $f(1)$

34. $f(5)$

35. $f(0)$

36. $f(-2)$

37. $f(a + h)$

38. $f(a + h) - f(a)$

In Exercises 39–54, graph the given relation. Use the vertical line test to determine whether the relation is a function.

39. $y = 3x - 2$

40. $y = -2x + 3$

41. $3x + 4y = 12$

42. $2x + 3y = 30$

43. $y = x^2 + 2x + 1$

44. $y = -x^2 - 2x + 3$

45. $y^2 - 6y + 9 = x$

46. $y^2 + 4x = x - 4$

47. $xy = 12$

48. $xy = -6$

49. $x^2 + y^2 = 25$

50. $x^2 - y^2 = 1$

51. $x^2 - y^2 = 25$　　　　　　　　　**52.** $x^2 + y^2 = 1$

53. $2x^2 + 5y^2 = 50$　　　　　　　　**54.** $4x^2 + 3y^2 = 48$

55. Use the distance formula to show that $x^2 + y^2 = r^2$ is the equation of a circle, centered at the origin, and with a radius of $|r|$.

1.3　INVERSES OF RELATIONS AND FUNCTIONS

If the components of each ordered pair in a given relation are interchanged, the new relation formed is called the **inverse relation** of the given relation.

Example 1　Find the inverse relation of $G = \{(1, 2), (2, 4), (3, 6), (4, 8)\}$.

Solution　Interchange the components of each ordered pair in relation G to find its inverse relation. The inverse relation of G, denoted as G^{-1}, is

$$G^{-1} = \{(2, 1), (4, 2), (6, 3), (8, 4)\}$$

The inverse relation of G is denoted as G^{-1}, but do not think of the "-1" as an exponent; it is not. The symbol G^{-1} is read as "the inverse relation of G." ■

Note that in Example 1, the domain of G is the range of G^{-1} and that the range of G is the domain of G^{-1}. We formalize this concept of inverse relation in a definition.

(1.8)

> **Definition.**　If G is any relation, and G^{-1} is the relation obtained from G by interchanging the components of each ordered pair in G, then G^{-1} is called the **inverse relation of G.**
> 　The domain of G^{-1} is the range of G, and the range of G^{-1} is the domain of G.

Example 2　Is the inverse relation of $\{(1, 2), (2, 3), (3, 4)\}$ a function?

Solution　The given set of ordered pairs is a function because to each first component there corresponds a single second component. The inverse relation is found by interchanging the components of each ordered pair:

$$\{(2, 1), (3, 2), (4, 3)\}$$

This, too, is a function. ■

Example 3　Is the inverse relation of $\{(1, 5), (2, 5), (3, 5)\}$ a function?

Solution　The given set of ordered pairs is a function because to each first component there corresponds a single second component. However, the inverse relation is *not* a function. After interchanging the components, three different second components correspond to the number 5.

$$\{(5, 1), (5, 2), (5, 3)\}$$

Thus, the inverse relation is not a function. ■

Example 4 Is the inverse relation of $y = 3x + 2$ a function?

Solution The equation $y = 3x + 2$ determines a function because it defines a set of ordered pairs and each x will give a unique y. To find the inverse relation of the given function $y = 3x + 2$, interchange the x and y in the equation to obtain $x = 3y + 2$. To decide whether $x = 3y + 2$ is a function, solve it for y to see if a single x will give a single y.

$$x = 3y + 2$$
$$x - 2 = 3y \qquad \text{Add} -2 \text{ to both sides.}$$
$$y = \frac{x - 2}{3} \qquad \text{Divide both sides by 3.}$$

Each x does determine just one y, so the inverse relation is a function. ■

Example 5 Is the inverse relation of $y = x^2$ a function?

Solution The equation $y = x^2$ determines a function, but the inverse relation does not. The inverse relation is found by reversing the roles of x and y. Thus, it is defined by the equation $x = y^2$ or $y = \pm \sqrt{x}$. Because the inverse relation contains pairs such as $(4, 2)$ and $(4, -2)$, in which two y values correspond to one number x, it is not a function. ■

Example 6 Is the inverse relation of $\{(x, y) \mid y = x^2 \text{ and } x \geqslant 0\}^*$ a function?

Solution The inverse relation is found by reversing the roles of x and y. The inverse relation is

$$\{(x, y) \mid x = y^2 \text{ and } y \geqslant 0\}$$

Although pairs such as $(4, 2)$ remain as elements of the inverse relation, pairs such as $(4, -2)$ do not. Because one value of x now gives only one value of y, the inverse relation is a function. ■

Example 7 On a single coordinate axes system, graph the function $y = x^2$ and its inverse relation $x = y^2$.

Solution The equation $y = x^2$ graphs as a parabola with vertex at the origin and opening upward. The inverse relation $x = y^2$ also graphs as a parabola with its vertex at the origin. However, the graph of the inverse relation opens to the right. See Figure 1-9. Note that the two graphs are symmetric about the line $y = x$. The

---

*This is read as "the set of all (x, y) such that $y = x^2$ and $x \geqslant 0$."

graph of an inverse relation will always be a reflection of the graph of the original relation. The line $y = x$ will always be the line of symmetry.

$y = x^2$			$x = y^2$	
x	y		x	y
0	0		0	0
1	1		1	1
-1	1		1	-1
2	4		4	2
-2	4		4	-2

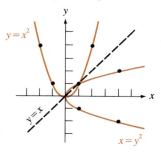

Figure 1-9

If a set of ordered pairs is a function, and if the inverse relation of that set of ordered pairs is also a function, the function is called **one-to-one.** We formalize this idea with a definition.

(1.9)

> **Definition.**   A function determined by pairs (x, y) is called **one-to-one** if and only if each value of y corresponds to exactly one number x.

Stated another way, a function f is one-to-one if and only if $f(a) = f(b)$ guarantees that $a = b$.

The definition of a one-to-one function implies this important fact.

(1.10)

> **The inverse relation of a one-to-one function is also a function.**

Example 8   Determine whether the function defined by $y = \dfrac{x + 2}{x}$ is a one-to-one function.

Solution   Note that, because to each number in the domain (x cannot be 0) there corresponds exactly one value of y, the given equation determines a function. This function will be one-to-one if each permissible value of y corresponds to exactly one number x. To see whether this is true, solve the equation for x as follows.

$$y = \frac{x + 2}{x}$$

$$xy = x + 2 \qquad \text{Multiply both sides by } x.$$

$$xy - x = 2 \qquad \text{Add } -x \text{ to both sides.}$$

$$x(y - 1) = 2 \qquad \text{Factor out } x.$$

$$x = \frac{2}{y - 1} \qquad \text{Divide both sides by } y - 1.$$

You can now see that as long as y does not equal 1, each value of y does determine exactly one number x. Thus, the function determined by the equation $y = \frac{x + 2}{x}$ is a one-to-one function. ■

Note that a relation is a correspondence from one set to another. For example, Figure 1-10 represents the relation $G = \{(2, 3), (2, 4), (3, 5), (4, 8)\}$. The

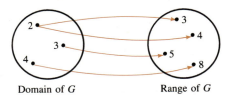

Domain of G Range of G

Figure 1-10

numbers in the left circle are the elements of the domain; the numbers in the right circle are the elements of the range. G is not a function because to the number 2 there corresponds both 3 and 4 in the range. The inverse relation of G is $G^{-1} = \{(3, 2), (4, 2), (5, 3), (8, 4)\}$, and it is represented in Figure 1-11. The relation G^{-1} is a function because to each element in the domain of G^{-1} there corresponds a single element in the range.

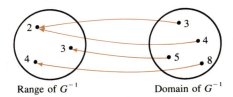

Range of G^{-1} Domain of G^{-1}

Figure 1-11

Figures 1-10 and 1-11 illustrate that the domain of a relation is the range of its inverse relation and that the range of a relation is the domain of its inverse relation.

Figure 1-12 illustrates a fundamental property of a one-to-one function. If a function $y = f(x)$ changes some number x to some number y, the inverse function f^{-1} will change y back to x. Function f pairs x with y, and function f^{-1} pairs y back with x.

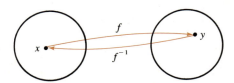

Figure 1-12

Exercise 1.3

In Exercises 1–16, find the inverse relation of the given relation and tell whether the inverse relation is a function.

1. $\{(2, 1), (3, 2), (4, 3)\}$

2. $\{(1, 1), (4, 2), (9, 3)\}$

3. $\{(2, 3), (2, 4), (2, 5)\}$

4. $\{(3, 2), (4, 2), (8, 3)\}$

5. $\{(5, 1), (4, 1), (3, 1), (2, 1)\}$

6. $\{(3, 5)\}$

7. $\{(1, 1), (2, 8), (3, 27), (4, 64)\}$

8. $\{(8, 8), (9, 9), (10, 10), (11, 11)\}$

9. $y = -2x - 2$

10. $y = 7x + 1$

11. $x + y = 3$

12. $2x - y = 4$

13. $y = x^2 + 2$

14. $y = 3x^2 - 1$

15. $y = x^2 - 1$ and $x \geq 0$

16. $2y - 1 - x^2 = 0$ and $x \geq 0$

In Exercises 17–32, graph the given relation and its inverse relation on the same set of axes. Note the line of symmetry.

17. $\{(2, 1), (3, 2), (4, 3)\}$

18. $\{(1, 1), (4, 2), (9, 3)\}$

19. $\{(2, 3), (2, 4), (2, 5)\}$

20. $\{(3, 2), (4, 2), (8, 3)\}$

21. $\{(5, 1), (4, 1), (3, 1), (2, 1)\}$

22. $\{(3, 5)\}$

23. $\{(1, 1), (2, 4), (3, 9)\}$

24. $\{(8, 8), (9, 9), (10, 10), (11, 11)\}$

25. $y = -2x - 2$

26. $y = 7x + 1$

27. $x + y = 3$

28. $2x - y = 4$

29. $y = x^2 + 2$

30. $y = 3x^2 - 1$

31. $y = x^2 - 1$ and $x \geq 0$

32. $2y - 1 - x^2 = 0$ and $x \geq 0$

In Exercises 33–46, tell whether the given functions are one-to-one.

33. $\{(1, 2), (2, 3), (3, 4)\}$

34. $\{(1, 1), (-1, 1), (2, 4), (-2, 4)\}$

35. $y = x^2$

36. $y = (x + 3)^2$

37. $y = x^3$

38. $y = (x - 5)^3$

39. $y = x$

40. $y = 3$

41. $y = \sqrt{x}$

42. $y = \sqrt[3]{x}$

43. $y = \dfrac{x}{x - 1}$

44. $y = \dfrac{x + 2}{x - 1}$

45. $y = x^2 + x^3$

46. $y = x + x^2$

In Exercises 47–50, find the range of the function by determining the domain of the function's inverse.

47. $y = \dfrac{x + 3}{x}$

48. $y = \dfrac{x}{x + 3}$

49. $y = \dfrac{3x - 2}{x + 2}$

50. $y = \dfrac{2x + 1}{x - 2}$

CHAPTER SUMMARY

Key Words

coordinate system (1.1)
distance between two points (1.1)
domain (1.2)
function (1.2)
inverse relation (1.3)
one-to-one function (1.3)

origin (1.1)
range (1.2)
relation (1.2)
vertical line test (1.2)
x- and y-axes (1.1)

Key Ideas

(1.1)   If $P(x_P, y_P)$ and $Q(x_Q, y_Q)$ are points on a horizontal line, then the distance between P and Q is $d(PQ) = |x_Q - x_P|$.

If $P(x_P, y_P)$ and $Q(x_Q, y_Q)$ are points on a vertical line, then the distance between P and Q is $d(PQ) = |y_Q - y_P|$.

The distance formula: $d(PQ) = \sqrt{(x_2 - x_1)^2 + (y_2 - y_1)^2}$

(1.2)   Finding the domain and range of relations and functions. Graphing relations and functions.

If any vertical line can intersect a graph more than once, that graph cannot represent a function.

(1.3)   If the components of each ordered pair in a relation G are interchanged, the inverse relation of G is formed.

The graph of an inverse relation is always the reflection of the graph of the original relation. The line $y = x$ is the axis of symmetry.

A function is one-to-one if and only if ordered pairs with different first components also have different second components.

The inverse relation of a one-to-one function is also a function.

REVIEW EXERCISES

In Review Exercises 1–12, find the distance between points P and Q.

1. $P(2, 5)$ and $Q(2, -2)$
2. $P(-2, -5)$ and $Q(-12, -5)$
3. $Q(5, 7)$ and $P(-12, 7)$
4. $Q(-5, -13)$ and $P(-5, 22)$
5. $P(0, 0)$ and $Q(3, 4)$
6. $P(0, 0)$ and $Q(12, 5)$
7. $P(2, 3)$ and $Q(5, 7)$
8. $P(-7, -6)$ and $Q(-4, -2)$
9. $Q(2, 8)$ and $P(7, 13)$
10. $Q(8, 2)$ and $P(13, 7)$

11. $Q(-2, -4)$ and $P(-9, 7)$　　　　**12.** $Q(5, -4)$ and $P(-3, -2)$

13. Determine whether a triangle with vertices at points $A(3, 1)$, $B(6, 2)$, and $C(4, 3)$ is an isosceles triangle.

14. Use the distance formula to show that points $A(0, -2)$, $B(2, 6)$, and $C(5, 18)$ all lie on the same line.

In Review Exercises 15–20, give the domain and range of the given relation. Assume that x and y represent real numbers in each ordered pair (x, y).

15. $\{(2, 1), (4, 2), (5, 6), (6, 8), (7, 7)\}$

16. $\{(1, 2), (-2, -2), (3, 3), (3, 4), (3, 5)\}$

17. $3y - 7x = 12$　　　　　　　　**18.** $y = 2x^2 - 10$

19. $2y = \sqrt{x - 10}$　　　　　　　**20.** $y = x^2 + x^4$

In Review Exercises 21–26, tell whether the given relation is a function. Assume that x and y represent real numbers in each ordered pair (x, y).

21. $\{(1, 1), (1, 2), (2, 3), (2, 4)\}$　　　**22.** $\{(1, 1), (2, 1), (3, 1), (4, 1)\}$

23. $x^2 - y^2 = 17$　　　　　　　　**24.** $y = x^4 + 2$

25. $x^2 + y^2 = 25$　　　　　　　　**26.** $x^2 - y^2 = 1$

In Review Exercises 27–30, find the required value if $f(x) = x^2 + 2x - 4$.

27. $f(0)$　　　　　　　　　　　**28.** $f(3)$

29. $f(h)$　　　　　　　　　　　**30.** $f(a + h) - f(a)$

In Review Exercises 31–34, graph the relation determined by the ordered pairs (x, y) defined by the given equation. Use the vertical line test to determine whether the relation is a function.

31. $y = \frac{1}{2}x - 3$　　　　　　　　**32.** $4x - y^2 = 3$

33. $x^2 - y = 4$　　　　　　　　　**34.** $xy = 6$

In Review Exercises 35–42, find the inverse relation of the given relation and tell whether the inverse relation is a function.

35. $\{(2, 1), (3, 1), (4, 1)\}$　　　　**36.** $\{(1, 10), (2, 20), (3, 30), (4, 40)\}$

37. $y = 7x + 12$　　　　　　　　**38.** $y = \dfrac{x + 17}{4}$

39. $3y = x^2$　　　　　　　　　　**40.** $xy = 36$

41. $y = \sqrt{2x - 1}$　　　　　　　**42.** $\sqrt{y - 4} = x$

In Review Exercises 43–46, graph the given relation and its inverse relation on the same set of axes. Note the line of symmetry.

43. $y = 4x + 2$　　　　　　　　**44.** $3x - y + 1 = 0$

45. $y = \dfrac{x^2}{2}$　　　　　　　　　**46.** $xy = 4$

In Review Exercises 47–52, tell whether the given function is one-to-one. Assume that x and y represent real numbers.

47. $3y = x - 2$ **48.** $y = x^2 - 3$

49. $y = x^4$ **50.** $y = \sqrt{x - 3}$

51. $y = x^3$ **52.** $y^3 = x$

CHAPTER TWO

THE TRIGONOMETRIC FUNCTIONS

The study of trigonometry is important for two reasons, one theoretical and one practical. The theoretical aspects of trigonometry appear in the development of many areas of mathematics, physics, biology, and chemistry. The practical aspects of trigonometry are necessary in the fields of surveying, navigation, mechanical engineering and machine design, electrical engineering, and even carpentry.

2.1   FUNCTIONS OF TRIGONOMETRIC ANGLES

An angle in trigonometry is an extension of the concept of angle that was defined in plane geometry.* There, an angle was defined as a figure formed by two rays originating from a common point called the **vertex.** A trigonometric angle is defined differently.

(2.1)

> **Definition.**   Consider two rays with common initial point O. A **trigono-metric angle** is the rotation required to move one ray, called the **initial side** of the angle, into coincidence with the other, called the **terminal side.** Angles that are rotations in a counterclockwise direction are considered to be positive. Angles that are rotations in a clockwise direction are considered to be negative.

This definition of a trigonometric angle is an extension of the definition of a geometric angle because one geometric angle can represent many, indeed infinitely many, different trigonometric angles. For example, the two rays forming the geometric angle in Figure 2-1 could represent any of the trigonometric angles represented in Figure 2-2.

Figure 2-1

Figure 2-2

To draw a trigonometric angle, we must include an arrow curved from the angle's initial side to its terminal side to indicate the rotation intended.

(2.2)

> **Definition.**   A trigonometric angle is in **standard position** if and only if it is drawn in a Cartesian coordinate system with its vertex at the origin, and its initial side along the positive x-axis. The angle is called a **first-, second-, third-,** or **fourth-quadrant angle** according to whether its terminal side lies in the first, second, third, or fourth quadrant. If the terminal side lies on the x- or y- axis, then the angle is called a **quadrantal angle.**

---

*For a brief review of plane geometry, refer to Appendix I.

The angles in Figure 2-3 are in standard position. Note that each initial side is along the positive *x*-axis and that each vertex is at the origin.

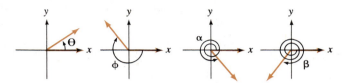

Figure 2-3

Angles in standard position are usually denoted by a single letter, often a Greek letter, next to the curved arrow. Figure 2-3 illustrates angles θ (theta), φ (phi), α (alpha), and β (beta). Note that angles θ and α are positive angles, and that angles φ and β are negative angles.

The angles in Figure 2-4 are not in standard position. Note that in 2-4a, the initial side is not on the positive *x*-axis. In 2-4b, the vertex is not at the origin; and in 2-4c, the initial side is not on the positive *x*-axis.

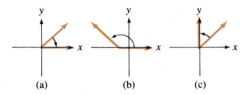

Figure 2-4

(2.3)   **Definition.**   If the terminal sides of two trigonometric angles in standard position coincide, the angles are called **coterminal.**

The three trigonometric angles shown in Figure 2-5 are coterminal. Note that these angles of 30°, 390°, and −330° are not equal angles. However, they are coterminal.

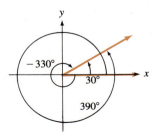

Figure 2-5

Example 1   Find three positive and three negative angles that are coterminal with an angle in standard position that measures 100°.

Solution   A trigonometric angle of 100° appears in Figure 2-6. A trigonometric angle of 100° plus 360° (100° plus one complete revolution) is coterminal with an angle of 100°; one positive answer is 460°. See Figure 2-7.

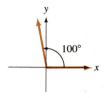

Figure 2-6

Another possible answer is the positive angle 820°, which is 100° plus 720° (100° plus two complete revolutions). A third positive angle is 1180°, which is 100° plus 1080°. Thus, 460°, 820°, and 1180° are three positive angles that are coterminal with 100°.

The angle of −260° is also coterminal with 100° because it is equal to 100° minus 360°. Similarly, two other negative angles that are coterminal with 100° are 100° − 720°, or −620°, and 100° − 1080°, or −980°. In fact, any angle of 100° ± n · 360°, where n is any whole number, is coterminal with an angle of 100°.

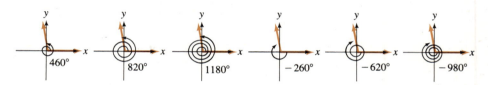

Figure 2-7

You have studied the concept of function. It is impossible to overestimate the importance of functions in mathematics, for they relate to virtually every concept studied. Most of the work in trigonometry is based on six trigonometric functions.

Domain and range of the trigonometric functions

Recall that a function assigns to each element of one set (called the **domain** of the function) a unique element of a second set (called the **range** of the function). The domain of each of the six trigonometric functions is a set of trigonometric angles. The values assigned by the functions (the range) are real numbers. Because the definitions of the six functions of a trigonometric angle are basic to all of trigonometry, they should be memorized.

(2.4)

Definition.   Let θ be a trigonometric angle in standard position as shown in Figure 2-8. Let $P(x, y)$ be any point (except the origin) on the terminal side of angle θ. Then $r = \sqrt{x^2 + y^2}$ is the length of line segment OP. The **six trigonometric functions of the angle θ** are:

$$\sin \theta = \frac{y}{r} \quad \text{read as ``sine theta''}$$

$$\cos \theta = \frac{x}{r} \quad \text{read as ``cosine theta''}$$

$$\tan \theta = \frac{y}{x} \quad \text{read as "tangent theta"}$$

$$\csc \theta = \frac{r}{y} \quad \text{read as "cosecant theta"}$$

$$\sec \theta = \frac{r}{x} \quad \text{read as "secant theta"}$$

$$\cot \theta = \frac{x}{y} \quad \text{read as "cotangent theta"}$$

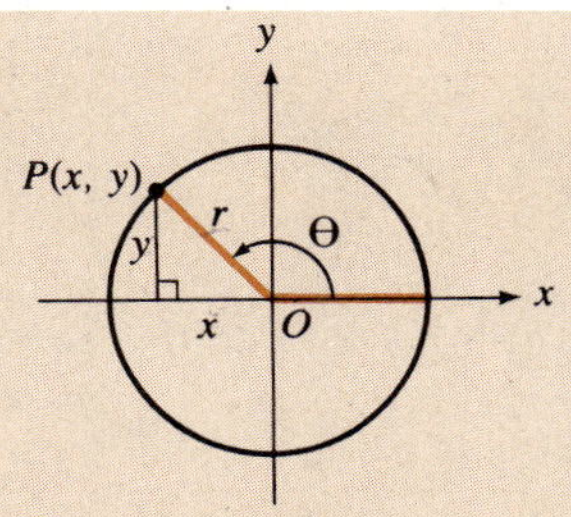

Figure 2-8

Remember that division by 0 is not permitted. If $x = 0$, $\tan \theta$ and $\sec \theta$ are undefined. If $y = 0$, $\csc \theta$ and $\cot \theta$ are undefined.

Definition 2.4 implies that the values of trigonometric functions of coterminal angles are equal.

Example 2 The point $(-3, 4)$ is on the terminal side of angle ϕ. Draw angle ϕ in standard position, and calculate the six trigonometric functions of ϕ.

Solution Refer to Figure 2-9. The point $P(-3, 4)$ lies at a distance $r = \sqrt{(-3)^2 + 4^2} = \sqrt{25} = 5$ units from point O, and ϕ is a second-quadrant angle. From the definitions, you obtain

$$\sin \phi = \frac{y}{r} = \frac{4}{5}$$

$$\cos \phi = \frac{x}{r} = \frac{-3}{5} = -\frac{3}{5}$$

$$\tan \phi = \frac{y}{x} = \frac{4}{-3} = -\frac{4}{3}$$

$$\csc \phi = \frac{r}{y} = \frac{5}{4}$$

$$\sec \phi = \frac{r}{x} = \frac{5}{-3} = -\frac{5}{3}$$

$$\cot \phi = \frac{x}{y} = \frac{-3}{4} = -\frac{3}{4}$$

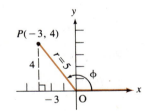

Figure 2-9

Any angle coterminal with angle ϕ will produce the same values for each of the six trigonometric functions.

The following theorem guarantees that any point on the terminal side of angle θ (in standard position) can be used to compute the six trigonometric functions of θ.

(2.5)

> **Theorem.** If angle θ is in standard position, and if P is some arbitrary point on the terminal side of angle θ, then the values of the trigonometric functions of θ are independent of the choice of point P.

Proof Let $P(x,\ y)$ and $P'(x',\ y')$ be two points on the terminal side of angle θ. Let $OP = r$ and $OP' = r'$. Points P,O,Q and P',O,Q' determine two triangles: triangle OPQ and triangle $OP'Q'$. See Figure 2-10. Since these triangles are similar, all ratios of corresponding sides are equal. In particular, we have

$$\sin \theta = \frac{y}{r} = \frac{y'}{r'}$$

and

$$\cos \theta = \frac{x}{r} = \frac{x'}{r'}$$

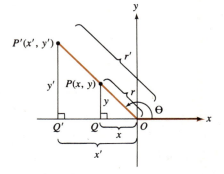

Figure 2-10

This shows that the value of $\sin \theta$ and $\cos \theta$ are independent of the choice of point P. It can be shown in a similar manner that the values of the remaining four trigonometric functions are also independent of the choice of point P. □

Example 3 Assume that $\sin \alpha = \frac{12}{13}$, and α is in quadrant II. Find the values of the other five trigonometric functions.

Solution Because any point P on the terminal side of α can be used, choose a point with y coordinate of 12 and $r = 13$, for then $\sin \alpha = \frac{y}{r} = \frac{12}{13}$ as required. To find the x coordinate, make use of the relationship $r^2 = x^2 + y^2$, and solve for x.

$$x^2 = r^2 - y^2$$
$$x = \pm \sqrt{r^2 - y^2}$$

Now substitute **12** for y and **13** for r in this equation.

$$x = \pm \sqrt{13^2 - 12^2}$$
$$x = \pm \sqrt{25}$$
$$x = \pm 5$$

Because angle α is in quadrant II, the x coordinate must be negative, so $x = -5$. See Figure 2-11. Write the values of the trigonometric functions by substituting the appropriate values of x, y, and r.

$$\cos \alpha = \frac{x}{r} = \frac{-5}{13} = -\frac{5}{13}$$

$$\tan \alpha = \frac{y}{x} = \frac{12}{-5} = -\frac{12}{5}$$

$$\csc \alpha = \frac{r}{y} = \frac{13}{12}$$

$$\sec \alpha = \frac{r}{x} = \frac{13}{-5} = -\frac{13}{5}$$

$$\cot \alpha = \frac{x}{y} = \frac{-5}{12} = -\frac{5}{12}$$

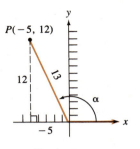

Figure 2-11

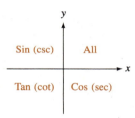

Figure 2-12

It is worthwhile to note in which quadrants each of the trigonometric functions is positive. Because r is always positive, the sine and cosecant functions are positive whenever y is positive. This occurs in quadrants I and II. Because x is positive in QI (quadrant I) and QIV (quadrant IV), the cosine and secant functions are positive in these quadrants also. The tangent and cotangent functions are positive when x and y agree in sign, and this happens in QI and QIII. Figure 2-12 will help you remember this information. The letters A, S, T, and C indicate the functions that are positive in the various quadrants. Some mathematics teachers remember these letters by fantasizing that **All Students Take Calculus.**

Example 4   Assume that angle β is *not* in QII and $\cos \beta = -\frac{3}{4}$. Find $\sin \beta$ and $\tan \beta$.

Solution   Because $\cos \beta$ is negative, β must be in QII or QIII. You are told that β is not in QII, so β is a third-quadrant angle. See Figure 2-13. Choose point $P(x, y)$ so that $x = -3$ and $r = 4$ (remember that r is always positive). Then, you have

$$y = -\sqrt{r^2 - x^2}$$
$$y = -\sqrt{4^2 - (-3)^2}$$
$$y = -\sqrt{7}$$

The minus sign is determined by the quadrant in which β lies—y is negative in QIII. Once you know the values of x, y, and r, you can compute the required quantities.

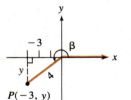

Figure 2-13

$$\sin \beta = \frac{y}{r} = \frac{-\sqrt{7}}{4} = -\frac{\sqrt{7}}{4}$$

$$\tan \beta = \frac{y}{x} = \frac{-\sqrt{7}}{-3} = \frac{\sqrt{7}}{3}$$

Note that the tangent is positive in QIII, as expected.

Example 5   Assume that angle θ is *not* in QII and $\tan \theta = -\frac{40}{9}$. Find $\sin \theta$ and $\sec \theta$.

Solution   The tangent function is negative in QII and QIV. Because θ is not in QII, θ is a fourth-quadrant angle. Choose point $P(9, -40)$ so that $\tan \theta = -\frac{40}{9}$. See Figure

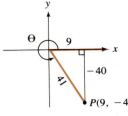

Figure 2-14

2-14. Use the Pythagorean Theorem to determine that $r = 41$. Use the values of x, y, and r to compute the required quantities.

$$\sin \theta = \frac{-40}{41} = -\frac{40}{41}$$

$$\sec \theta = \frac{41}{9}$$

Exercise 2.1

In Exercises 1–12, tell whether the angle is in standard position. Indicate whether the angle is positive or negative.

1.

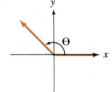

2.

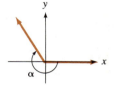

3.

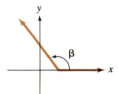

4.

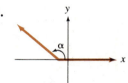

5.

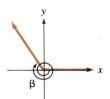

6.

7.

8.

9.

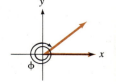

10.

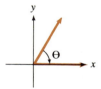

11. 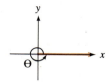   **12.**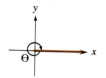

In Exercises 13–26, tell whether θ is a QI, QII, QIII, or QIV angle. Assume that θ is in standard position.

13.   **14.**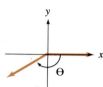

15.   **16.**

17. θ = 170°   **18.** θ = 25°

19. θ = 225°   **20.** θ = 350°

21. θ = 612°   **22.** θ = 460°

23. θ = −70°   **24.** θ = −212°

25. θ = −1190°   **26.** θ = −1470°

In Exercises 27–34, tell whether the given angles are coterminal. Assume that all angles are in standard position.

27. 40°, 400°   **28.** 90°, −270°

29. 135°, 270°   **30.** 135°, −135°

31. 740°, 380°   **32.** −340°, −700°

33. 1035°, −405°   **34.** 10°, −10°

In Exercises 35–50, point P is on the terminal side of angle θ, which is in standard position. Draw angle θ, with θ positive, and calculate the six trigonometric functions of angle θ.

35. P(3, 4)   **36.** P(5, 12)

37. P(−5, −12)   **38.** P(3, −4)

39. P(−9, 40)   **40.** P(9, −40)

41. P(1, 1)   **42.** P(−3, 3)

43. P(−3, 4)   **44.** P(−1, −1)

45. P(−4, 3)   **46.** P(24, 10)

47. P(3, 5)   **48.** P(−3, −4)

49. P(24, −10)   **50.** P(−5, 12)

In Exercises 51–60, find the remaining trigonometric functions of angle θ. Assume that θ is positive and in standard position.

51. $\sin \theta = \dfrac{3}{5}$; θ in QI

52. $\tan \theta = 1$; θ not in QI

53. $\cot \theta = \dfrac{5}{12}$; $\cos \theta = -\dfrac{5}{13}$

54. $\cos \theta = \dfrac{\sqrt{5}}{5}$; $\csc \theta = -\dfrac{\sqrt{5}}{2}$

55. $\sin \theta = -\dfrac{9}{41}$; θ in QIV

56. $\tan \theta = -1$; θ not in QII

57. $\sec \theta = -\dfrac{5}{3}$; $\csc \theta = \dfrac{5}{4}$

58. $\tan \theta = -\dfrac{3}{5}$; $\sin \theta = \dfrac{3\sqrt{34}}{34}$

59. $\tan \theta = -\dfrac{40}{9}$; $\cos \theta = \dfrac{9}{41}$

60. $\sin \theta = -\dfrac{\sqrt{7}}{4}$; $\cos \theta = -\dfrac{3}{4}$

2.2   THE EIGHT FUNDAMENTAL RELATIONSHIPS OF TRIGONOMETRIC FUNCTIONS

If the product of two numbers is 1, the numbers are called **reciprocals** of each other. For example, 2 and $\frac{1}{2}$ are reciprocals, as are $-\frac{3}{4}$ and $-\frac{4}{3}$. Note that $\sqrt{2} + 1$ and $\sqrt{2} - 1$ are reciprocals also, because

$$(\sqrt{2} + 1)(\sqrt{2} - 1) = 1$$

Thus, the reciprocal of N does not always *look* like $\frac{1}{N}$. Perhaps you have already noticed that the six trigonometric functions can be grouped in reciprocal pairs. For example, cos θ and sec θ are reciprocals because

$$\frac{x}{r} \quad \text{and} \quad \frac{r}{x}$$

are reciprocals. Similarly, sin θ and csc θ are reciprocals, and so are tan θ and cot θ. These results are summarized for easy reference.

(2.6)

> **Reciprocal Relationships.**   For any angle θ for which the functions are defined,
>
> $$\sin \theta \csc \theta = 1 \qquad \sin \theta = \frac{1}{\csc \theta} \qquad \csc \theta = \frac{1}{\sin \theta}$$
>
> $$\cos \theta \sec \theta = 1 \qquad \cos \theta = \frac{1}{\sec \theta} \qquad \sec \theta = \frac{1}{\cos \theta}$$
>
> $$\tan \theta \cot \theta = 1 \qquad \tan \theta = \frac{1}{\cot \theta} \qquad \cot \theta = \frac{1}{\tan \theta}$$

There are many other algebraic relationships among the six trigonometric functions. These relationships are true only for those angles for which the functions are defined.

Example 1   Show that $\dfrac{\sin \theta}{\cos \theta} = \tan \theta$.

Solution   Using the definitions of the sine and cosine functions, you have

$$\frac{\sin \theta}{\cos \theta} = \frac{\dfrac{y}{r}}{\dfrac{x}{r}} = \frac{y}{r} \cdot \frac{r}{x} = \frac{y}{x} = \tan \theta$$

Example 2   Show that $\dfrac{\cos \theta}{\sin \theta} = \cot \theta$.

Solution   Using the definitions of the cosine and sine functions, you have

$$\frac{\cos \theta}{\sin \theta} = \frac{\dfrac{x}{r}}{\dfrac{y}{r}} = \frac{x}{r} \cdot \frac{r}{y} = \frac{x}{y} = \cot \theta$$

The six trigonometric functions involve the quantities x, y, and r, but, because of the Pythagorean Theorem, these three quantities are related by the formula $x^2 + y^2 = r^2$. This formula can be used to establish three more important relationships.

By agreement, $\sin^2 \theta$ is standard notation for $(\sin \theta)^2$; therefore, $\sin^2 \theta = \sin \theta \cdot \sin \theta$. Powers of the other trigonometric functions are denoted in a similar way.

Example 3   Use the result $y^2 + x^2 = r^2$ to show that $\sin^2 \theta + \cos^2 \theta = 1$.

Solution   Divide $y^2 + x^2 = r^2$ by r^2 to get

$$\frac{y^2}{r^2} + \frac{x^2}{r^2} = \frac{r^2}{r^2}$$

or

$$\left(\frac{y}{r}\right)^2 + \left(\frac{x}{r}\right)^2 = 1$$

Then, use the definitions of the trigonometric functions to get
$$\sin^2 \theta + \cos^2 \theta = 1$$

Example 4   Use the result $y^2 + x^2 = r^2$ to show that $\tan^2 \theta + 1 = \sec^2 \theta$.

Solution   Divide $y^2 + x^2 = r^2$ by x^2 and use the definitions of the trigonometric functions to get

$$\frac{y^2}{x^2} + \frac{x^2}{x^2} = \frac{r^2}{x^2}$$

$$\left(\frac{y}{x}\right)^2 + 1 = \left(\frac{r}{x}\right)^2$$

or

$$\tan^2 \theta + 1 = \sec^2 \theta$$

Example 5   Use the result $x^2 + y^2 = r^2$ to show that $\cot^2 \theta + 1 = \csc^2 \theta$.

Solution   Divide $x^2 + y^2 = r^2$ by y^2. Then you have

$$\frac{x^2}{y^2} + \frac{y^2}{y^2} = \frac{r^2}{y^2}$$

$$\left(\frac{x}{y}\right)^2 + 1 = \left(\frac{r}{y}\right)^2$$

or

$$\cot^2 \theta + 1 = \csc^2 \theta$$

The relationships developed so far are summarized in the following chart.

(2.7)

> **The Eight Fundamental Relationships.**   For any trigonometric angle θ for which the functions are defined,
>
> $$\sin \theta = \frac{1}{\csc \theta} \qquad\qquad \cos \theta = \frac{1}{\sec \theta}$$
>
> $$\tan \theta = \frac{1}{\cot \theta} \qquad\qquad \tan \theta = \frac{\sin \theta}{\cos \theta}$$
>
> $$\cot \theta = \frac{\cos \theta}{\sin \theta} \qquad\qquad \sin^2 \theta + \cos^2 \theta = 1$$
>
> $$\tan^2 \theta + 1 = \sec^2 \theta \qquad \cot^2 \theta + 1 = \csc^2 \theta$$

Many other relationships can also be established.

Example 6   Show that $\tan \theta = \dfrac{\sec \theta}{\csc \theta}$.

Solution   Make use of the fundamental relationships of the trigonometric functions and proceed as follows.

$$\tan \theta = \frac{\sin \theta}{\cos \theta}$$

$$= \frac{\dfrac{1}{\csc \theta}}{\dfrac{1}{\sec \theta}}$$

$$= \frac{1}{\csc \theta} \cdot \frac{\sec \theta}{1}$$

$$= \frac{\sec \theta}{\csc \theta}$$

∎

Example 7 Assume that θ is in QII and that $\sin \theta = \frac{4}{5}$. Use the eight fundamental relationships of trigonometric functions to find the values of the other five trigonometric functions.

Solution Because the sine and cosecant of an angle are reciprocals, you have

$$\csc \theta = \frac{1}{\sin \theta}$$

$$= \frac{5}{4}$$

Because $\sin^2 \theta + \cos^2 \theta = 1$, you have

$$\cos^2 \theta = 1 - \sin^2 \theta$$
$$\cos \theta = \pm \sqrt{1 - \sin^2 \theta}$$

Because θ is in QII, $\cos \theta$ must be negative. Therefore, you have

$$\cos \theta = -\sqrt{1 - \left(\frac{4}{5}\right)^2}$$

$$= -\sqrt{\frac{9}{25}}$$

$$= -\frac{3}{5}$$

Because $\sec \theta$ is the reciprocal of $\cos \theta$, you have

$$\sec \theta = -\frac{5}{3}$$

Use the relationship $\tan \theta = \dfrac{\sin \theta}{\cos \theta}$ to find $\tan \theta$.

$$\tan \theta = \frac{\sin \theta}{\cos \theta}$$

$$= \frac{\dfrac{4}{5}}{-\dfrac{3}{5}}$$

$$= -\frac{4}{3}$$

Finally, cot θ is the reciprocal of tan θ.

$$\cot \theta = -\frac{3}{4}$$

Example 8   Given that $\cos \theta = \frac{1}{3}$ and that tan θ is negative, use the fundamental trigonometric relationships to find the values of the other five trigonometric functions.

Solution   You must first decide in which quadrant the terminal side of angle θ lies. Because cos θ is positive, θ must be a QI or QIV angle. Because tan θ is negative, θ must be a QII or QIV angle. Putting these facts together, you know that θ is a QIV angle. See Figure 2-15. From the eight basic trigonometric relationships, it follows that

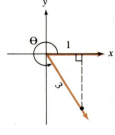

Figure 2-15

$$\sec \theta = \frac{1}{\cos \theta}$$

$$= \frac{1}{\frac{1}{3}}$$

$$= 3$$

Because $\tan^2 \theta = \sec^2 \theta - 1$, you have $\tan \theta = \pm \sqrt{\sec^2 \theta - 1}$. Because tan θ is negative, you have

$$\tan \theta = -\sqrt{3^2 - 1}$$
$$= -\sqrt{8}$$
$$= -2\sqrt{2}$$

Because $\dfrac{\sin \theta}{\cos \theta} = \tan \theta$, it follows that $\sin \theta = \tan \theta \cos \theta$, and

$$\sin \theta = -2\sqrt{2} \cdot \frac{1}{3}$$

$$= -\frac{2\sqrt{2}}{3}$$

Continuing, find csc θ.

$$\csc \theta = \frac{1}{\sin \theta}$$

$$= -\frac{3}{2\sqrt{2}}$$

$$= -\frac{3\sqrt{2}}{4} \qquad \text{Multiply numerator and denominator by } \sqrt{2}.$$

Finally, because $\cot\theta$ is the reciprocal of $\tan\theta$, you have

$$\cot\theta = \frac{1}{-2\sqrt{2}}$$

$$= -\frac{\sqrt{2}}{4} \qquad \text{Multiply numerator and denominator by } -\sqrt{2}.$$

Trigonometric functions of $(-\theta)$

We consider three more basic relationships. Consider angles θ and $-\theta$, drawn in standard position as shown in Figure 2-16. Points P and Q are on the terminal sides of the respective angles. Let (x, y) be coordinates of point P, and let $(x, -y)$ be the coordinates of point Q. The trigonometric functions of these angles are

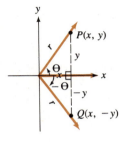

$$\begin{cases} \sin(-\theta) = \dfrac{-y}{r} \\[2mm] \sin\theta = \dfrac{y}{r} \end{cases} \qquad \begin{cases} \cos(-\theta) = \dfrac{x}{r} \\[2mm] \cos\theta = \dfrac{x}{r} \end{cases} \qquad \begin{cases} \tan(-\theta) = \dfrac{-y}{x} \\[2mm] \tan\theta = \dfrac{y}{x} \end{cases}$$

Figure 2-16   From these results, it follows that

(2.8)

$$\sin(-\theta) = -\sin\theta$$
$$\cos(-\theta) = \cos\theta$$
$$\tan(-\theta) = -\tan\theta$$

In the exercises you will be asked to verify these results for angles that are drawn in the second, third, and fourth quadrants.

(2.9)   **Definition.**   If $f(-x) = f(x)$ for all x, then f is called an **even function.** If $f(-x) = -f(x)$ for all x, then f is called an **odd function.**

Equations 2.8 show that the cosine function is an even function and that the sine and tangent functions are odd functions.

Example 9   Show that $y = f(x) = x^2 + x^6$ is an even function.

Solution   The function is an even function if and only if $f(-x) = f(x)$ for all x. Substitute $-x$ for x and simplify.

$$f(-x) = (-x)^2 + (-x)^6$$
$$= x^2 + x^6$$
$$= f(x) \quad \text{(for all } x\text{)}$$

Because $f(-x) = f(x)$ for all x, $f(x)$ is an even function.

Example 10   Show that $y = f(\theta) = \cos\theta - \sin\theta$ is neither an even function nor an odd function.

Solution   The function is an even function if and only if $f(-\theta) = f(\theta)$ for all θ. The function is an odd function if and only if $f(-\theta) = -f(\theta)$ for all θ. Calculate $f(-\theta)$ and $-f(\theta)$, and compare the results with $f(\theta)$. Note that

$$f(\theta) = \cos\theta - \sin\theta$$
$$f(-\theta) = \cos(-\theta) - \sin(-\theta)$$
$$= \cos\theta + \sin\theta$$

and

$$-f(\theta) = -(\cos\theta - \sin\theta)$$
$$= -\cos\theta + \sin\theta$$

Because $f(-\theta) \neq f(\theta)$, the function $f(\theta) = \cos\theta - \sin\theta$ cannot be an even function. Because $f(-\theta) \neq -f(\theta)$, the function $f(\theta)$ cannot be an odd function. Hence, it is neither even nor odd.   ∎

Exercise 2.2

In Exercises 1–8, use the letters x, y, and r and the definitions of the trigonometric functions to verify the following equations.

1. $\tan\theta \cos\theta = \sin\theta$

2. $\cot\theta \sin\theta = \cos\theta$

3. $\tan\theta = \dfrac{1}{\cos\theta \csc\theta}$

4. $\cot\theta = \cos\theta \csc\theta$

5. $\sin^2\theta + \sin^2\theta \cot^2\theta = 1$

6. $\cot\theta = \dfrac{\csc\theta}{\sec\theta}$

7. $\cot^2\theta + \sin^2\theta = \csc^2\theta - \cos^2\theta$

8. $\tan^2\theta + \cos^2\theta = \sec^2\theta - \sin^2\theta$

In Exercises 9–16, use one or more of the eight fundamental trigonometric relationships to verify the following equations.

9. $\tan\theta \cos\theta = \sin\theta$

10. $\cot\theta \sin\theta = \cos\theta$

11. $\tan\theta = \dfrac{1}{\cos\theta \csc\theta}$

12. $\cot\theta = \cos\theta \csc\theta$

13. $\sin^2\theta + \sin^2\theta \cot^2\theta = 1$

14. $\cot\theta = \dfrac{\csc\theta}{\sec\theta}$

15. $\cot^2\theta + \sin^2\theta = \csc^2\theta - \cos^2\theta$

16. $\tan^2\theta + \cos^2\theta = \sec^2\theta - \sin^2\theta$

In Exercises 17–26, use one or more of the eight fundamental trigonometric relationships to find the values of the remaining trigonometric functions.

17. θ in QI; $\sin\theta = \dfrac{4}{5}$

18. θ in QIII; $\sin\theta = -\dfrac{4}{5}$

19. θ in QII; $\cos\theta = -\dfrac{5}{13}$

20. θ in QIII; $\cos\theta = -\dfrac{5}{13}$

21. θ in QII; $\tan \theta = -\dfrac{4}{3}$

22. θ in QIV; $\tan \theta = -\dfrac{3}{4}$

23. θ in QI; $\cot \theta = \dfrac{9}{40}$

24. θ in QII; $\csc \theta = \dfrac{13}{5}$

25. $\sec \theta = -\dfrac{13}{5}$; θ in QIII

26. $\sec \theta = -\dfrac{13}{5}$; θ in QII

In Exercises 27–36, tell whether the given function is even, odd, or neither.

27. $y = x^3$

28. $y = x^2$

29. $y = x^4 + x^2$

30. $y = x^6 - 3x^4 + x$

31. $y = \cot \theta$

32. $y = \csc \theta + \sec \theta$

33. $y = \sin \theta + \cos \theta$

34. $y = \cos \theta + \sec \theta$

35. $y = 1 + x$

36. $y = -4 + \cos \theta$

37. Show that $y = \csc \theta$ is an odd function.

38. Show that $y = \cot \theta$ is an odd function.

39. Is the sum of two odd functions still odd?

40. Is the sum of two even functions still even?

41. Is the product of two odd functions still odd?

42. Is the product of two even functions still even?

43. Let θ be a second-quadrant angle. Draw a figure similar to Figure 2-16 and show that $\sin(-\theta) = -\sin \theta$, $\cos(-\theta) = \cos \theta$, and $\tan(-\theta) = -\tan \theta$.

44. Let θ be a third-quadrant angle. Draw a figure similar to Figure 2-16 and show that $\sin(-\theta) = -\sin \theta$, $\cos(-\theta) = \cos \theta$, and $\tan(-\theta) = -\tan \theta$.

45. Let θ be a fourth-quadrant angle. Draw a figure similar to Figure 2-16 and show that $\sin(-\theta) = -\sin \theta$, $\cos(-\theta) = \cos \theta$, and $\tan(-\theta) = -\tan \theta$.

46. If θ is a second-quadrant angle, is $\sin(-\theta)$ positive or negative?

47. If θ is a third-quadrant angle, is $\cos(-\theta)$ positive or negative?

48. Identify the product $\sin(-\theta) \cos(-\theta) \tan(-\theta)$ as positive or negative if θ is a second-quadrant angle.

49. Identify the product $\sin(-\theta) \cos(-\theta) \tan(-\theta)$ as positive or negative if θ is a third-quadrant angle.

50. Identify the product $\sin(-\theta) \cos(-\theta) \tan(-\theta)$ as positive or negative if θ is a fourth-quadrant angle.

2.3   TRIGONOMETRIC FUNCTIONS OF CERTAIN ANGLES

Certain angles are more common than others. For example, angles of 90° occur more often in geometric theorems than do angles of 89°. Angles of 30° and 45° receive more attention than angles of 31° and 43.5°. Let's find values for the trigonometric functions of some of these common angles.

An angle of 0° placed in standard position has both its initial side and its terminal side along the positive *x*-axis. See Figure 2-17. The point $P(1, 0)$ lies on the terminal side, and r, the distance of P from the origin, is 1 unit. Four of the six trigonometric functions of $0°$ are

$$\sin 0° = \frac{y}{r} = \frac{0}{1} = 0$$

$$\cos 0° = \frac{x}{r} = \frac{1}{1} = 1$$

$$\tan 0° = \frac{y}{x} = \frac{0}{1} = 0$$

$$\sec 0° = \frac{r}{x} = \frac{1}{1} = 1$$

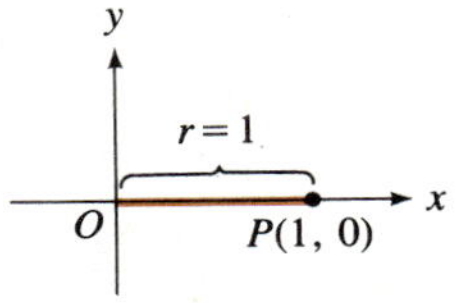

Figure 2-17

Values for $\cot 0°$ and $\csc 0°$ are undefined because division by 0 is not permitted.

In a similar manner, the point $P(0, 1)$ lies on the terminal side of a $90°$ angle in standard position and at a distance $r = 1$ from the origin. See Figure 2-18. Again, four functions may be computed.

$$\sin 90° = \frac{y}{r} = \frac{1}{1} = 1$$

$$\cos 90° = \frac{x}{r} = \frac{0}{1} = 0$$

$$\csc 90° = \frac{r}{y} = \frac{1}{1} = 1$$

$$\cot 90° = \frac{x}{y} = \frac{0}{1} = 0$$

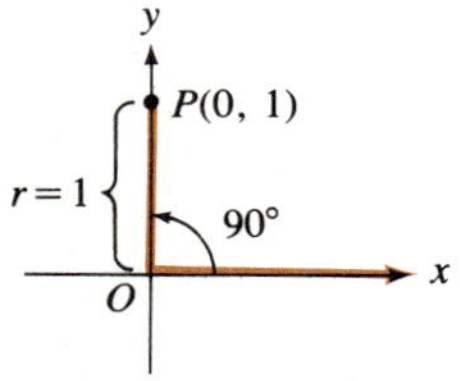

Figure 2-18

Values for $\tan 90°$ and $\sec 90°$ are undefined.

The point $P(-1, 0)$ lies on the terminal side of an angle of $180°$ that is placed in standard position. The distance of P from the origin is $r = +1$. Remember that r is always positive. See Figure 2-19. Four functions can be computed and the other two are undefined.

$$\sin 180° = \frac{y}{r} = \frac{0}{1} = 0$$

$$\cos 180° = \frac{x}{r} = \frac{-1}{1} = -1$$

$$\tan 180° = \frac{y}{x} = \frac{0}{-1} = 0$$

$$\sec 180° = \frac{r}{x} = \frac{1}{-1} = -1$$

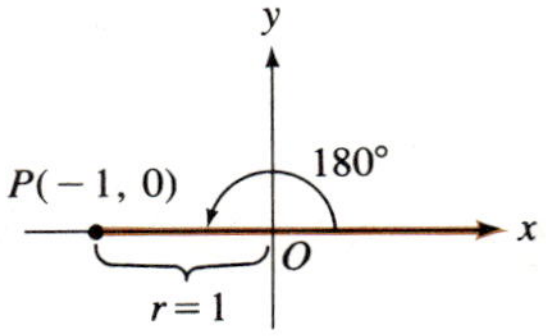

Figure 2-19

Finally, $P(0, -1)$ is a point on the terminal side of an angle of $270°$ placed in standard position. Again, four functions have values and two functions do not. See Figure 2-20.

$$\sin 270° = \frac{y}{r} = \frac{-1}{1} = -1$$

$$\cos 270° = \frac{x}{r} = \frac{0}{1} = 0$$

$$\cot 270° = \frac{x}{y} = \frac{0}{-1} = 0$$

$$\csc 270° = \frac{r}{y} = \frac{1}{-1} = -1$$

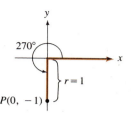

Figure 2-20

We summarize these results in a chart. This information should be memorized.

(2.10)

θ	sin θ	cos θ	tan θ	csc θ	sec θ	cot θ
0°	0	1	0	undefined	1	undefined
90°	1	0	undefined	1	undefined	0
180°	0	−1	0	undefined	−1	undefined
270°	−1	0	undefined	−1	undefined	0

Example 1   Find sin 990°.

Solution   Because 990° is greater than 360°, find an angle that is less than 360° but coterminal with 990°. This can be done by repeatedly subtracting 360° from 990° until an angle less than 360° is found. Then, find the sine of that angle.

$$\sin 990° = \sin[990° - 2(360°)]$$
$$= \sin 270°$$
$$= -1$$ ■

If $P(x, y)$ is a point on the terminal side of a 30° angle placed in standard position, and r is the distance from P to the origin, then triangle OPA is a 30°, 60° right triangle. See Figure 2-21. Recall that the leg opposite the 30° angle measures one-half of the length of the hypotenuse, and the leg opposite the 60° angle is $\sqrt{3}$ times the leg opposite the 30° angle. Therefore, $r = 2y$ and $x = \sqrt{3}\, y$.

$$\sin 30° = \frac{y}{r} = \frac{y}{2y} = \frac{1}{2}$$

$$\cos 30° = \frac{x}{r} = \frac{\sqrt{3}y}{2y} = \frac{\sqrt{3}}{2}$$

$$\tan 30° = \frac{y}{x} = \frac{y}{\sqrt{3}y} = \frac{1}{\sqrt{3}} = \frac{\sqrt{3}}{3}$$

$$\csc 30° = 2$$

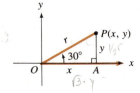

Figure 2-21

$$\sec 30° = \frac{2\sqrt{3}}{3}$$

$$\cot 30° = \sqrt{3}$$

The cosecant, secant, and cotangent functions were found by taking the respective reciprocals of the sine, cosine, and tangent functions.

A similar argument determines the trigonometric functions of a 60° angle. See Figure 2-22. Triangle OPA is again a 30°, 60° right triangle with $r = 2x$ and $y = \sqrt{3}x$. Thus, we have

$$\sin 60° = \frac{y}{r} = \frac{\sqrt{3}x}{2x} = \frac{\sqrt{3}}{2}$$

$$\cos 60° = \frac{x}{r} = \frac{x}{2x} = \frac{1}{2}$$

$$\tan 60° = \frac{y}{x} = \frac{\sqrt{3}x}{x} = \sqrt{3}$$

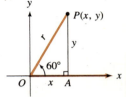

Figure 2-22

The following functions have values that are the reciprocals of the previous three.

$$\csc 60° = \frac{2}{\sqrt{3}} = \frac{2\sqrt{3}}{3}$$

$$\sec 60° = 2$$

$$\cot 60° = \frac{1}{\sqrt{3}} = \frac{\sqrt{3}}{3}$$

Because an angle of 45° is a base angle of the isosceles right triangle in Figure 2-23, it follows that $x = y$ and that $r = \sqrt{2}x = \sqrt{2}y$. Thus, we have

$$\sin 45° = \frac{y}{r} = \frac{y}{\sqrt{2}y} = \frac{1}{\sqrt{2}} = \frac{\sqrt{2}}{2}$$

$$\cos 45° = \frac{x}{r} = \frac{x}{\sqrt{2}x} = \frac{1}{\sqrt{2}} = \frac{\sqrt{2}}{2}$$

$$\tan 45° = \frac{y}{x} = 1$$

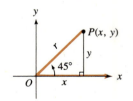

Figure 2-23

Taking the reciprocals of these values gives

$$\csc 45° = \sqrt{2}$$

$$\sec 45° = \sqrt{2}$$

$$\cot 45° = 1$$

We summarize all of the results of this section in the following chart.

(2.11)

θ	$\sin\theta$	$\cos\theta$	$\tan\theta$	$\csc\theta$	$\sec\theta$	$\cot\theta$
0°	0	1	0	undefined	1	undefined
30°	$\dfrac{1}{2}$	$\dfrac{\sqrt{3}}{2}$	$\dfrac{\sqrt{3}}{3}$	2	$\dfrac{2\sqrt{3}}{3}$	$\sqrt{3}$
45°	$\dfrac{\sqrt{2}}{2}$	$\dfrac{\sqrt{2}}{2}$	1	$\sqrt{2}$	$\sqrt{2}$	1
60°	$\dfrac{\sqrt{3}}{2}$	$\dfrac{1}{2}$	$\sqrt{3}$	$\dfrac{2\sqrt{3}}{3}$	2	$\dfrac{\sqrt{3}}{3}$
90°	1	0	undefined	1	undefined	0
180°	0	-1	0	undefined	-1	undefined
270°	-1	0	undefined	-1	undefined	0

It is now possible to determine the values of trigonometric functions of certain angles in any quadrant.

Example 2 Find the value of the six trigonometric functions of $\theta = 120°$.

Solution Draw an angle of 120° in standard position and mark a point $P(x, y)$ on the terminal side at a distance r from the origin. See Figure 2-24. Draw segment PA so it is perpendicular to the x-axis. This forms triangle OPA, which is called **Reference** the **reference triangle.** Because P is in the second quadrant, x is negative and **triangle** y is positive. And r is always positive. Because the reference triangle OPA is a 30°, 60° right triangle, $r = -2x$ and $y = -\sqrt{3}x$. (Remember that if x is negative, $-2x$ and $-\sqrt{3}x$ are positive.) Thus, you have

$$\sin 120° = \frac{y}{r} = \frac{-\sqrt{3}x}{-2x} = \frac{\sqrt{3}}{2}$$

$$\cos 120° = \frac{x}{r} = \frac{x}{-2x} = -\frac{1}{2}$$

$$\tan 120° = \frac{y}{x} = \frac{-\sqrt{3}x}{x} = -\sqrt{3}$$

$$\csc 120° = \frac{1}{\sin 120°} = \frac{2}{\sqrt{3}} = \frac{2\sqrt{3}}{3}$$

$$\sec 120° = \frac{1}{\cos 120°} = -2$$

$$\cot 120° = \frac{1}{\tan 120°} = \frac{1}{-\sqrt{3}} = -\frac{\sqrt{3}}{3}$$

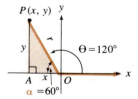

Figure 2-24

Note that angle α (not in standard position) formed by the terminal side of θ and the x-axis is 60°. Also note that the values of the functions of 120° equal the values of the functions of a 60° angle, except for an occasional minus sign.

Example 3   Find the values of the six trigonometric functions of $\theta = 225°$.

Solution   Draw angle θ in standard position. Place point $P(x, y)$ on the terminal side at a distance r from the origin and draw segment PA perpendicular to the x-axis to form the reference triangle OPA. See Figure 2-25. Because point P is in the third quadrant, x and y are both negative (and r is, of course, positive). Triangle OPA is an isosceles right triangle, so $x = y$ and $r = -\sqrt{2}x = -\sqrt{2}y$. Thus, you have

$$\sin 225° = \frac{y}{r} = \frac{y}{-\sqrt{2}y} = -\frac{1}{\sqrt{2}} = -\frac{\sqrt{2}}{2}$$

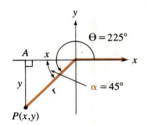

Figure 2-25

$$\cos 225° = \frac{x}{r} = \frac{x}{-\sqrt{2}x} = -\frac{1}{\sqrt{2}} = -\frac{\sqrt{2}}{2}$$

$$\tan 225° = \frac{y}{x} = 1$$

$$\csc 225° = \frac{1}{\sin 225°} = -\sqrt{2}$$

$$\sec 225° = \frac{1}{\cos 225°} = -\sqrt{2}$$

$$\cot 225° = \frac{1}{\tan 225°} = 1$$

Note that the acute angle, angle α, between the terminal side of angle θ and the x-axis is 45°. Also note that the six values of the functions of 225° equal the six values of the functions of 45°, except for sign.

Reference angle   A pattern emerges from Examples 2 and 3. Associated with any angle θ is an acute angle in the reference triangle called the **reference angle.** The reference angle is that acute angle formed by the terminal side of angle θ and the x-axis. The reference angle is not always in standard position. However, if it were, it would be a first-quadrant angle with six positive values for the six trigonometric functions. The useful fact is this:

(2.12) | **The six trigonometric functions of any angle θ are equal to those of the reference angle of θ, except possibly for sign.**

The appropriate sign can be determined independently by considering the quadrant in which the terminal side of θ lies.

Example 4   Find the values of the six trigonometric functions of $\theta = 210°$.

Solution   Sketch an angle of 210° in standard position to determine in which quadrant the terminal side of θ lies. See Figure 2-26. Because $\theta = 210°$, the reference angle must equal 30°. Because 210° is a third-quadrant angle, only its tangent and its cotangent are positive. The other trigonometric functions of 210° are negative.

$$\sin 210° = -\sin 30° = -\frac{1}{2}$$

$$\cos 210° = -\cos 30° = \frac{-\sqrt{3}}{2}$$

$$\tan 210° = +\tan 30° = \frac{\sqrt{3}}{3}$$

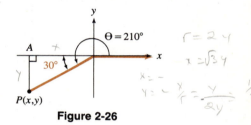

Figure 2-26

As usual, taking the reciprocals gives the values of the other functions.

$$\csc 210° = -2$$

$$\sec 210° = -\frac{2}{\sqrt{3}} = -\frac{2\sqrt{3}}{3}$$

$$\cot 210° = +\sqrt{3}$$

■

Example 5   Find the values of the six trigonometric functions of $\theta = -405°$.

Solution   A sketch helps determine the quadrant in which the angle θ terminates and also its reference angle. See Figure 2-27. Recall that a negative angle has its rotation in a clockwise direction. Because an angle of $-405°$ is a fourth-quadrant angle, only its cosine and secant are positive.

$$\sin(-405°) = -\sin 45° = -\frac{\sqrt{2}}{2}$$

$$\cos(-405°) = +\cos 45° = +\frac{\sqrt{2}}{2}$$

$$\tan(-405°) = -\tan 45° = -1$$

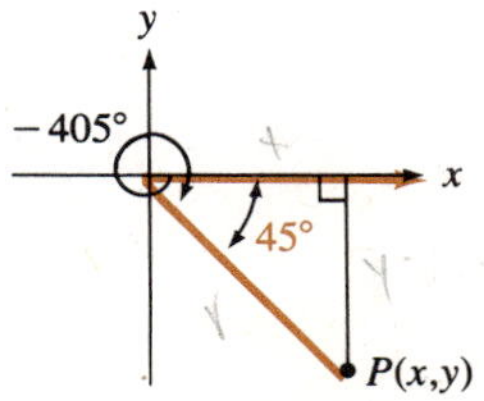

Figure 2-27

Take reciprocals to find the values of the other three functions.

$$\csc(-405°) = -\sqrt{2}$$
$$\sec(-405°) = +\sqrt{2}$$
$$\cot(-405°) = -1$$

Example 6   If $\sin\theta = -\dfrac{1}{2}$ and $\cos\theta = \dfrac{\sqrt{3}}{2}$, find θ.

Solution   Because $\sin\theta$ is negative and $\cos\theta$ is positive, θ must be a QIV angle. See Figure 2-28. Because

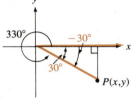

$$\sin 30° = \frac{1}{2} \quad \text{and} \quad \cos 30° = \frac{\sqrt{3}}{2}$$

the reference angle must be 30°. What fourth-quadrant angle has a reference angle of 30°? There are infinitely many. Two examples are $\theta = -30°$ and $\theta = 330°$. *All* the values of θ can be expressed as $330° \pm n \cdot 360°$ where n is any whole number.

Figure 2-28

Exercise 2.3

In Exercises 1–6 draw each angle in standard position and find the value of the sine, cosine, and tangent functions. **Do not use a calculator.**

1. 135°   **2.** 225°

3. 240°   **4.** 300° $-40 \ -\frac{\sqrt{3}}{2}, \frac{V}{2} \ r \ V3$

5. 315° $(-45) \ -\frac{\sqrt{2}}{2}, \frac{\sqrt{2}}{2}, -1$ **6.** 330°

In Exercises 7–12, draw each angle in standard position and find the value of the cosecant, secant, and cotangent functions. **Do not use a calculator.**

7. 225°   **8.** 135°

9. 300°   **10.** 240°

11. 330°   **12.** 315°

In Exercises 13–18, find the value of the sine, cosine, and tangent of the given angle. **Do not use a calculator.**

13. 390°   **14.** 480°

15. 510°   **16.** $-690°$

17. $-45°$ **18.** $-150°$

In Exercises 19–34, evaluate each expression. **Do not use a calculator.**

19. $\sin 0° + \cos 0° \tan 45°$ **20.** $\sin^2 90° + \cos 180° \tan 0°$

21. $\cos^2 90° + \cos 90° \sin^2 180°$ **22.** $\cos^2 0° + \sin^2 90° + \cot^2 90°$

23. $\sin^2 270° + \csc^2 270° + \cot^2 270°$ **24.** $\cos 180° \sin 180° - \tan^2 180°$

25. $\sin 30° \cos 60° - 2 \tan^2 60°$

26. $\sin^2 120° \cos 45° + \tan 45° \sin 90°$

27. $\sin 45° \cos 330° - \tan 150° \tan 60°$

28. $\cos 30° \tan 60° + \cos^3 45° \tan 45°$

29. $\csc^2 210° \sec 30° - \sec 315° \cot 60°$

30. $\csc 90° \csc 210° + \csc 45° \sin 135°$

31. $\sec 0° \tan 0° - \cot 30° \cot 45°$

32. $\cot^2 270° \csc 90° + \sec 60° \cot 30°$

33. $\csc^2 60° \sin^2 240°$

34. $\sec^2 45° \cos^2 315°$

In Exercises 35–48, use the given information to find values of θ, where $0° \leq \theta < 360°$.
Do not use a calculator.

35. $\tan \theta = \dfrac{\sqrt{3}}{3}$; $\sin \theta = \dfrac{1}{2}$

36. $\tan \theta = -1$; $\cos \theta = \dfrac{-\sqrt{2}}{2}$

37. $\tan \theta = -\sqrt{3}$; $\cos \theta = \dfrac{1}{2}$

38. $\cot \theta = \sqrt{3}$; $\cos \theta = \dfrac{-\sqrt{3}}{2}$

39. $\sin \theta = -\dfrac{1}{2}$; $\sec \theta = \dfrac{-2\sqrt{3}}{3}$

40. $\cos \theta = \dfrac{\sqrt{3}}{2}$; $\csc \theta = 2$

41. $\tan \theta = -1$; $\sec \theta = \sqrt{2}$

42. $\tan \theta = -\sqrt{3}$; $\cos \theta = -\dfrac{1}{2}$

43. $\sec \theta = -\sqrt{2}$; $\cot \theta = -1$

44. $\sin \theta = -1$

45. $\tan \theta$ is undefined.

46. $\csc \theta = -\sqrt{2}$; $\cot \theta = -1$

47. $\cos \theta = \dfrac{\sqrt{3}}{2}$; $\sin \theta = -\dfrac{1}{2}$

48. $\sin \theta = -\dfrac{\sqrt{3}}{2}$; $\cos \theta = \dfrac{1}{2}$

In Exercises 49–60, use the given information to find all values of α, where possible, if $0° \leq \alpha < 360°$.

49. $\sin \alpha = \dfrac{1}{2}$

50. $\cos \alpha = -\dfrac{\sqrt{3}}{2}$

51. $\tan \alpha = \dfrac{\sqrt{3}}{3}$

52. $\cot \alpha = -\sqrt{3}$

53. $\sec \alpha = -2$

54. $\csc \alpha = -2$

55. $\sin \alpha = -\dfrac{1}{2}$; α in QII

56. $\tan \alpha = -1$; α not in QII

57. $\cot \alpha = \sqrt{3}$; α not in QI

58. $\cos \alpha = \dfrac{1}{2}$; α in QIII

59. $\sin \alpha \cos \alpha = \dfrac{\sqrt{3}}{4}$

60. $\sin \alpha \cos \alpha = -\dfrac{\sqrt{3}}{4}$

2.4 TRIGONOMETRIC FUNCTIONS OF ANY ANGLE

Not every angle is one of the common angles considered in the previous section. We now discuss how to find values such as $\sin 23°$, $\cos 313.27°$, or $\tan 107°$. One method requires a protractor, ruler, and compass to construct the **unit circle** (a circle with radius of 1, centered at the origin). While this method is not very accurate, it does reinforce the definitions of the trigonometric functions. It might also prove useful if the batteries in your calculator die. A second method utilizes tables of values of trigonometric functions, such as Table A in Appendix IV in the back of this book. A third method is by far the easiest and most efficient. Use a calculator.

Example 1   Use a protractor, compass, and ruler to find approximate values for the sine, cosine, and tangent of an angle of 107°.

Solution   Extend the compass to a unit length as measured on the ruler and construct a circle with radius of 1. Use the protractor to draw an angle of 107° in standard position as in Figure 2-29. Point P is the intersection of the circle and the terminal side of the 107° angle. Draw PA perpendicular to the x-axis. Because $OP = r = 1$,

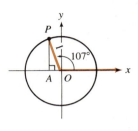

$$\sin 107° = \frac{AP}{OP} = \frac{AP}{1} = AP$$

Figure 2-29

Carefully measure AP. That value, 0.96, is approximately equal to sin 107°. To find the cosine of 107°, also refer to Figure 2-29, and note that

$$\cos 107° = \frac{OA}{OP} = \frac{OA}{1} = OA$$

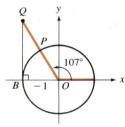

Carefully measure OA and note that, because an angle of 107° is a second-quadrant angle, the directed distance from O to A is negative. That value, -0.29, is approximately equal to cos 107°.

To find the tangent of 107°, draw a tangent line to the unit circle at point B and extend it until it intersects with the extension of segment OP. See Figure 2-30. The directed distance from O to B is -1, so we have

Figure 2-30 $$\tan 107° = \frac{BQ}{OB} = \frac{BQ}{-1} = -BQ$$

Carefully measure BQ. The negative of this measure, -3.27, is approximately the value of tan 107°. ■

Example 2   Use Table A in Appendix IV to find the sine, cosine, tangent, and cotangent of 43.5°.

Solution   In the **Degrees** column on the left side of the table, locate the number 43.5. Move to the right in that row and read the entries in the columns headed by **Sin, Cos, Tan,** and **Cot.**

Radians	Degrees	Sin	Cos	Tan	Cot		
			· · ·				
.7575	43.4	.6871	.7266	.9457	1.057	46.6	.8133
.7592	43.5	.6884	.7254	.9490	1.054	46.5	.8116
.7610	43.6	.6896	.7242	.9523	1.050	46.4	.8098
		Cos	**Sin**	**Cot**	**Tan**	**Degrees**	**Radians**

$$\sin 43.5° \approx 0.6884$$
$$\cos 43.5° \approx 0.7254$$
$$\tan 43.5° \approx 0.9490$$
$$\cot 43.5° \approx 1.054$$

Example 3   Use Table A in Appendix IV to find the sine, cosine, tangent, and cotangent of an angle of 46.5°.

Solution   When searching the **Degrees** column on the left side of the table, you find that it ends at 45°. However, 46.5° can be found on the right side in a column which is footed by **Degrees**. See the table in Example 2. Move to the left in the row containing 46.5° and read the entries in the columns footed by **Cos, Sin, Cot,** and **Tan.**

$$\sin 46.5° \approx 0.7254$$
$$\cos 46.5° \approx 0.6884$$
$$\tan 46.5° \approx 1.054$$
$$\cot 46.5° \approx 0.9490$$

Note that the same row of the table was used in both Examples 2 and 3; it did double duty for the complementary angles 43.5° and 46.5°. (Remember that if the sum of two acute angles is 90°, they are called **complementary angles.**) Note also that

$$\sin 43.5° = \cos 46.5° \quad \text{and} \quad \tan 43.5° = \cot 46.5°$$

are true statements. Because $\sin 43.5° = \cos 46.5°$, their reciprocals must be equal also. Therefore, we have

$$\csc 43.5° = \sec 46.5°$$

(2.13)

> **Definition.**   The trigonometric functions of sine and cosine are called **cofunctions.** The tangent and cotangent functions are cofunctions, as are the secant and cosecant functions.

Trigonometric functions of complementary angles   It's always true that the sine of an acute angle θ is equal to the cosine of the complement of θ. In like manner, the tangent of an acute θ is equal to the cotangent of the complement of θ, and the secant of an acute angle θ is equal to the cosecant of the complement of θ. These facts are summarized in the following theorem.

(2.14)

> **Theorem.**   If θ is any acute angle, any trigonometric function of θ is equal to the cofunction of the complement of θ.
>
> $$\sin \theta = \cos(90° - \theta) \quad \text{and} \quad \cos \theta = \sin(90° - \theta)$$
> $$\tan \theta = \cot(90° - \theta) \quad \text{and} \quad \cot \theta = \tan(90° - \theta)$$
> $$\csc \theta = \sec(90° - \theta) \quad \text{and} \quad \sec \theta = \csc(90° - \theta)$$

You will be asked to prove these cofunction relationships in the exercises. In a later chapter, you will see that these relationships can be extended to all values of θ.

Example 4   Use Table A in Appendix IV to find the values of the sine, cosine, tangent, and cotangent of an angle of 107°.

Solution   Although an angle of 107° does not appear in the table, its reference angle, 73°, does. Remember that any trigonometric function of an angle in standard position can differ only in sign from that same trigonometric function of its reference angle. Because a 107° angle is a second-quadrant angle, only the sine and cosecant functions are positive. The rest are negative.

$$\sin 107° = +\sin 73° \approx 0.9563$$
$$\cos 107° = -\cos 73° \approx -0.2924$$
$$\tan 107° = -\tan 73° \approx -3.271$$
$$\cot 107° = -\cot 73° \approx -0.3057$$

Using Calculators

Only a few years ago, commercially published tables provided the only practical way of finding values of trigonometric functions. However, the inexpensive pocket calculator has changed all that. There is no question that the easiest and most efficient way to evaluate the trigonometric functions is to use a calculator.

Example 5   Use a calculator to find the values of **a.** sin 23°, **b.** cos 313.27°, **c.** cos(−28.2°), and **d.** tan 90°.

Solution   Set your calculator for degree measure of angles.
   a.   To find sin 23°, enter the number 23 and press the ⎡SIN⎤ key. The display should read .3907311285. To the nearest ten-thousandth, sin 23° = 0.3907.
   b.   To evaluate cos 313.27°, enter the number 313.27 and press the ⎡COS⎤ key. The display should read 0.685437198. To the nearest ten-thousandth, cos 313.27° = 0.6854.
   c.   To evaluate cos(−28.2°), enter the number 28.2. Press the ⎡+/−⎤ and ⎡COS⎤ keys. The display should read .8813034521. To the nearest ten-thousandth, cos(−28.2°) = 0.8813.
   d.   Finally, to attempt to evaluate tan 90°, enter 90 and press the ⎡TAN⎤ key. The display will either blink 9's at you or read ERROR. Either way, it is telling you that tan 90° is undefined.

Example 6   Find the value of sec 43°.

Solution   Neither tables nor calculators will allow you to evaluate the secant function directly. You must use the property that sec 43° is the reciprocal of cos 43°. To find sec 43° on a calculator, set your calculator for degrees, enter the number 43, and then press in order the ⎡COS⎤ and ⎡1/x⎤ keys. This gives the reciprocal of cos 43°, which is sec 43° ≈ 1.367327461. To the nearest ten-thousandth, sec 43° = 1.3673.   ◼

Example 7 If θ is an acute angle and cos θ = 0.7660, find angle θ.

Solution Angle θ can be found by using either a calculator or Table A.

   If you use a calculator, be sure it is set for degrees. Enter the number .7660 and press the INV and COS keys. If your calculator does not have an INV key, consult your owner's manual. To the nearest tenth, θ = 40.0°.

   If you use Table A, find the column headed by **Cos** at the top of the page. Run your finger down the column, moving to successive pages if necessary, until you find the number .7660. Move to the left in that row to find the value of 40.0° in the degree column.   ◼

Example 8 If α is an acute angle and tan α = 5.671, find angle α.

Solution Be sure your calculator is set for degrees. Enter the number 5.671 and press the INV and TAN keys in that order. To the nearest tenth of a degree, α = 80.0°.

   To use Table A, find the column footed by **Tan** at the bottom of the page. Run your finger up the column, moving to successive pages if necessary, to find the number 5.671. Move to the right to find the value of 80.0° in the degree column.   ◼

Example 9 If θ is between 180° and 270° and sin θ = −0.9397, find angle θ.

Solution See Figure 2-31. Any trigonometric function of θ has the same value as that same trigonometric function of θ's reference angle α, except possibly for sign. Enter the number .9397 in your calculator and press the INV and SIN keys. The display will give the value of the acute reference angle α. To the nearest tenth, α = 70.0°. Because θ is a third-quadrant angle, add 180° to α to find angle θ. Thus, θ = 180° + 70.0° = 250.0°. Table A can be used to find angle α if you do not have a calculator.

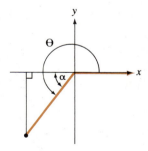

Figure 2-31

◼

Exercise 2.4

In Exercises 1–4, use the unit circle in Illustration 1 on the following page to compute the sine, cosine, and tangent ratios of the given angles.

1. 50°   **2.** 170°

3. 235°   **4.** 340°

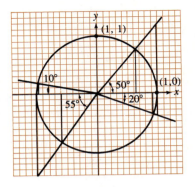

Illustration 1

In Exercises 5–10, draw the unit circle and use it to estimate the sine, cosine, and tangent ratios of the given angles.

5. 12°   **6.** 72°

7. 115°   **8.** 175°

9. −260°   **10.** −310°

In Exercises 11–20, use Table A in Appendix IV to find the values of the sine, cosine, and tangent of the given angles.

11. 17°   **12.** 29°

13. 62°   **14.** 82°

15. 119°   **16.** 121°

17. −233°   **18.** −182°

19. 1723°   **20.** 2811°

In Exercises 21–26, use a calculator to find the values of the given functions. Note that in each exercise, the second function is the cofunction of the first and that the second angle is the complement of the first.

21. $\sin 20°$; $\cos 70°$ **22.** $\tan 9°$; $\cot 81°$

23. $\cos 5°$; $\sin 85°$ **24.** $\cot 64°$; $\tan 26°$

25. $\sec 84°$; $\csc 6°$ **26.** $\sec 12°$; $\csc 78°$

In Exercises 27–36, tell whether the given statement is true for all acute angles A. If it is not, answer "false."

27. $\sin A = \sin(90° - A)$ **28.** $\cos A = \dfrac{1}{\sec A}$

29. $\sin^2 A + \sin^2(90° - A) = 1$ **30.** $\sin A = \cos(90° - A)$

31. $\tan A = \cot(90° - A)$ **32.** $\sin A = \dfrac{1}{\sec(90° - A)}$

33. $\cos A = \dfrac{1}{\csc(90° - A)}$

34. $\sin(90° - A) + \sin A = 1$

35. $\cot A = \dfrac{\cos(90° - A)}{\sin(90° - A)}$

36. $\cot A = \dfrac{\sin(90° - A)}{\cos(90° - A)}$

In Exercises 37–48, use a calculator to compute the required values to four decimal places.

37. $\sin 23.1°$

38. $\sin 57.8°$

39. $\cos 133.7°$

40. $\cos 211.7°$

41. $\tan 223.5°$

42. $\tan(-223.5°)$

43. $\csc 312.4°$

44. $\csc 129.2°$

45. $\sec(-47.4°)$

46. $\sec 11.3°$

47. $\cot 640.6°$

48. $\cot 302.2°$

In Exercises 49–60, angle θ ($0° \le \theta < 360°$) is in a given quadrant and the value of a trigonometric function is known. Use a calculator to find angle θ to the nearest tenth of a degree.

49. QI; $\tan \theta = 0.2493$

50. QI; $\sin \theta = 0.9986$

51. QII; $\cos \theta = -0.3420$

52. QII; $\cos \theta = -0.9063$

53. QIII; $\sin \theta = -0.4540$

54. QIII; $\cos \theta = -0.7193$

55. QIV; $\tan \theta = -5.6713$

56. QIV; $\sin \theta = -0.1908$

57. QI; $\csc \theta = 1.3250$

58. QII; $\sec \theta = -57.2987$

59. QIII; $\cot \theta = 1.1918$

60. QIV; $\csc \theta = -11.4737$

61. Devise a method to use the unit circle to find the secant, cosecant, and cotangent of an angle θ.

62. In Illustration 2, angle A is complementary to angle B. Prove that $\sin A = \cos B$, $\tan A = \cot B$, and $\csc A = \sec B$.

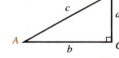

Illustration 2

2.5   RIGHT TRIANGLE TRIGONOMETRY

The word *trigonometry* means "measuring triangles (trigons)." The possibility of solving triangles—determining all sides and all angles when only some are known—has made trigonometry indispensible in astronomy, navigation, and surveying. In this section, we consider several applications of right triangle trigonometry.

Example 1   A right triangle ABC has an acute angle A that equals $27°$ and a hypotenuse c of length 14 ft. Solve the triangle by finding the other acute angle and the lengths of the two unknown sides.

Solution   Let the lengths of the two unknown sides of right triangle ABC be represented by a and b. Then, triangle ABC may be drawn in QI of a coordinate system as in Figure 2-32. In reference triangle ABC, $27°$ is the reference angle, $r = c = $

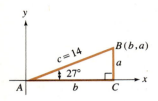

Figure 2-32

14, and point $B(x, y) = B(b, a)$. Because the triangle is a right triangle, angle A and angle B are complementary. Thus, angle $B = 63°$. The values of a and b may be found by using two trigonometric functions.

$$\sin 27° = \frac{a}{c} = \frac{a}{14}$$

$$\cos 27° = \frac{b}{c} = \frac{b}{14}$$

Use a calculator and solve these equations for a and b.

$$a = 14 \sin 27°$$
$$\approx 14(0.4539904997)$$
$$\approx 6.355866996$$
$$\approx 6.4$$

Similarly, we have

$$b = 14 \cos 27°$$
$$\approx 14(0.8910065242)$$
$$\approx 12.47409134$$
$$\approx 12$$

The remaining two sides of the triangle are 6.4 ft and 12 ft. ∎

Accuracy and significant digits   You may be curious why the answers in Example 1 were rounded off to two digits (called **significant digits**). If the hypotenuse were *exactly* 14 ft and the acute angle *exactly* 27°, rounding off would be unnecessary. More likely, however, 14 ft and 27° are not exact, but are measurements that are only approximate. In that case, the impressive strings of decimal digits are unwarranted.

Example 2   In Example 1, the hypotenuse of the triangle was measured to two significant digits and the angle to the nearest degree. How many significant digits are appropriate in the answers?

Solution   Because the hypotenuse is measured to two significant digits as 14 ft, the actual value lies between 13.5 ft and 14.5 ft. The angle of 27° lies between 26.5° and 27.5°. Calculate the values of a for the worst possible cases. Using the smallest possible values for the hypotenuse and angle gives

$$a = 13.5 \sin 26.5° \approx 6.023670477$$

The complete answer from Example 1 is

$$a = 14 \sin 27° \approx 6.355866996$$

Using the largest possible values of the hypotenuse and angle gives

$$a = 14.5 \sin 27.5° \approx 6.695354892$$

Note that the previous results agree only in the units digit of 6. In spite of how impressive they might appear, all of the digits to the right of the decimal points are meaningless. However, because the hypotenuse is measured to two significant digits, convention allows rounding all sides to two significant digits also.   ■

Calculators routinely provide answers to 8, 10, or 12 figures. You must decide how many of these are significant. Resist the temptation of thinking "If I have them, I'll use them." A good rule of thumb for determining acceptable accuracy is provided in the following table.

(2.15)

Accuracy in measurements of sides	Accuracy in angles
Two significant digits	Nearest degree
Three significant digits	Nearest tenth of a degree
Four significant digits	Nearest hundredth of a degree

When solving triangles, remember that answers can be only as accurate as the least accurate of the given data. However, if a calculation requires several inter-mediate steps, do not round off until you have the final answer.

It is not always easy to decide how many significant digits a number has. For example, is the number 140 accurate to two or three significant digits? If the number is rounded to the nearest ten, the zero is merely a placeholder and 140 has two significant digits. On the other hand, if the number has been rounded to the nearest unit, the zero is significant and 140 has three significant digits. In this book, we will assume the greatest possible number of significant digits unless stated otherwise—140 has three significant digits.

3234 has four significant digits.
104 has three significant digits.
140.00 has five significant digits.
0.00012 has two significant digits.
0.000120 has three significant digits.
0.0003 has one significant digit.
1.0003 has five significant digits.

Right
triangle
trigonometry

To solve right triangles, it is helpful to view the definitions of the trigonometric functions in a different light. We place right triangle ABC on a coordinate system so it is the reference triangle for the acute angle A. See Figure 2-33. Let a be the length of BC, the side opposite angle A. Let b be the length of AC, the side adjacent to angle A. Finally, let the hypotenuse have length c. The six trigono-metric functions of the acute angle A can be defined as ratios involving sides of the right triangle ABC.

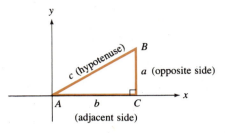

Figure 2-33

(2.16)

> **Definition.**   If angle A is an acute angle in right triangle ABC, then
>
> $$\sin A = \frac{\text{opposite side}}{\text{hypotenuse}} \qquad \csc A = \frac{\text{hypotenuse}}{\text{opposite side}}$$
>
> $$\cos A = \frac{\text{adjacent side}}{\text{hypotenuse}} \qquad \sec A = \frac{\text{hypotenuse}}{\text{adjacent side}}$$
>
> $$\tan A = \frac{\text{opposite side}}{\text{adjacent side}} \qquad \cot A = \frac{\text{adjacent side}}{\text{opposite side}}$$

Remember that the terminology of adjacent and opposite sides is relative to acute angles of right triangles only. Note that a side adjacent to one acute angle of a right triangle is opposite the other acute angle. Also note that the above definitions are simply a result of the earlier definitions of the six basic trigonometric functions.

In many trigonometry problems, various angles are often given by schemes that may be unfamiliar to you. Some terminology may require explanation.

Angle of elevation and depression

If an observer looks up at an object (an airplane, the sun, or a cloud, perhaps), the angle that the observer's line of sight makes with the horizontal is called the **angle of elevation.** If the observer looks down to see the object, the angle made with the horizontal is called the **angle of depression.** See Figure 2-34.

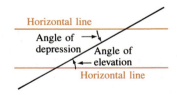

Figure 2-34

In nautical navigation and surveying, the concept of bearing is used. See Figure 2-35. The **bearing** of point A from point O (the observer) is the acute angle measured from the north-south line to the line segment OA. This bearing is denoted as N 30° E and is read as "north 30° east" (or "30° east of north"). The bearing of point B from O is N 75° W, the bearing of point C from O is S 20° W, and the bearing of point D from O is S 80° E.

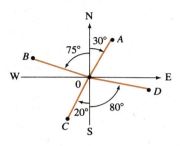

Figure 2-35

Example 3   From a location 23.0 meters from the base of a flagpole, the angle of elevation to the top of the flagpole is 37.5°. How tall is the flagpole?

Solution   Consider the right triangle in Figure 2-36. The flagpole is the side opposite the 37.5° angle, and 23.0 m is the length of the adjacent side. Because the tangent of the angle involves the opposite and the adjacent sides, you have

$$\tan 37.5° = \frac{\text{opposite side}}{\text{adjacent side}}$$

$$= \frac{h}{23.0}$$

Solve for h.

$$h = 23.0 \tan 37.5°$$
$$\approx (23.0)(0.7673)$$
$$\approx 17.6479$$
$$\approx 17.6$$

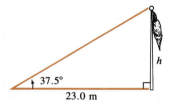

Figure 2-36

The flagpole is 17.6 m tall. Note that the answer has been properly rounded to three significant digits, which is consistent with the accuracy of the data. ■

Example 4   A circus tightrope walker ascends from the ground to a platform 75 ft above the arena by walking a taut cable that is 92 ft long. At what angle is the cable from the horizontal?

Solution   Figure 2-37 aids in determining the various parts of a triangle. You are given the lengths of the hypotenuse and a side. You must find angle θ. First, use the sine ratio to find the value of $\sin \theta$.

$$\sin \theta = \frac{\text{opposite side}}{\text{hypotenuse}}$$

$$= \frac{75}{92}$$

$$\approx 0.8152$$

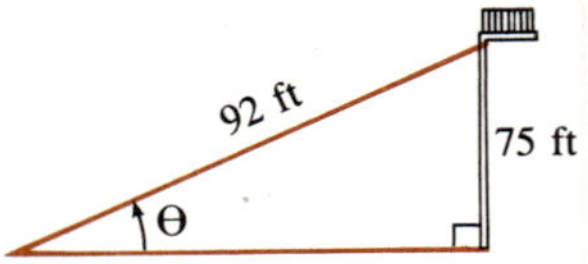

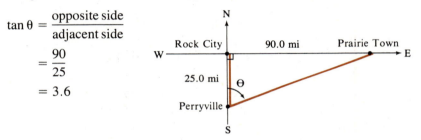

Figure 2-37

Now if you look up .8152 in the body of Table A in Appendix IV, or enter .8152 and press ⃞INV⃞ ⃞SIN⃞ on a calculator, you will find that $\theta \approx 55°$. This result is accurate to the nearest degree and is consistent with the accuracy of the given sides.

Example 5   Perryville is 25.0 mi due south of Rock City, and Prairie Town is 90.0 mi due east of Rock City. What is the bearing of Prairie Town from Perryville?

Solution   See Figure 2-38. To find the bearing, find angle θ by using the tangent ratio.

$$\tan \theta = \frac{\text{opposite side}}{\text{adjacent side}}$$

$$= \frac{90}{25}$$

$$= 3.6$$

Figure 2-38

The angle with tangent of 3.6 is 74.5° (from either a calculator or the tables). Because the distances are accurate to three significant digits, the angle is given to the nearest tenth of a degree. The bearing of Prairie Town from Perryville is N 74.5° E. Incidentally, the bearing of Perryville from Prairie Town is S 74.5° W.

Example 6   Two forest fire lookouts are on a north-south line, 17.2 mi apart. The bearing of a fire from lookout point A is S 27.0° W and the bearing of the fire from point B is N 63.0° W. How far from B is the fire?

Solution   See Figure 2-39. Because the sum of the three angles of any triangle must be 180°, the angle at point F is 90°, so triangle AFB is a right triangle with hypotenuse AB. Find the distance that B is from the fire by using the cosine ratio.

$$\cos 63.0° = \frac{\text{adjacent side}}{\text{hypotenuse}}$$

$$= \frac{a}{17.2}$$

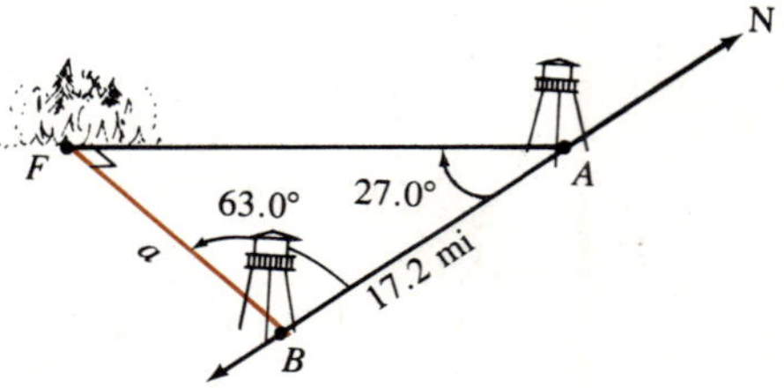

Figure 2-39

Solve for *a*.

$$a = 17.2\cos 63.0°$$
$$\approx 7.81$$

The fire is 7.81 mi from lookout point *B*. If the angle at point *F* had not been 90°, this technique would not have worked. The problem would have been much more difficult.   ■

Example 7   A television tower stands on top of a building. From a point 75.3 ft from the base of the building, the angles of elevation to the top and the base of the tower are 60.1° and 47.4°, respectively. How tall is the tower?

Solution   See Figure 2-40. You can find the total height *H* of the structure by using the tangent ratio.

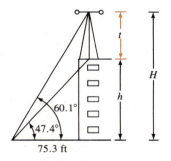

Figure 2-40

$$\tan 60.1° = \frac{H}{75.3}$$
$$75.3(\tan 60.1°) = H$$
$$75.3(1.7391) \approx H$$
$$130.95 \approx H$$

The total height of the structure is approximately 130.95 ft.
    You can find the height *h* of the building by using the tangent ratio.

$$\tan 47.4° = \frac{h}{75.3}$$
$$75.3(\tan 47.4°) = h$$
$$75.3(1.0875) \approx h$$
$$81.89 \approx h$$

The height of the building is approximately 81.89 ft.
    The height *t* of the tower is the difference between the total height of the structure and the height of the building. Thus, you have

$$t = H - h$$
$$\approx 130.95 - 81.89$$
$$\approx 49.06$$

The height of the tower is approximately 49.1 ft. Note that the final answer is rounded off to the proper degree of accuracy. ■

Exercise 2.5

In Exercises 1–4, solve each triangle by finding the unknown sides and angles.

✓1.

2.

3.

4.

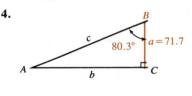

5. If the angle of elevation to the top of a flagpole from a point 40 ft from its base is 20°, find the height of the flagpole.

6. A person on the edge of a cliff looks down at a boat on a lake. The angle of depression of the person's line of sight is 10.0°, and the line-of-sight distance from the person to the boat is 555 ft. How high is the cliff?

7. A car drives up a long hill on a road that makes an angle of 7.5° with the horizontal. How far will the car travel to reach the top if the horizontal distance traveled is 715 ft?

8. On an approach to an airport a plane is descending at an angle of 3.5°. How much altitude is lost as the plane travels a horizontal distance of 21.2 mi?

9. A plane loses 2750 ft in altitude as it travels a horizontal distance of 39,300 ft. What is its angle of descent to the nearest tenth of a degree?

10. A train rises 230 ft as it travels 1 mi up a steep grade. What is its angle of ascent? (Hint: 5280 ft = 1 mi.)

11. The angle of depression from a point in the top of a tree to a point on the ground is 57°. Find the height of the tree if the line-of-sight distance from the top of the tree to the point on the ground is 34 ft.

12. An observer noted that the angle of elevation to a plane passing over a landmark was 32°. If the landmark was 1500 m from the observer, what was the altitude of the plane?

13. A ship leaves from a port of call on a bearing of S 12.7° E. How far south has the ship traveled during a trip of 327 mi?

14. The bearing of Madison, Wisconsin, from Stevens Point, Wisconsin, is S 4.1° E. The distance between the cities is 108 mi. How much farther east is Madison than Stevens Point?

15. A ship leaves port and sails 8800 km due west. It then sails 4500 km due south. To the nearest tenth of a degree, what is the ship's bearing from its port?

16. A ship is 3.3 mi from a lighthouse. It is also due north of a buoy that is 2.5 mi due east of the lighthouse. Find the bearing of the ship from the lighthouse.

17. Two lighthouses are on an east–west line. The bearing of a ship from one lighthouse is N 59° E, and the bearing from the other lighthouse is N 31° W. How far apart are the lighthouses if the ship is 5.0 mi from the first lighthouse?

18. Two lookout stations are on a north–south line. The bearing of a forest fire from one lookout is S 67° E, and the bearing of the fire from the second lookout is N 23° E. If the fire is 3.5 km from the second lookout station, how far is the fire from the other lookout station?

19. A plane flying at 18,100 ft passes directly over an observer. Thirty seconds later, the observer notes that the plane's angle of elevation is 31.0°. How fast is the plane going in miles per hour? (Hint: 5280 ft = 1 mi.)

20. A plane flying horizontally at 650 mph passes directly over a small city. One minute later the pilot notes that the angle of depression to that city is 13°. What is the plane's altitude in feet?

21. Use the information in Illustration 1 to compute the height of George Washington's face on Mount Rushmore.

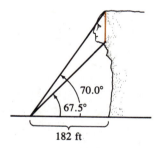

70.0°

67.5°

182 ft

Illustration 1

22. A boat is 537 m from a lighthouse and has a bearing from the lighthouse of N 33.7° W. A second boat is 212 m from the same lighthouse and has a bearing from the lighthouse of S 20.1° W. How many meters north of the second boat is the first?

23. A plane is flying at an altitude of 5120 ft. As it approaches an island, the navigator determines the angles of depression as in Illustration 2 on page 58. What is the length of the island in feet and in miles?

24. From an observation point 6500 m from a launch site, an observer watches the vertical flight of a rocket. At one instant, the angle of elevation of the rocket is 15°. How far will the rocket ascend in the time it takes the angle of elevation to increase by 57°?

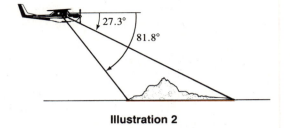

Illustration 2

25. Refer to Illustration 3 and find θ.

26. Refer to Illustration 3 and find φ.

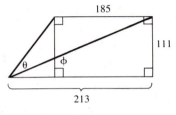

Illustration 3

2.6 MORE RIGHT TRIANGLE TRIGONOMETRY (OPTIONAL)

We must work with two triangles simultaneously to solve many right-triangle trigonometry problems.

Example 1   An observer A notices that the Space Needle of Seattle is due north, and the angle of elevation to its top is 44.4°. A second observer, B, 706 ft due east of A, notices that the bearing of the Space Needle is N 48.8° W. How tall is the Space Needle?

Solution   This problem is harder than the previous examples because two triangles must be solved. Triangle PAB lies on the ground, while triangle APQ sits on its edge with the Space Needle as one of its sides. See Figure 2-41. Not enough information is given about triangle APQ to determine the height, h, from that triangle alone. If the distance x were known, however, h could be computed. Note that

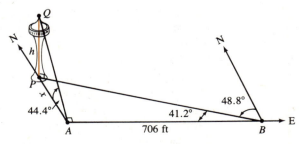

Figure 2-41

$$\tan 44.4° = \frac{h}{x}$$

or

$$h = x \tan 44.4°$$

The distance x may be found by working with the triangle on the ground. Triangle *PAB* is a right triangle with a right angle at A. Angle *PBA* is $90° - 48.8° = 41.2°$. Form the equation

$$\tan 41.2° = \frac{x}{706}$$

or

$$x = 706 \tan 41.2°$$

Putting these facts together gives

$$h = 706 \tan 41.2° \tan 44.4°$$
$$\approx 706(0.8754)(0.9793)$$
$$\approx 605.2391293$$
$$\approx 605$$

Seattle's Space Needle is approximately 605 ft tall.

Example 2   Two Coast Guard lookouts, A and B, are on an east–west line, 21.3 km apart. The bearing of a ship from lookout A is N 23.0° W, and the bearing of the ship from lookout B is N 65.0° E. How far is the ship from lookout B?

Solution   See Figure 2-42 and note that triangle *BAS* is not a right triangle. However, if the perpendicular h is drawn from point S to side BA, two right triangles are formed. If you let y represent the length of BD, then $21.3 - y$ will represent the length of DA. Thus, you have

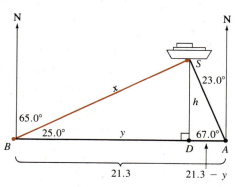

Figure 2-42

$$\tan 25.0° = \frac{h}{y} \qquad \text{or} \qquad h = y\,(\tan 25.0°)$$

and

$$\tan 67.0° = \frac{h}{21.3 - y} \quad \text{or} \quad h = (21.3 - y)\tan 67.0°$$

Setting the two values of h equal to each other and solving for y gives

$$y \tan 25.0° = (21.3 - y)\tan 67.0°$$
$$y \tan 25.0° = 21.3(\tan 67.0°) - y \tan 67.0°$$
$$y \tan 25.0° + y \tan 67.0° = 21.3 \tan 67.0°$$
$$y(\tan 25.0° + \tan 67.0°) = 21.3 \tan 67.0°$$
$$y = \frac{21.3 \tan 67.0°}{\tan 25.0° + \tan 67.0°}$$
$$\approx \frac{21.3(2.3559)}{0.4663 + 2.3559}$$
$$\approx 17.78$$

You can now use the cosine ratio to find x.

$$\cos 25.0° = \frac{17.78}{x}$$
$$x = \frac{17.78}{\cos 25.0°}$$
$$\approx \frac{17.78}{0.9063}$$
$$\approx 19.618$$

The ship is approximately 19.6 km from lookout B.

Trigonometric functions can be used to relate the various sides and angles of geometric figures.

Example 3   An isosceles triangle has equal sides of k units and a vertex angle of α degrees. Express the length b of its base in terms of k and α.

Solution   See Figure 2-43, in which CD is the perpendicular drawn from the vertex of the vertex angle to the base. Because this perpendicular bisects the vertex angle, you have

$$\sin\frac{\alpha}{2} = \frac{x}{k}$$
$$x = k \sin\frac{\alpha}{2}$$

Because the perpendicular CD also bisects the base, you have

Figure 2-43 $b = 2x$

Substitute the previously obtained value for x to get

$$b = 2k \, \sin \frac{\alpha}{2}$$

■

Example 4   A regular polygon has n equal sides, each of length a. The radius of the circumscribed circle is R. Express a as a function of n and R.

Solution   See Figure 2-44, in which point O is the center of an n-sided regular polygon with each side a units long. Because the number of degrees in one complete revolution is $360°$, and because an n-sided regular polygon has n equal central angles, angle $BOC = \frac{360°}{n}$. The perpendicular from point O to side BC bisects angle BOC and side BC. Thus, you have

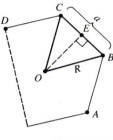

Figure 2-44

$$BE = \frac{a}{2}$$

angle $BEO = 90°$

and

$$\text{angle } BOE = \frac{1}{2}\left(\frac{360°}{n}\right) = \frac{180°}{n}$$

Because triangle BOE is a right triangle, you can write

$$\sin(\text{angle } BOE) = \frac{BE}{OB}$$

After substituting values for the above quantities, you have

$$\sin\left(\frac{180°}{n}\right) = \frac{\frac{a}{2}}{R}$$

or

$$a = 2R \sin\left(\frac{180°}{n}\right)$$

■

Exercise 2.6

1. Compute the height of the Sears Tower using the information given in Illustration 1.

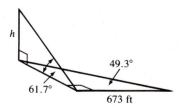

h

$49.3°$

$61.7°$

673 ft

Illustration 1

2. Compute the height of the Empire State Building using the information given in Illustration 2.

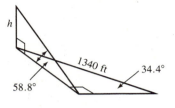

Illustration 2

3. Bill and Paula, standing on the same side of and in line with the Washington Monument, are looking at its top. The angle of elevation from Bill's position is 34.1°, and the angle of elevation from Paula's position is 60.0°. If Bill and Paula stand on level ground and are 500 ft apart, how tall is the monument?

4. Use the information given in Illustration 3 to compute the height of the figure part of the Statue of Liberty.

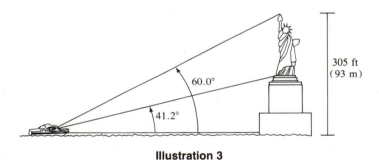

Illustration 3

5. Use the information given in Illustration 4 to compute the height of the Gateway Arch in St. Louis.

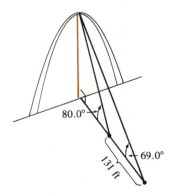

Illustration 4

6. The bearing of point C from point A is N 25.0° E, and the bearing of point C from point B is N 40.0° W. If A and B are on an east–west line and 200 km apart, how far is B from C?

7. Find the height of the triangle shown in Illustration 5.

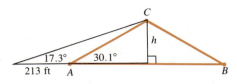

Illustration 5

8. Find the length of side *AC* shown in Illustration 5.

9. Gail is standing 152 ft due east of a television tower. She then travels along a bearing of S 60.0° W. How tall is the tower if the angle of elevation to its top from her closest position to the tower is 58.2°?

10. Jeffrey stands due south of the Eiffel Tower and notes that the angle of elevation to its top is 60.0°. His brother Grant, standing 298 ft due east of Jeff, notes an angle of elevation of 56.9°. How tall is the tower? (Hint: Consider a triangle on the ground and use the Pythagorean Theorem.)

11. Two tangents are drawn from a point *P* to a circle of radius *r* as in Illustration 6. The angle between the tangents is θ. How long is each tangent?

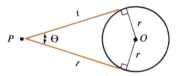

Illustration 6

12. A regular polygon has *n* equal sides, each of length *a*. The radius of the inscribed circle is *r*. Express *a* as a function of *n* and *r*.

13. An isosceles triangle has a base angle of α and a side of *k* cm. Express the length *b* of its base in terms of α and *k*.

14. At noon a ship left port on a bearing of N α E and steamed *k* km. At the same time another ship left port on a bearing of N α W and steamed *k* km. Express the distance *d* between the two ships in terms of α and *k*.

15. Point *A* is *k* meters from a building. From point *A* the angle of elevation to the top of the building is α, and the angle of depression to the base of the building is β. Express the height *H* of the building in terms of *k*, α, and β.

16. A ladder *k* ft long reaches a height of *H* ft on the side of a building, and the ladder makes an angle of α with the horizontal. The ladder slips so that it reaches a height of only *h* ft on the side of the building. It then makes an angle of β with the horizontal. Express the distance *d* that the ladder has come down the building in terms of *k*, α, and β.

17. From an observation point *D* meters from a launch site, an observer watches the vertical flight of a rocket. At one instant, the angle of elevation of the rocket is θ, and at a later instant, ϕ. How far has the rocket flown during that time?

18. Two helicopters are hovering at an altitude of 5000 ft. The pilot of one helicopter observes a crate being dropped from the other plane. At one instant the pilot measures the angle of depression of the crate to be α. When the crate hits the ground, the pilot measures the angle of depression to be β. How far has the crate fallen in that time?

19. Two tall buildings are separated by a distance of d meters. From the top of the shorter building, the angle of elevation to the top of the taller building is α, and the angle of depression to the base of the taller building is β. Express the height a of the shorter building in terms of d and β.

20. Express the height b of the taller building described in Exercise 19 in terms of d, α, and a.

2.7   INTRODUCTION TO VECTORS

In many applications of mathematics, there is a need for quantities that have both magnitude and direction. The enrollment in an art class, for example, is a quantity that possesses magnitude only, but the flight of an airplane must be described by both a speed and a direction. "Thirty-seven students" adequately describes the enrollment, whereas "350 miles per hour north-east" describes the flight.

Quantities with both magnitude and direction are called **vector quantities** and are represented by mathematical entities called **vectors.**

(2.17)

> **Definition.**   A **vector** is a directed line segment. The direction of the vector is indicated by the angle it makes with some convenient reference line. The **magnitude** of the vector is the length of the line segment. If a vector is denoted by **v,** then the magnitude of the vector is denoted by $|\mathbf{v}|$.

A vector quantity has both magnitude and direction but not location; any two directed line segments with the same length and the same direction are regarded as **equal vectors.** It is usual to position a vector, when drawing diagrams, in the most convenient location. A force of 30 lb, for example, exerted in a north-westerly direction may be represented by the directed line segment in Figure 2-45. The length of the arrow is 30 units and because it is the terminal side of a 135° angle, it "points" northwest.

A 20-mph wind blowing from the east may be represented by the vector in Figure 2-46.

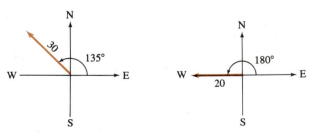

Figure 2-45   **Figure 2-46**

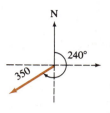

Figure 2-47

An airplane flying 350 mph on a **heading** (an intended direction of travel measured clockwise from the north line) of 240° can be represented by the vector in Figure 2-47.

Vector quantities, such as forces, can be added, but the process must take into account both their magnitudes and directions. Two forces of 40 lb, for example, exerted on the same object might not combine to be an 80-lb force. If they acted in opposite directions, the net force would be zero.

The sum of two vectors is another vector, called the **resultant.** Vector quantities may be added by using a parallelogram law.

Adding two vectors

If two vectors originating at a common point are adjacent sides of a parallelogram, their resultant vector (or vector sum) is that parallelogram's diagonal drawn from the common point. In Figure 2-48, the sum of vectors **AB** and **AD** is the vector **AC,** given by the diagonal of the parallelogram *ABCD*. Note that vector **DC** has the same magnitude and direction as vector **AB.**

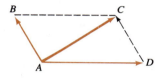

Figure 2-48

Example 1   A boat capable of 8.0 mph in still water attempts to go directly across a river flowing at 3.0 mph. By what angle is the boat pushed off its intended path? What is the effective speed of the boat?

Solution   The two given velocities can be represented by vectors as in Figure 2-49. The direction the boat travels, called its **course,** is represented by a vector sum, the diagonal of the rectangle. Angle θ specifies the direction in which the boat is forced to travel, and the length of *OP* specifies the effective speed of the boat. Because *AP* is also 3.0 units, $\tan \theta = \frac{3}{8}$. From this relationship, you can calculate θ. You then find that the river pushes the boat approximately 21° off its intended path. Use the Pythagorean Theorem to find the length of *OP* and, hence, the speed of the boat.

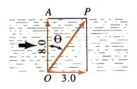

Figure 2-49

$$OP = \sqrt{8^2 + 3^2}$$
$$= \sqrt{64 + 9}$$
$$= \sqrt{73}$$
$$\approx 8.5$$

The effective speed of the boat is approximately 8.5 mph. ■

Example 2   An airplane capable of a speed of 270 mph in still air sets a heading of 75°. A very strong wind is blowing in the direction of 165° and forces the plane onto a course that is due east. What is the velocity of the wind, and what is the ground speed (the speed relative to the ground) of the plane?

Solution   The velocities involved are represented in Figure 2-50. Vector **w** represents the wind velocity and vector **v** represents the resultant velocity, or **ground speed,** of the plane. Because $165° - 75° = 90°$, the vector parallelogram is a rectangle with angle α equal to $90°$. Because each triangle formed by the diagonal is a right triangle, you can write

$$\tan 15° = \frac{|\mathbf{w}|}{270}$$

Multiplying both sides by 270 and simplifying gives

$$|\mathbf{w}| = 270 \tan 15°$$
$$|\mathbf{w}| \approx 72$$

Also, you have

$$\cos 15° = \frac{270}{|\mathbf{v}|}$$

$|\mathbf{v}| \cos 15° = 270$		Multiply both sides by $|\mathbf{v}|$.
$|\mathbf{v}| = \dfrac{270}{\cos 15°}$		Divide both sides by $\cos 15°$.
$|\mathbf{v}| \approx 280$		Simplify.

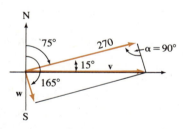

Figure 2-50

The wind speed is approximately 72 mph, and the ground speed of the plane (the resultant of the plane's air speed and the wind speed) is approximately 280 mph.   ■

Instead of two vectors adding to form a single resultant, it is possible to separate a single vector into several components. Suppose that a car weighing 3000 lb is parked on a hill. It is pulled directly downward by gravity with a force of 3000 lb. Part of this force appears as a tendency to roll the car down the hill, and another part presses the car against the road. Just how the 3000 lb is apportioned depends on the angle of the hill. If there were no hill, there would be no tendency to roll; if the hill were very steep, only a small force would hold the car to the road. The weight of the car is said to be **resolved** into two components—one directed down the hill, and the other directed into the hill.

These vectors obey the parallelogram law. See Figure 2-51. Note that the angle of the hill, α, is also the angle between two of the vectors because both of these angles are complementary to angle β.

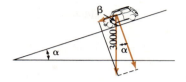

Figure 2-51

Example 3 A 3000-lb car sits on a 23.0° incline. What force is required to prevent the car from rolling down the hill? With what force is it held to the roadway?

Solution You must find the magnitudes of vectors **t** and **n** in Figure 2-52. Because the figure *OACB* is a rectangle, the opposite sides are equal and angle $B = 90°$. Hence, you have

$$\sin 23.0° = \frac{|\mathbf{t}|}{3000}$$
$$|\mathbf{t}| = 3000 \sin 23.0°$$
$$|\mathbf{t}| \approx 1170$$

Also, you have

$$\cos 23.0° = \frac{|\mathbf{n}|}{3000}$$
$$|\mathbf{n}| = 3000 \cos 23.0°$$
$$|\mathbf{n}| \approx 2760$$

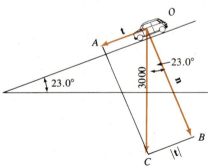

Figure 2-52

A force of approximately 1170 lb is required to prevent rolling, and a force of approximately 2760 lb keeps the car on the hill. Note that, because the angles are to the nearest tenth of a degree, only three-place accuracy is permitted for the sides. In this example, the lengths of the vectors have only three significant digits.

Exercise 2.7

1. A boat capable of a speed of 6 mph in still water attempts to go directly across a river. As the boat crosses the river, it drifts 30° from its intended path. How strong is the current? What is the effective speed of the boat?

2. A boat capable of a speed of 11 mph in still water attempts to go directly across a river with a current of 5.6 mph. By what angle is the boat pushed off its intended path? What is the effective speed of the boat?

3. Laura can row a boat $\frac{1}{2}$ mph in still water. She attempts to row straight across a river that has a current of 1 mph. If she must row for 2 hours to cross the river, by what angle is the current pushing her off her intended path? Give your answer to the nearest degree.

4. A boat attempts to go directly across a river with a current of 3.7 mph. The current causes the boat to drift 23° from its intended path. How far will the boat travel if the trip takes 10 minutes?

5. A plane has a heading of 260.0° and is flying at 357 mph. If a southerly wind causes the plane's course (the direction it is actually going) to be due west, find the ground speed of the plane (its speed relative to the ground).

6. A plane has an airspeed of 411 mph and a heading of 90.0°. A wind from the north is blowing at 31.0 mph. By how many degrees is the plane blown off its heading? Find the ground speed of the plane (its speed relative to the ground).

7. A plane leaves an airport with a heading of 45.0° and an airspeed of 201 mph. At the same time, another plane leaves the same airport with a heading of 135.0° and an airspeed of 305 mph. At the end of 2 hours, what is the bearing of the first plane from the second?

8. A rifle with a muzzle velocity of 4100 feet per second is fixed at an angle of elevation of 32°. What is the horizontal component of the bullet's velocity?

9. A 317-lb weight is hanging from the ceiling on a long rope. A man pushes horizontally against the weight to rotate the rope through an angle of 11.2°. What resultant force is being counteracted by the rope?

10. A plane leaves an airport with a heading of 170°. At the same time, a second plane leaves the same airport with a heading of 260°. One hour later, the first plane is 1300 miles directly southeast of the second plane. How fast is the first plane going?

11. What force is required to keep a 2210-lb car from rolling down a ramp that makes a 10.0° angle with the horizontal?

12. A force of 25 lb is necessary to hold a barrel in place on a ramp that makes an angle of 7° with the horizontal. How much does the barrel weigh?

13. A vehicle presses against a roadway with a force of 1100 lb. How much does the vehicle weigh if the roadway has an 18.0° grade?

14. A board will break if it is subjected to a force greater than 350 lb. Will the board hold a 450-lb piano supported by a single dolly as it slides up the board and into a truck? Assume that the board makes an angle of 35.0° with the horizontal.

15. A garden tractor weighing 351 lb is being driven up a ramp onto a trailer. If the tractor presses against the ramp with a 341-lb force, what angle does the ramp make with the horizontal?

16. If a force of 21.3 lb is necessary to keep a 50.1-lb barrel from rolling down an inclined plane, what angle does the inclined plane make with the horizontal?

17. A 201-lb force is directed due east. What force, directed due north, is needed to produce a resultant force of 301 lb? What angle is formed by the vectors representing the 201-lb and the 301-lb force?

18. A 312-lb force is directed due west. What force, directed exactly southeast, would cause the resultant force to be directed due south?

19. It requires 35 lb of force to keep two children from sliding down the chute shown in Illustration 1. If one child weighs 110 lb and the other 51 lb, what angle does the chute make with the horizontal?

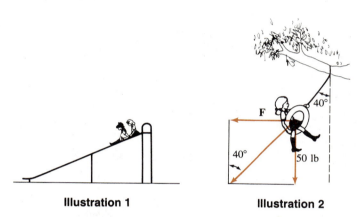

Illustration 1   **Illustration 2**

20. A 50-lb girl is playing in a tire swing hanging from the limb of a large tree. See Illustration 2. To get her started, a friend pulls the swing backward until it makes an angle of 40° with the vertical. What horizontal force F is required to hold the swing in this position?

21. A weight of 160 lb is supported by a string as in Illustration 3. Find the vertical and horizontal components of force F_1. (Hint: The vertical component of F_1 supports half the weight.)

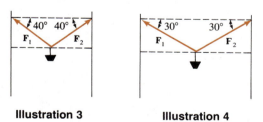

Illustration 3   **Illustration 4**

22. A weight of 220 lb is supported by a cable as shown in Illustration 4. Find the vertical and horizontal components of force F_1.

In Exercises 23–24, you must draw some lines to create right triangles.

23. A plane leaves an airport at 12:00 noon with a heading of 60.0° and an airspeed of 451 mph. One hour later, another plane leaves the same airport with a heading of 290.0° and an airspeed of 611 mph. What is the bearing of the first plane from the second at 2:00 P.M.? Assume no wind.

24. A plane leaves an airport at 3:00 P.M. with a heading of 70.0° and an airspeed of 512 mph. Two hours later, another plane leaves the same airport with a heading of 100.0° and an airspeed of 621 mph. How far apart are the two planes at 6:00 P.M.? Assume no wind.

In Exercises 25–26, you must recall some theorems about rhombuses.

25. Two forces of 30 lb make an angle of 30° with each other. What is the magnitude of their resultant?

26. Two equal forces of *f* lb each make an angle of θ with each other. What is the magnitude of their resultant?

2.8   MORE APPLICATIONS OF THE TRIGONOMETRIC FUNCTIONS

In this chapter, you have learned how to use trigonometry to calculate forces, directions, velocities, and distances. These are important applications of trigonometry to navigation, mechanics, and surveying, but they represent only a small fraction of the possible uses of trigonometry. Here are four more.

1. Slope of a straight line

If a line passes through points $P(x_1, y_1)$ and $Q(x_2, y_2)$, the **slope** of the line is defined as

$$\text{slope of } PQ = \frac{\text{rise}}{\text{run}} = \frac{y_2 - y_1}{x_2 - x_1}$$

In the right triangle *PQR* in Figure 2-53, $y_2 - y_1$ is the length of the side opposite angle α, and $x_2 - x_1$ is the length of the side adjacent to angle α. The slope, therefore, is the ratio of the opposite side to the adjacent side, and

slope of *PQ* = tan α

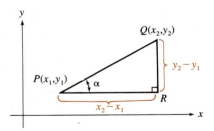

Figure 2-53

Angle α, measured counterclockwise from the horizontal line segment, is called the **angle of inclination** of line *PQ*. When α = 90°, tan α is undefined. This indicates that a vertical line has no slope.

2. Optics

Light travels more rapidly in air than in water or glass. When light passes from one substance into another in which it slows down, its path is bent. This is called **refraction** and is the principle behind camera lenses, eye glasses, and the sparkle of a diamond.

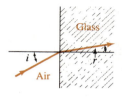

Figure 2-54

Suppose, in Figure 2-54, that a beam of light is passing from air into glass. Angle i is called the **angle of incidence,** and angle r the **angle of refraction.** The two are related by a simple formula named after the Dutch astronomer Willebrord Snell (1591–1629):

$$\frac{\sin i}{\sin r} = n$$

where n is a constant, called the **refractive index,** dependent upon the optical properties of the glass. If, for example, light striking a piece of glass at a $10°$ angle is refracted to an angle of $6°$, the refractive index is

$$
\begin{aligned}
n &= \frac{\sin i}{\sin r} \\
&= \frac{\sin 10°}{\sin 6°} \\
&\approx \frac{0.1736}{0.1045} \\
&\approx 1.7
\end{aligned}
$$

The refractive index of glass ranges from 1.5 to 1.9, whereas the refractive index of diamond is approximately 2.4. This is the reason why diamonds sparkle more than rhinestones.

3. Electrical engineering

Certain components of an electronic circuit behave in ways that are conveniently analyzed by using trigonometric functions. For example, a resistor and an inductor connected as a series network will oppose the flow of an alternating current. A measure of this opposition is called the network's **impedance.** The impedance, z, is determined by the *reactance, X,* of the inductor, and the *resistance, R,* of the resistor. The three quantities z, R, and X represent the sides of the right triangle in Figure 2-55. These quantities are related to each other by the Pythagorean Theorem as follows.

$$z^2 = X^2 + R^2$$

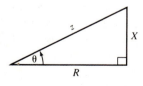

Figure 2-55

The angle θ shown in the figure is called the **phase angle,** and from the figure we can see that

$$\tan \theta = \frac{X}{R}$$

Example 1   An alternating current generator G and an impedance z, consisting of a resistance R and an inductive reactance X, are connected as in Figure 2-56. If $R = 300$ ohms and $X = 400$ ohms, find the impedance z and the phase angle θ.

Solution   The circuit and the impedance triangle shown in the figure will help you determine both z and θ. By the Pythagorean Theorem, you have

$$z^2 = X^2 + R^2$$

By substituting **300** for R and **400** for X and simplifying, you get

$$z = \sqrt{\mathbf{300}^2 + \mathbf{400}^2}$$
$$= \sqrt{250{,}000}$$
$$= 500$$

Thus, the impedance is 500 ohms.
   To find the phase angle θ, use the fact that

$$\tan \theta = \frac{X}{R}$$
$$= \frac{400}{300}$$
$$\approx 1.3333$$

Thus, angle θ is approximately $53.1°$.

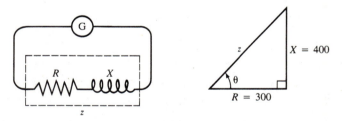

Figure 2-56

4. The seasons

In answer to the question "Why is summer hotter than winter?" many people respond "Because the earth is closer to the sun in the summer." However, this is not true. In fact, the earth is closest to the sun in December. Summers and winters are caused by the angle at which the sun's energy hits the earth. This varies as the earth moves along its orbit because the axis of the earth is tilted by approximately $23\frac{1}{2}°$. In the summer, sunlight approaches the earth at angles close to $90°$; but in the winter, the sunlight is not nearly as direct.
   In the northern hemisphere at noon on June 21 (the summer solstice), the sun is as far north as it will ever get; this is the first day of summer. Six months later, at noon on December 22, the sun appears $47°$ farther south in the sky; this is the beginning of winter. The heat energy from the sun that reaches the earth

is proportional to the cosine of the angle between the sun's rays and the vertical. Because this angle is greater during the winter months, less energy reaches us and the days are cold.

Example 2   At noon on the first day of summer in Havana, Cuba, the sun is directly overhead. How much less energy reaches Havana at noon on the first day of winter? See Figure 2-57.

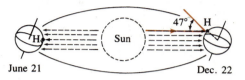

Figure 2-57

Solution   The energy E that reaches the earth is proportional to the cosine of the angle of incidence. In June, $E = k \cos 0° = k$, while in December, $E = k \cos 47° \approx 0.68k$. Havana receives about $1 - 0.68 = 0.32$, or 32% less energy on the first day of winter than on the first day of summer.   ■

Exercise 2.8

1. A line passes through the points $(1, 3)$ and $(3, 5)$. What is its angle of inclination?
2. A line passes through the points $(0, -1)$ and $(-2\sqrt{3}, 1)$. What is its angle of inclination?
3. A beam of light enters a block of ice at the angle of incidence of $8.0°$ and is refracted to $6.1°$. What is the index of refraction of the ice?
4. A clear liquid, tetrachloroethylene (used in dry cleaning), has an index of refraction equal to that of a certain brand of glass. Would you be able to see a piece of that glass submerged in the liquid? How could you distinguish real diamonds from a handful of fakes?
5. An impedance consists of a resistance of 120 ohms and an inductive reactance of 50 ohms. Find the impedance.
6. Use the information in Exercise 5 to find the phase angle.
7. An impedance consists of a resistance of 200 ohms in series with an unknown inductance. If the impedance measures 290 ohms, what is the phase angle?
8. Use the information in Exercise 7 to find the inductive reactance.
9. The impedance of a series resister–inductor network is measured as 240 ohms, and the phase angle as $60°$. Find the resistance R.
10. Use the information in Exercise 10 to find the inductive reactance X.

In Exercises 11–12, assume that the amount of the sun's energy reaching the earth is directly proportional to the cosine of the angle of incidence.

11. Find the amount of the sun's energy that reaches earth if the angle of incidence is $30°$

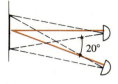

Illustration 1

12. If only 90% of the sun's energy reaches the earth, what is the angle of incidence?

13. A photographer aims his lights directly at a painting he wishes to photograph. How much less light will hit the painting if he lowers his lights by 20°? See Illustration 1. Assume that the light remains the same distance from the painting.

CHAPTER SUMMARY

Key Words

angle   (2.1)
angle of depression   (2.5)
angle of elevation   (2.5)
angle of incidence   (2.8)
angle of refraction   (2.8)
bearing   (2.5)
cofunctions   (2.4)
complementary angles   (2.4)
cosecant θ   (2.1)
cosine θ   (2.1)
cotangent θ   (2.1)
coterminal angles   (2.1)
equal vectors   (2.7)
even functions   (2.2)
ground speed   (2.7)
heading   (2.7)
impedance   (2.8)
initial side of a
   trigonometric angle   (2.1)
odd functions   (2.2)
phase angle   (2.8)

quadrant   (2.1)
quadrantal angle   (2.1)
reciprocals   (2.2)
reference angle   (2.3)
reference triangle   (2.3)
refraction   (2.8)
refractive index   (2.8)
resultant   (2.7)
secant θ   (2.1)
significant digits   (2.5)
sine θ   (2.1)
slope of a line   (2.8)
standard position of a
   trigonometric angle   (2.1)
tangent θ   (2.1)
terminal side of a
   trigonometric angle   (2.1)
trigonometric angle   (2.1)
unit circle   (2.4)
vectors   (2.7)
vertex of an angle   (2.1)

Key Ideas

(2.1)   If $P(x, y)$ is any point on the terminal side of an angle θ in standard position and at a distance r from the origin, then

$$\sin \theta = \frac{y}{r} \qquad \cos \theta = \frac{x}{r} \qquad \tan \theta = \frac{y}{x}$$

$$\csc \theta = \frac{r}{y} \qquad \sec \theta = \frac{r}{x} \qquad \cot \theta = \frac{x}{y}$$

(2.2)   The eight fundamental relationships.

$$\sin \theta = \frac{1}{\csc \theta} \qquad\qquad \cos \theta = \frac{1}{\sec \theta}$$

$$\tan \theta = \frac{1}{\cot \theta} \qquad\qquad \tan \theta = \frac{\sin \theta}{\cos \theta}$$

$$\cot \theta = \frac{\cos \theta}{\sin \theta} \qquad\qquad \sin^2 \theta + \cos^2 \theta = 1$$

$$\tan^2 \theta + 1 = \sec^2 \theta \qquad\qquad \cot^2 \theta + 1 = \csc^2 \theta$$

Three other important relationships.

$$\sin(-\theta) = -\sin \theta$$
$$\cos(-\theta) = \cos \theta$$
$$\tan(-\theta) = -\tan \theta$$

(2.3)   The six trigonometric functions of any angle θ are equal to those of the reference angle of θ, except possibly for sign.

	0°	30°	45°	60°	90°	180°	270°
sin θ	0	$\dfrac{1}{2}$	$\dfrac{\sqrt{2}}{2}$	$\dfrac{\sqrt{3}}{2}$	1	0	-1
cos θ	1	$\dfrac{\sqrt{3}}{2}$	$\dfrac{\sqrt{2}}{2}$	$\dfrac{1}{2}$	0	-1	0
tan θ	0	$\dfrac{\sqrt{3}}{3}$	1	$\sqrt{3}$	—	0	—

(2.4)   If θ is any acute angle, then any trigonometric function of θ is equal to the cofunction of the complement of θ.

(2.5)   When solving right triangles, your answers can be only as accurate as the least accurate of the given data.

If θ is an acute angle in a right triangle, then

$$\sin \theta = \frac{\text{opposite side}}{\text{hypotenuse}} \qquad\qquad \csc \theta = \frac{\text{hypotenuse}}{\text{opposite side}}$$

$$\cos \theta = \frac{\text{adjacent side}}{\text{hypotenuse}} \qquad\qquad \sec \theta = \frac{\text{hypotenuse}}{\text{adjacent side}}$$

$$\tan \theta = \frac{\text{opposite side}}{\text{adjacent side}} \qquad\qquad \cot \theta = \frac{\text{adjacent side}}{\text{opposite side}}$$

(2.6)   Many times, auxiliary lines can be drawn to form right triangles. These right
(Optional)   triangles can then be solved by using the methods of Section 2.5.

(2.7)   Vectors can be used to describe quantities that have both magnitude and direction.

Vector quantities are added by using a parallelogram law.

REVIEW EXERCISES

In Review Exercises 1–4, tell whether the given angle is in standard position.

1.

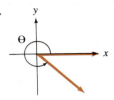

2.

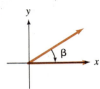

3.

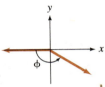

4.

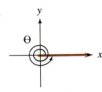

5. Are angles of 360° and 410° coterminal?

6. Are angles of 190° and 820° coterminal?

In Review Exercises 7–10, find the values of the remaining trigonometric functions of angle θ, which is in standard position.

7. $\sin \theta = \dfrac{-7}{10}$; θ in QIII

8. $\tan \theta = \dfrac{7}{9}$; θ not in QI

9. $\cos \theta = \dfrac{-7}{10}$; θ in QII

10. $\cot \theta = \dfrac{-9}{8}$; θ in QIV

11. Use the eight fundamental relationships to show that

$$\frac{1}{\sec \theta} = \sin \theta \cot \theta$$

12. Use the eight fundamental relationships to show that $\cos \theta \csc \theta = \cot \theta$.

In Review Exercises 13–16, evaluate each trigonometric expression. **Do not use a calculator.**

13. $\sin 45° \cos 30°$

14. $\cos 120° \tan 135°$

15. $\tan^2 225° \cos^2 30° \sin^2 300°$

16. $\sec 30° \csc 30° + \sec 330° \csc 330°$

In Review Exercises 17–20, find the value of the sine, cosine, and tangent of the given angle. **Do not use a calculator.**

17. 930°

18. 1380°

19. −300°

20. −585°

In Review Exercises 21–24, draw the unit circle and use it to estimate the sine, cosine, and tangent ratios of the given angles.

21. 15°

22. 160°

23. 265°

24. 340°

In Review Exercises 25–30, use Table A in Appendix IV to evaluate the sine, cosine, and tangent of the given angle. Check your work with a calculator.

25. 15°   **26.** 160°

27. 265°   **28.** 340°

29. − 160°   **30.** − 340°

In Review Exercises 31–36, use Table A in Appendix IV to find angle α ($0° \leq \alpha < 360°$), which is in the given quadrant. Check your work with a calculator.

31. QII; sin α = 0.8746   **32.** QIII; tan α = 0.6009

33. QIV; cos α = 0.7314   **34.** QI; sec α = 1.871

35. QII; cot α = −0.1763   **36.** QIV; csc α = −1.046

37. From a location 32.1 m from the base of a flagpole, the angle of elevation to its top is α. Find α if the flagpole is 10.0 m tall.

38. The angle of depression from a window in a building to a point on the ground is 17.7°. If the point on the ground is 187 ft from the base of the building, how high is the observer?

39. Owatonna, Minnesota, is approximately 55 mi due south of Minneapolis, and the bearing of Winona, Minnesota, from Minneapolis is about S 45° E. How far is Winona from Owatonna if Owatonna is due west of Winona?

40. Assume that the bearing of South Bend, Indiana, from Fort Wayne, Indiana, is N 48° W and that the distance between the cities is about 71 mi. Further assume that South Bend is due north of Indianapolis and that the bearing of Fort Wayne from Indianapolis is N 21° E. How far is Fort Wayne from Indianapolis?

41. A barrel weighing 60 lb rests on a ramp that makes an angle of 10° with the horizontal. How much force is necessary to keep the barrel from rolling down the ramp?

42. A 2500-lb car rests on a hill. A force of 500 lb is required to keep the car from rolling down the hill. How steep is the grade?

CHAPTER THREE

RADIAN MEASURE AND THE CIRCULAR FUNCTIONS

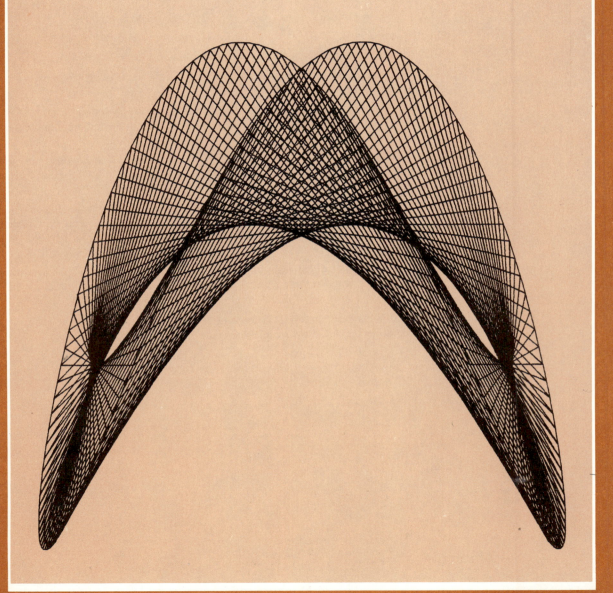

In geometry, angles are measured in degrees. In this chapter, we will introduce and use a different unit of angle measurement: the radian.

3.1 RADIAN MEASURE

People of ancient times noticed that the annual progression of the seasons appeared to repeat in 360-day cycles. Accordingly, they divided their calendar of the "year's circle" into 360 days. It was reasonable, then, to divide the circle itself into 360 equal parts, and degree measure was born. Although calendars were revised when closer observation showed that a year contains approximately 365 days, the 360-degree circle persists to this day. While still very common, degree measure is not convenient in higher mathematics. A different unit, called a **radian,** is used to express angular measure.

Radian measure

(3.1)

> **Definition.** Consider the circle with radius r in Figure 3-1. The measure of central angle θ is **1 radian** if and only if θ intercepts an arc whose length is r.

Figure 3-1

Because the circumference of a circle is $2\pi r$ and each arc with length r that is marked off on the circumference determines an angle of 1 radian,

$$\frac{2\pi r}{r} = 2\pi$$

is the number of radians in one complete revolution. Thus, we have

(3.2)   **2π radians $= 360°$**

Dividing both sides of this equation by 2 leads to the fundamental relationship between radians and degrees.

(3.3)   **π radians $= 180°$**

Equation (3.3) is very important and should be memorized.

Example 1   Change **a.** 30°, **b.** 45°, **c.** 60°, **d.** 225°, **e.** $-315°$, and **f.** $D°$ to radians.

Solution   **a.** $180° = \pi$ radians   **b.** $180° = \pi$ radians

$\dfrac{1}{6}(180°) = \dfrac{1}{6}(\pi \text{ radians})$ $\dfrac{1}{4}(180°) = \dfrac{1}{4}(\pi \text{ radians})$

$30° = \dfrac{\pi}{6} \text{ radians}$ $45° = \dfrac{\pi}{4} \text{ radians}$

c. $180° = \pi \text{ radians}$

$\dfrac{1}{3}(180°) = \dfrac{1}{3}(\pi \text{ radians})$

$60° = \dfrac{\pi}{3} \text{ radians}$

d. $180° = \pi \text{ radians}$

$1° = \dfrac{\pi}{180} \text{ radians}$

$225(1°) = 225\left(\dfrac{\pi}{180} \text{ radians}\right)$

$225° = \dfrac{5}{4}\pi \text{ radians}$

e. $180° = \pi \text{ radians}$

$1° = \dfrac{\pi}{180} \text{ radians}$

$-315(1°) = -315\left(\dfrac{\pi}{180} \text{ radians}\right)$

$-315° = -\dfrac{7\pi}{4} \text{ radians}$

f. $180° = \pi \text{ radians}$

$1° = \dfrac{\pi}{180} \text{ radians}$

$D(1°) = D\left(\dfrac{\pi}{180} \text{ radians}\right)$

$D° = \dfrac{\pi D}{180} \text{ radians}$

To find the number of degrees in 1 radian, we proceed as follows.

$$\pi \text{ radians} = 180°$$
$$1 \text{ radian} = \dfrac{180°}{\pi}$$
$$\approx 57.3°$$

Hence, 1 radian is approximately equal to $57.3°$.

Example 2   Change to degrees:

a. $\dfrac{2}{3}\pi$ radians, **b.** $\dfrac{5}{7}\pi$ radians, **c.** -7 radians, and **d.** R radians.

Solution   **a.** $\pi \text{ radians} = 180°$

$\dfrac{2}{3}(\pi \text{ radians}) = \dfrac{2}{3}(180°)$

$\dfrac{2\pi}{3} \text{ radians} = 120°$

b. $\pi \text{ radians} = 180°$

$\dfrac{5}{7}(\pi \text{ radians}) = \dfrac{5}{7}(180°)$

$\dfrac{5\pi}{7} \text{ radians} = \dfrac{900°}{7}$

c. $\pi \text{ radians} = 180°$

$\dfrac{1}{\pi}(\pi \text{ radians}) = \dfrac{1}{\pi}(180°)$

$1 \text{ radian} = \dfrac{180°}{\pi}$

$-7(1 \text{ radian}) = -7\left(\dfrac{180°}{\pi}\right)$

$-7 \text{ radians} = -\dfrac{1260°}{\pi}$

d. $\pi \text{ radians} = 180°$

$1 \text{ radian} = \dfrac{180°}{\pi}$

$R(1 \text{ radian}) = R\left(\dfrac{180°}{\pi}\right)$

$R \text{ radians} = \dfrac{R180°}{\pi}$

The following table gives the degree measure and the corresponding radian measure of five common angles.

Degree measure	0°	30°	45°	60°	90°
Radian measure	0	$\dfrac{\pi}{6}$	$\dfrac{\pi}{4}$	$\dfrac{\pi}{3}$	$\dfrac{\pi}{2}$

The information given in this table helps us to convert angles such as 120° and 225° from degree measure to radian measure in the following way.

$$120° = 2(60°) = 2\left(\frac{\pi}{3}\right) = \frac{2\pi}{3}$$

$$225° = 5(45°) = 5\left(\frac{\pi}{4}\right) = \frac{5\pi}{4}$$

We can also use this information to change angles such as $\frac{7\pi}{4}$ and $\frac{11\pi}{6}$ from radian measure to degree measure.

$$\frac{7\pi}{4} = 7\left(\frac{\pi}{4}\right) = 7(45°) = 315°$$

$$\frac{11\pi}{6} = 11\left(\frac{\pi}{6}\right) = 11(30°) = 330°$$

Arc length   The use of radians permits easy calculation of arc length in a circle. Suppose that the central angle in Figure 3-2 is in radians. Because the arc length s is the same fractional part of the circle's circumference that the angle θ is of one complete revolution, we can set up the proportion

$$\frac{s}{2\pi r} = \frac{\theta}{2\pi}$$

Figure 3-2   and solve for s.

$$2\pi s = 2\pi r\theta$$
$$s = r\theta$$

We summarize this important result.

(3.4)

> **Formula for Arc Length.**   If a central angle θ in a circle with radius r is in radians, the length s of the intercepted arc is given by the formula
>
> $$s = r\theta$$

Note that the radian measure of a central angle θ of a circle is the ratio of its intercepted arc to the circle's radius. In a circle of radius 5 in., a central angle that intercepts an arc of 10 in. is $\frac{10 \text{ in.}}{5 \text{ in.}}$, or 2 radians. In a circle of radius 8 cm, a central angle that intercepts an arc of 12 cm is $\frac{12\ cm}{8\ cm}$, or $\frac{3}{2}$ radians. In a circle of radius r, a central angle that intercepts an arc of length s is $\frac{s}{r}$ radians. In general, if θ is a central angle in radians, we have

(3.5) $$\theta = \frac{s}{r}$$

where s is the length of the intercepted arc and r is the radius of the circle.

Example 3   If a central angle θ of circle with radius of 5 cm is 80°, find the length of the intercepted arc.

Solution   First change θ to radians.

$$180° = \pi\ \text{radians}$$

$$1° = \frac{\pi}{180}\ \text{radians}$$

$$80° = \frac{4\pi}{9}\ \text{radians}$$

Because the radius of the circle is 5 cm and $\theta = \frac{4\pi}{9}$, substitute **5** for r and $\frac{4\pi}{9}$ for θ in the formula $s = r\theta$, and simplify.

$$s = 5\left(\frac{4\pi}{9}\right)\ \text{cm} \approx 5(1.39626)\ \text{cm} \approx 6.981\ \text{cm}$$

Figure 3-3

It is easy to find the area of a **sector** of a circle if the central angle of the sector is in radians. See Figure 3-3. We set up the proportion that arc s is to the circumference as the area A of the sector is to the area of the circle.

$$\frac{s}{2\pi r} = \frac{A}{\pi r^2}$$

We solve this proportion for A.

$$2\pi r A = s\pi r^2$$

$$A = \frac{s\pi r^2}{2\pi r}$$

$$A = \frac{1}{2} rs$$

Because $s = r\theta$, we can substitute $r\theta$ for s.

$$A = \frac{1}{2} r^2 \theta$$

(3.6)

> **Formula for the Area of a Sector.**   If a sector of a circle with radius r has a central angle θ in radians, then the area of the sector is given by the formula
>
> $$A = \frac{1}{2}r^2\theta$$

Example 4   A sector of a circle has a central angle of 50.0° and an area of 605 sq cm. Find the radius of the circle.

Solution   Change the angle of 50.0° to radians.

$$180° = \pi \text{ radians}$$

$$1° = \frac{\pi}{180} \text{ radians}$$

$$50.0° = \frac{5\pi}{18} \text{ radians}$$

Because the values of A and θ are known, substitute those values into the formula $A = \frac{1}{2}r^2\theta$ and solve for r.

$$A = \frac{1}{2}r^2\theta$$

$$605 = \frac{1}{2}r^2\frac{5\pi}{18}$$

$$\frac{605(2)(18)}{5\pi} = r^2$$

$$1386.56 \approx r^2$$

$$\sqrt{1386.56} \approx r$$

$$37.2 \approx r$$

The radius of the circle is approximately 37.2 cm.

Example 5   What is the length of the arc intercepted by a central angle of $\frac{5\pi}{6}$ radians in a circle of 18-meter radius?

Solution   By the formula $s = r\theta$ and the known values $r = 18$ and $\theta = \frac{5\pi}{6}$, it follows that

$$s = 18\left(\frac{5\pi}{6}\right)$$

$$= 15\pi$$

$$\approx 47$$

The arc length is approximately 47 m.

Example 6   An automobile is traveling north as it enters a curve. After it travels 1530 ft around the curve, the road straightens and heads northwest. What is the radius of the curve?

Solution   From Figure 3-4, the change in direction of the automobile is 45° (from north to northwest). Because angle α is supplementary to both the 45° angle and to θ, angle θ must be 45°. Before using the formula $s = r\theta$, θ must be expressed in radians: $\theta = \frac{\pi}{4}$.

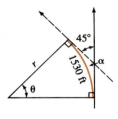

$$s = r\theta$$

$$1530 = r\left(\frac{\pi}{4}\right)$$

$$r = \frac{4(1530)}{\pi}$$

$$r \approx 1948$$

Figure 3-4

The radius of the curve is approximately 1948 ft. ■

Example 7   A pulley with a diameter of exactly 8 in. drives another pulley with diameter of exactly 6 in. If the larger pulley turns through one revolution, through what angle does the smaller pulley turn?

Solution   The radii of the two pulleys are exactly 4 in. and 3 in. When the larger pulley turns through one revolution (2π radians), a point on the edge of the pulley moves $s = r\theta = 4(2\pi) = 8\pi$ in. If the belt does not slip, that motion is transferred to the smaller pulley, which turns through an angle

$$\theta = \frac{s}{r} = \frac{8\pi}{3} \approx 8.38 \text{ radians}$$ ■

Example 8   Scott knows that the moon is about 237,000 mi from the earth, but he has forgotten its diameter. If the angle between his lines of sight to either side of the moon is 0.52°, how can Scott estimate its diameter? Give the answer to three significant digits.

Solution   Because the moon is so far from the earth, the length of the arc $\overset{\frown}{AB}$ (with the earth as its center, and intercepted by a diameter of the moon) is a good estimate of the length of the diameter. See Figure 3-5. Because the moon is approximately 237,000 mi from the earth, side EB of triangle EAB is approximately 237,000 mi long as well. Change 0.52° to radians and use the formula $s = r\theta$.

$$\pi \text{ radians} = 180°$$

$$\frac{\pi}{180} \text{ radians} = 1°$$

$$\mathbf{0.52}\left(\frac{\pi}{180} \text{ radians}\right) = \mathbf{0.52}(1°)$$

$$0.0091 \text{ radians} \approx 0.52°$$

Substitute **237,000** for r and **0.0091** for θ into the formula $s = r\theta$ to get

$$s = r\theta \approx 237{,}000(0.0091) \approx 2160$$

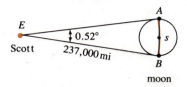

Figure 3-5

Scott's estimate of the diameter of the moon is 2160 mi.

Exercise 3.1

In Exercises 1–12, change each angle to radians.

1. $15°$ **2.** $75°$ **3.** $120°$ **4.** $150°$

5. $210°$ **6.** $240°$ **7.** $300°$ **8.** $330°$

9. $780°$ **10.** $660°$ **11.** $-520°$ **12.** $-880°$

In Exercises 13–24, each angle is expressed in radians. Change each angle to degrees.

13. $\dfrac{3}{4}\pi$ **14.** 3π **15.** $\dfrac{5}{2}\pi$ **16.** $\dfrac{7}{3}\pi$

17. $\dfrac{4}{3}\pi$ **18.** $\dfrac{11}{6}\pi$ **19.** 6 **20.** 8

21. -10 **22.** -5 **23.** $12\dfrac{1}{2}$ **24.** -15.3

In Exercises 25–36, find the values of the trigonometric functions. All angles are measured in radians. **Do not use a calculator or tables.**

25. $\sin\dfrac{\pi}{3}$ **26.** $\cos\dfrac{\pi}{6}$ **27.** $\tan\dfrac{\pi}{4}$ **28.** $\sin\pi$

29. $\cos\dfrac{3\pi}{4}$ **30.** $\tan\dfrac{5\pi}{4}$ **31.** $\sin\dfrac{7\pi}{6}$ **32.** $\sin\dfrac{10\pi}{3}$

33. $\csc\dfrac{11}{6}\pi$ **34.** $\sec\dfrac{7}{4}\pi$ **35.** $\cot\left(-\dfrac{5}{3}\pi\right)$ **36.** $\csc\left(-\dfrac{5}{3}\pi\right)$

37. Find the radius of a circle if a central angle of $25°$ intercepts an arc of 17 cm.

38. Find the central angle in radians that intercepts an arc of 10 cm in a circle with diameter of 10 cm.

39. Find the central angle in degrees that intercepts a 10-in. arc on a 15-in. diameter wheel.

40. If the radius of a circle is 3 m, find the central angle in degrees that intercepts an arc of 6 m.

In Exercises 41–44, use 3960 miles for the radius of the earth. Consider this estimate to be accurate to three figures.

41. The latitude of Manchester, New Hampshire, is 43.0° N. How far is Manchester from the equator? See Illustration 1.

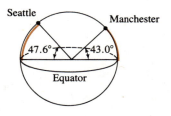

Illustration 1

42. The latitude of Seattle, Washington, is 47.6° N. How far is Seattle from the equator? See Illustration 1.

43. St. Louis, Missouri, is 2670 mi north of the equator. Find its latitude.

44. Pittsburgh, Pennsylvania, is 2800 mi north of the equator. Find its latitude.

45. Find the area, to the nearest hundredth, of a sector of a circle if the sector has a central angle of exactly 30° and the circle has a radius of exactly 20 units.

46. If a circle contains a sector with central angle of $\frac{2\pi}{3}$ and area of 30 sq m, find the diameter of the circle.

47. The diameter of the moon is approximately 2160 mi. How far is the center of the moon from earth if the central angle from a point on earth intercepting a diameter on the moon is 0.56°?

48. A regular octagon is inscribed in a circle 30 cm in diameter. How long is an arc intercepted by one of the sides of the octagon?

49. A railroad track curves along a 17° arc of a circle. If the radius of the circle is 250 m, how long is the track?

50. The earth is approximately 93 million miles from the sun. In one day, the earth moves through an arc of 0.986°. How many miles does the earth travel through space in one week?

51. A **nautical mile** is defined as $\frac{1}{60}$° of arc on the equator. If the radius of the earth is 3960 statute miles, find the number of statute miles contained in one nautical mile. Give your answer to the nearest hundredth.

52. A ship steams 500 nautical miles. How far has the ship gone in statute miles? Give your answer to the nearest hundredth. (Hint: See Exercise 51.)

3.2 LINEAR AND ANGULAR VELOCITY

The question "How fast is it moving?" may be answered in two ways, depending on what "it" is. "How fast is that train moving?" might be answered "70 miles per hour" or "90 feet per second." These answers indicate the train's **linear velocity,*** a measure of how far (miles, feet, meters, and so on) the train will travel in one unit of time (hour, minute, second, and so on). It is not necessary that linear velocity be in a straight line; a car could travel at 55 mph around a curve.

The question "How fast is that phonograph record turning?" might be answered "$33\frac{1}{3}$ revolutions per minute" or, if you enjoy the "oldies," "78 revolutions per minute." These answers indicate the record's **angular velocity,** a measure of the angle (radian, degree, or revolutions) through which it rotates in one unit of time (hour, minute, second, and so on).

Example 1   What is the angular velocity of the earth about its axis?

Solution   Because the earth rotates once in 24 hours, its angular velocity is one revolution per 24 hours, or 1/24 revolution per hour. Since one revolution is 2π radians, the earth's angular velocity is also 2π radians per 24 hours, or $\pi/12$ radians per hour. Other possible answers are one revolution per day, 15 degrees per hour, or 2π radians per day. ■

There are situations in which linear velocity and angular velocity are related. For example, the question "How fast is the tire of that car moving?" might be answered in two ways. Because the tire moves with the car, it has linear velocity. Because the tire is spinning, it has angular velocity as well. Since for a given linear velocity, small motorcycle tires spin faster than the bigger tires of a semi-trailer, one suspects that linear velocity, angular velocity, and the radius of the wheel are all related.

If a wheel of radius r rolls a distance s without slipping, any point on the circumference of the wheel also moves a distance s, measured along the arc of the wheel. See Figure 3-6. The wheel rotates through an angle θ, and

$$s = r\theta \quad (\theta \text{ in radians})$$

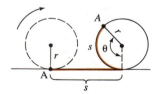

Figure 3-6

---

*The words *linear velocity* and *speed* are used interchangeably; the word *velocity* alone denotes a vector quantity, as used in Section 2.6.

If this movement is accomplished in a length of time, t, then $\frac{s}{t}$ is the linear velocity of the wheel, and $\frac{\theta}{t}$ is the angular velocity. Dividing both sides of the equation $s = r\theta$ by t gives a formula that relates linear and angular velocity.

$$\frac{s}{t} = r\left(\frac{\theta}{t}\right)$$

Denote the linear velocity $\frac{s}{t}$ by v, and the angular velocity $\frac{\theta}{t}$ by ω (the Greek letter omega). The relation between linear and angular velocity is

(3.7) $v = r\omega$

where ω is in units of radians per unit time.

Example 2   What is the linear velocity of a point on the earth's equator in mph?

Solution   Assume that the radius of the earth is 4000 mi. From Example 1, the angular velocity of the earth can be expressed as $\omega = \frac{\pi}{12}$ radians/hour. To find the linear velocity, use the formula $v = r\omega$, substitute **4000** for r and $\frac{\pi}{12}$ for ω, and simplify.

$$v = r\omega$$

$$= 4000\left(\frac{\pi}{12}\right)$$

$$\approx 1047$$

Because 4000 mi is the radius of the earth to two-place accuracy, round the answer to two-place accuracy also. The linear velocity of a point on the earth's equator is approximately 1000 mph.   ■

Example 3   A bicycle with 24-in. wheels is traveling down a road at 10 mph. Find the angular velocity of the wheels in rpm (revolutions per minute).

Solution   Because the radius of each wheel is 12 in. and the angular velocity is to be given in revolutions per minute, change 10 mi/hr to units of in./min.

$$10\,\frac{mi}{hr} \cdot 63360\,\frac{in.}{mi} \cdot \frac{1}{60}\,\frac{hr}{min} = 10560\,\frac{in.}{min}$$

Substitute **10560** for v and **12** for r in the formula $v = r\omega$, and solve for ω.

$$v = r\omega$$

$$10560 = 12\omega$$

$$880\,\frac{rad}{min} = \omega$$

To find ω in rpm, multiply 880 rad/min by $\frac{1}{2\pi}$ rev/rad.

$$\omega = 880\,\frac{rad}{min} \cdot \frac{1}{2\pi}\,\frac{rev}{rad}$$

$$\omega = \frac{440}{\pi} \frac{\text{rev}}{\text{min}}$$

$$\omega \approx 140 \text{ rpm}$$

Example 4   An 8-in. diameter pulley drives a 6-in. diameter pulley. The larger pulley makes 15 revolutions per second. What is the angular velocity of the smaller pulley in revolutions per second?

Solution   The angular velocity of the drive pulley is

$$15 \frac{\text{rev}}{\text{sec}} \cdot 2\pi \frac{\text{rad}}{\text{rev}} = 30\pi \frac{\text{rad}}{\text{sec}}$$

If you assume that the belt that connects the two pulleys does not slip, the linear velocities of points on either circumference are the same—the product $r_1\omega_1$ for one pulley is equal to the product $r_2\omega_2$ for the second pulley. Thus, you have

$$r_1\omega_1 = v = r_2\omega_2$$

$$4(30\pi) = 3(\omega_2) \qquad \text{Substitute 4 for } r_1, 30\pi \text{ for } \omega_1, \text{ and 3 for } r_2.$$

$$\frac{4(30\pi)}{3} = \omega_2$$

$$40\pi = \omega_2$$

The angular velocity of the smaller pulley is 40π radians per second. To convert to revolutions per second, multiply by $\frac{1}{2\pi} \frac{\text{rev}}{\text{rad}}$ and simplify.

$$40\pi \frac{\text{rad}}{\text{sec}} \cdot \frac{1}{2\pi} \frac{\text{rev}}{\text{rad}} = 20 \frac{\text{rev}}{\text{sec}}$$

The angular velocity of the smaller pulley is 20 rev/sec.

Exercise 3.2

In Exercises 1–6, find the angular velocity of the object in the unit specified.

1. The minute hand of a clock in rad/hr.

2. The second hand of a clock in rad/sec.

3. The minute hand of a clock in rad/sec.

4. The earth in its orbit in rad/month.

5. The moon in its orbit in rad/day. Assume that the moon circles the earth in 29.5 days.

6. A phonograph record turning at $33\frac{1}{3}$ rpm in rad/sec.

7. A car is traveling 60 mph (approximately 88 ft/sec). How fast are its 30-in. diameter tires spinning in rev/sec?

8. A Volkswagen is traveling 30 mph. Find the angular velocity of its 22-in. diameter tires in rpm.

9. A large truck is traveling 65 mph. How fast are its 20-in. radius tires spinning in rpm?

10. The 30-in. diameter tires of a bus are turning at 400 rpm. What is the linear velocity of the bus in mph?

11. The 27-in. diameter tires of a bicycle are turning at the rate of 125 rpm. What is the linear velocity of the bicycle in ft/sec?

12. A wheel that is driven by a belt is making 1 rev/sec. If the wheel is 6 in. in diameter, what is the linear velocity of the belt in ft/sec?

13. A wheel that is 8.0 cm in diameter is driven by a belt with a linear velocity of 40 cm/sec. How many rpm is the wheel making?

14. An idler pulley 3 in. in diameter is making 95 rpm. What is the linear velocity of the belt driving the pulley in in./sec?

15. A belt drives two pulleys. One pulley is 10 in. in diameter and the other is 12 in. in diameter. How many rpm is the large pulley making if the small pulley is turning at 20 rpm?

16. A belt drives two wheels that are 15 and 20 in. in diameter. If the 20-in. wheel is turning at 10 rpm, how fast is the other wheel turning in rpm?

17. The 10-in. diameter sprocket of a ten-speed bike drives a 3-in. diameter sprocket that is attached to the 27-in. diameter rear wheel. If the pedals make 30 rpm, how fast is the bike traveling in mph?

18. What is the linear velocity due to the rotation of the earth of Green Bay, Wisconsin (latitude of 44.5° N)? Assume the radius of the earth to be 3960 mi. See Illustration 1.

19. What is the linear velocity due to the rotation of the earth of Miami, Florida (latitude of 25.8° N)? Assume the radius of the earth to be 3960 mi. See Illustration 2.

20. What is the latitude of Pocatello, Idaho, if its linear velocity due to the rotation of the earth is 759 mph? Assume the radius of the earth to be 3960 mi.

21. What is the latitude of Salem, Oregon, if its linear velocity due to the rotation of the earth is 734 mph? Assume the radius of the earth to be 3960 mi.

22. If a gear of radius R_1 drives a gear of radius R_2, show that the angular velocity of the driven gear is R_1/R_2 times that of the driving gear. See Illustration 3.

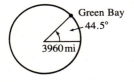

Illustration 1

Illustration 2

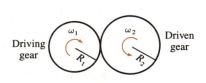

Illustration 3

23. Show that, if the gears in Illustration 3 are separated by an idler gear of any radius, the angular velocity of the driven gear is still R_1/R_2 times that of the driving gear. See Illustration 4.

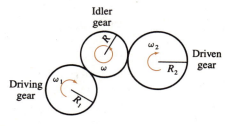

Illustration 4

3.3   THE CIRCULAR FUNCTIONS

In Chapter 2, we defined the six trigonometric functions as functions of an angle in standard position. We now present an alternative definition that makes use of the **unit circle,** the circle that is centered at the origin, and has a radius of 1 unit.

(3.8)

> **Definition.**   Let θ be a central angle of the unit circle shown in Figure 3-7. Let $P(x, y)$ be the point where the terminal side of angle θ intersects the circle. Then, $r = OP = 1$. The **six trigonometric functions of angle θ** are
>
> $\sin \theta = y$
>
> $\cos \theta = x$
>
> $\tan \theta = \dfrac{y}{x}$ $(x \neq 0)$
>
> $\csc \theta = \dfrac{1}{y}$ $(y \neq 0)$
>
> $\sec \theta = \dfrac{1}{x}$ $(x \neq 0)$
>
> $\cot \theta = \dfrac{x}{y}$ $(y \neq 0)$
>
> **Figure 3-7**

Note that the unit-circle definition of the trigonometric functions is a special case of the angle-in-standard-position definition. In the unit-circle definition, the radius r is required to be 1 unit in length. In the angle-in-standard-position definition, r can be any positive number.

Recall that the domain of the function defined by the equation $y = f(x)$ is the set of all admissible values of x, and the range is the set of all values of y. For

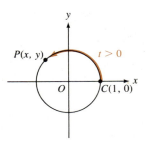

Figure 3-8

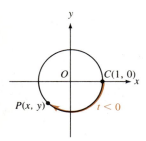

Figure 3-9

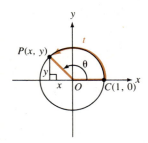

Figure 3-10

virtually all functions previously studied, both the domain and the range of the function have been subsets of the real number system. A real number x was fed into the function, and the function produced a corresponding real number y. Thus far, however, the trigonometric functions have been a different breed: the domain of $y = \sin \theta$, for example, has been a set of angles, rather than a set of real numbers. One of the advantages of the unit-circle definition is subtle but important. The unit-circle approach allows us to bring the trigonometric functions into conformity with other functions—that is, to show that the trigonometric functions can be thought of as functions with real number domains.

The argument begins by noting that any real number can represent the length of exactly one arc on the unit circle. If t is a positive number, we can find the arc of length t by measuring a distance t in a counterclockwise direction along an arc of the unit circle beginning at the point $C(1, 0)$. This determines arc CP of length t. See Figure 3-8. If t is a negative number, we can find the arc of length t by measuring a distance $|t|$ in a clockwise direction along an arc of the unit circle beginning at the point $C(1, 0)$. This determines arc CP of length $|t|$. See Figure 3-9.

In each case, marking off the arc determines the unique point P with coordinates (x, y) that corresponds to the real number t. Now suppose that we let t be any real number and place an arc of length $|t|$ on the unit circle as in Figure 3-10. From the previous section we know that, if s is an arc intercepted by a central angle θ and θ is measured in radians, we have

$$s = r\theta$$

We can substitute t for s and 1 for r in the formula $s = r\theta$ to obtain

$$t = \theta$$

or

$$\theta = t$$

Thus, when the measure of an arc on the unit circle is the real number t, then t is also the radian measure of the central angle determined by that arc. This gives

$\sin \theta = \sin t$	$\csc \theta = \csc t$
$\cos \theta = \cos t$	$\sec \theta = \sec t$
$\tan \theta = \tan t$	$\cot \theta = \cot t$

where θ is an angle measured in radians and t is a real number. Hence, we can think of each trigonometric expression as being either a trigonometric function of an angle measured in radians or as a trigonometric function of a real number. The important point is this: *The trigonometric functions can now be thought of as functions that have domains and ranges that are subsets of the real numbers*.

Example 1   Find **a.** $\sin\left(\dfrac{\pi}{3}\right)$ and **b.** $\cos\left(\dfrac{5\pi}{2}\right)$.

Solution   **a.** $\sin\left(\dfrac{\pi}{3}\right) = \sin\left(\dfrac{\pi}{3}\ \text{rad}\right) = \dfrac{\sqrt{3}}{2}$

b. $\cos\left(\dfrac{5\pi}{2}\right) = \cos\left(\dfrac{5\pi}{2}\,\text{rad}\right) = \cos\left(\dfrac{\pi}{2}\,\text{rad}\right) = 0$ ■

Example 2   Find **a.** $\tan\left(\dfrac{3\pi}{4}\right)$ and **b.** $\csc\left(\dfrac{7\pi}{6}\right)$.

Solution   **a.** $\tan\left(\dfrac{3\pi}{4}\right) = \tan\left(\dfrac{3\pi}{4}\,\text{rad}\right) = -\tan\left(\dfrac{\pi}{4}\,\text{rad}\right) = -1$

b. $\csc\left(\dfrac{7\pi}{6}\right) = \csc\left(\dfrac{7\pi}{6}\,\text{rad}\right) = \dfrac{1}{\sin\left(\dfrac{7\pi}{6}\,\text{rad}\right)} = -2$ ■

Example 3   Find the coordinates (x, y) of the point P on the unit circle that correspond to the real numbers **a.** π,   **b.** $\dfrac{\pi}{6}$, and   **c.** $\dfrac{-5\pi}{4}$.

Solution   **a.**   From Definition (3.8), you have

$$x = \cos\theta$$
$$y = \sin\theta$$

But, if θ is measured in radians, you also have

$$x = \cos t$$
$$y = \sin t$$

where t is the real number that is the radian measure of central angle θ. Hence, the coordinates (x, y) of the point P that correspond to the real number π are

$$x = \cos\pi = \cos(\pi\,\text{rad}) = -1$$
$$y = \sin\pi = \sin(\pi\,\text{rad}) = 0$$

Thus, point P has coordinates $(-1, 0)$.

Note that because the circumference of the unit circle is 2π, the number π is the measure of half that circumference. From Figure 3-11, you can see that the coordinates of the point P must be $(-1, 0)$.

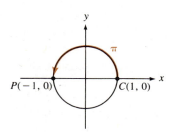

Figure 3-11

b. The point P on the unit circle that corresponds to $\dfrac{\pi}{6}$ has coordinates of

$$x = \cos\left(\frac{\pi}{6}\right) = \cos\left(\frac{\pi}{6}\text{rad}\right) = \frac{\sqrt{3}}{2}$$

$$y = \sin\left(\frac{\pi}{6}\right) = \sin\left(\frac{\pi}{6}\text{rad}\right) = \frac{1}{2}$$

Thus, the point P has coordinates $\left(\dfrac{\sqrt{3}}{2}, \dfrac{1}{2}\right)$.

c. The point P on the unit circle that corresponds to $\dfrac{-5\pi}{4}$ has coordinates of

$$x = \cos\left(\frac{-5\pi}{4}\right) = \cos\left(\frac{-5\pi}{4}\text{rad}\right) = -\frac{\sqrt{2}}{2}$$

$$y = \sin\left(\frac{-5\pi}{4}\right) = \sin\left(\frac{-5\pi}{4}\text{rad}\right) = \frac{\sqrt{2}}{2}$$

Thus, the point P has coordinates $\left(-\dfrac{\sqrt{2}}{2}, \dfrac{\sqrt{2}}{2}\right)$.

Example 4   Find the coordinates of the point P on the unit circle that correspond to the real numbers **a.** 7 and **b.** -1.37.

Solution   **a.**   The coordinates of the point P are

$$x = \cos(7) = \cos\,(7\,\text{rad})$$
$$y = \sin(7) = \sin(7\,\text{rad})$$

Use your calculator, set in radian mode, to determine that

$$x \approx 0.7539$$

and

$$y \approx 0.6570$$

Thus, point P has approximate coordinates of $(0.7539, 0.6570)$.

b.   The coordinates of the point P are

$$x = \cos(-1.37) = \cos(-1.37\,\text{rad}) \approx 0.1994$$
$$y = \sin(-1.37) = \sin(-1.37\,\text{rad}) \approx -0.9799$$

Thus, point P has approximate coordinates of $(0.1994, -0.9799)$.

Exercise 3.3

In Exercises 1–12, evaluate each given expression. **Do not use a calculator.**

1. $\sin \dfrac{\pi}{6}$ **2.** $\cos \dfrac{\pi}{6}$ **3.** $\cos\left(-\dfrac{5\pi}{6}\right)$ **4.** $\sin\left(-\dfrac{5\pi}{6}\right)$

5. $\tan \dfrac{5\pi}{4}$ **6.** $\cot \dfrac{5\pi}{4}$ **7.** $\csc \dfrac{5\pi}{6}$ **8.** $\sec \dfrac{5\pi}{6}$

9. $\sin\left(-\dfrac{8\pi}{3}\right)$ **10.** $\cos \dfrac{8\pi}{3}$ **11.** $\tan \dfrac{15\pi}{4}$ **12.** $\csc \dfrac{7\pi}{3}$

In Exercises 13–24, use a calculator to evaluate each given expression to four decimal places. Remember to set your calculator in radian mode.

13. $\sin 2$ **14.** $\sin 3$ **15.** $\cos 8$ **16.** $\tan 5$

17. $\sin \dfrac{3}{\pi}$ **18.** $\cos\left(-\dfrac{3}{\pi}\right)$ **19.** $\sec 5$ **20.** $\csc 4$

21. $\cot 1$ **22.** $\cot(-1)$ **23.** $\sin(3+\pi)$ **24.** $\cos(3-\pi)$

In Exercises 25–48, find the coordinates of the point P on the unit circle that correspond to each given real number. **Do not use a calculator.**

25. $\dfrac{3\pi}{2}$ **26.** $\dfrac{-\pi}{2}$ **27.** $-\pi$ **28.** 2π

29. 3π **30.** 4π **31.** $\dfrac{-7\pi}{2}$ **32.** $\dfrac{9\pi}{2}$

33. $\dfrac{\pi}{4}$ **34.** $\dfrac{-\pi}{4}$ **35.** $\dfrac{-3\pi}{4}$ **36.** $\dfrac{5\pi}{4}$

37. $\dfrac{\pi}{3}$ **38.** $\dfrac{-\pi}{6}$ **39.** $\dfrac{-2\pi}{3}$ **40.** $\dfrac{2\pi}{3}$

41. $\dfrac{4\pi}{3}$ **42.** $\dfrac{-4\pi}{3}$ **43.** $\dfrac{-5\pi}{3}$ **44.** $\dfrac{5\pi}{6}$

45. $\dfrac{11\pi}{6}$ **46.** $\dfrac{17\pi}{6}$ **47.** $\dfrac{23\pi}{6}$ **48.** $\dfrac{-17\pi}{6}$

49. Show that the area of a sector of the unit circle is given by the formula $A = \frac{1}{2}\theta$, where θ is the central angle of the sector measured in radians.

50. Show that the area of triangle *OAR* in Illustration 1 is given by the formula
$A = \frac{1}{2}\cos\theta\sin\theta$

51. Show that the area of triangle *OTP* in Illustration 1 is given by the formula
$A = \frac{1}{2}\tan\theta$

52. Use the inequality

area of $\triangle OAR \leqslant$ area of sector $OTR \leqslant$ area of $\triangle OTP$

to show that the ratio $\frac{\sin\theta}{\theta}$ approaches 1 as θ approaches 0. Note that θ must be a real number for this ratio to have meaning. This fact is extremely important in calculus. (*Hint:* Refer to Exercises 49–51 and to Illustration 1.)

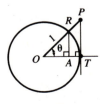

Illustration 1

3.4 GRAPHS OF FUNCTIONS INVOLVING SIN *X* AND COS *X*

In this section we will discuss the graphs of the sine and cosine functions. Because a picture is often worth a thousand words, a study of these graphs will emphasize some properties of the sine and cosine functions that are not obvious from their equations.

We begin by noting that the sine of an angle θ is equal to the sine of any angle that is coterminal with angle θ. Similarly, the cosine of an angle θ is equal to the cosine of any angle that is coterminal with angle θ. Therefore, for any real number x, we have

$$\sin x = \sin(x \pm 2\pi) = \sin(x \pm 4\pi) = \cdots$$
$$\cos x = \cos(x \pm 2\pi) = \cos(x \pm 4\pi) = \cdots$$

The sine and cosine functions are called **periodic functions** because, as x increases, the values of sin x and cos x repeat in a predictable way.

(3.9)

> **Definition.**   A function f is said to be **periodic** with period p if p is the smallest positive number for which
>
> $$f(x) = f(x + p)$$
>
> for all x in the domain of f.

The sine function has a period of 2π because sin $x = \sin(x + 2\pi)$ for all x, and sin $x = \sin(x + p)$ is true for no positive number p that is less than 2π. The cosine function also has a period of 2π because cos $x = \cos(x + 2\pi)$ for all x, and cos $x = \cos(x + p)$ is true for no positive number p that is less than 2π.

By plotting several points (Table A in Appendix IV or a calculator will help), we can graph the function defined by $y = \sin x$. In an equation such as $y = \sin x$, the variable x is often called the **argument** of the function. Remember that the argument can be thought of either as an angle in radians or as a real number.

The table of values in Figure 3-12 shows that, as x increases from 0 to $\frac{\pi}{2}$, the values of sin x increase from 0 to 1. As x continues to increase from $\frac{\pi}{2}$ to π, the values of sin x begin to decrease, heading back to 0. As x goes from π to $\frac{3\pi}{2}$, the values of sin x continue to decrease until they reach a value of -1 at $x = \frac{3\pi}{2}$. As x continues to increase from $\frac{3\pi}{2}$ to 2π, the values of sin x begin to increase, reaching the value of 0 when $x = 2\pi$. If we plot the pairs of values $(x, \sin x)$ shown in the table of values, we can draw the graph of $y = \sin x$.

$y = \sin x$

x	y
0	0
$\frac{\pi}{6}$	$.5$
$\frac{\pi}{3}$	$.87$
$\frac{\pi}{2}$	1
$\frac{2\pi}{3}$	$.87$
$\frac{5\pi}{6}$	$.5$
π	0
$\frac{7\pi}{6}$	$-.5$
$\frac{4\pi}{3}$	$-.87$
$\frac{3\pi}{2}$	-1
$\frac{5\pi}{3}$	$-.87$
$\frac{11\pi}{6}$	$-.5$
2π	0

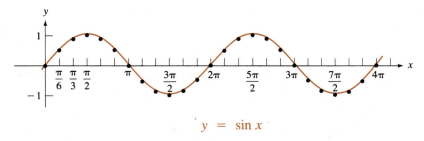

$y = \sin x$

Figure 3-12

The graph of $y = \sin x$ doesn't start at 0, nor does it end at 2π. It continues to oscillate in both directions forever. Because of the periodic nature of this function, however, if we know its behavior through one period, we know its behavior everywhere. Note that the values of $\sin x$ are always between -1 and 1, inclusive.

The graph of the cosine function is similar to that of the sine function. Because $\cos 0 = 1$, we will draw its graph beginning at the point $(0, 1)$. As x increases from 0 to π, the values of $\cos x$ decrease, dropping to a value of -1 when $x = \pi$. As x increases from π to 2π, the values of $\cos x$ increase back to 1 when $x = 2\pi$. The graph of $y = \cos x$ appears in Figure 3-13. As with the values of $\sin x$, the values of $\cos x$ are always between -1 and 1, inclusive.

$y = \cos x$

x	y
0	1
$\frac{\pi}{6}$	$.87$
$\frac{\pi}{3}$	$.5$
$\frac{\pi}{2}$	0
$\frac{2\pi}{3}$	$-.5$
$\frac{5\pi}{6}$	$-.87$
π	-1
$\frac{7\pi}{6}$	$-.87$
$\frac{4\pi}{3}$	$-.5$
$\frac{3\pi}{2}$	0
$\frac{5\pi}{3}$	$.5$
$\frac{11\pi}{6}$	$.87$
2π	1

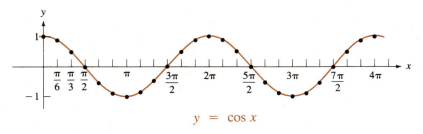

$y = \cos x$

Figure 3-13

Example 1   Graph the function defined by $y = 3 \sin x$.

Solution   Because $\sin x$ has values between 1 and -1 as x increases from 0 to 2π, the values of $3 \sin x$ must be between 3 and -3 in that same interval. As x increases from 0 to $\frac{\pi}{2}$, the values of $3 \sin x$ increase from 0 to 3. As x increases from $\frac{\pi}{2}$ to $\frac{3\pi}{2}$, the values of $3 \sin x$ decrease from 3 to -3. As x increases from $\frac{3\pi}{2}$ to 2π, the values of $\sin x$ increase back to their starting value of 0. Thus, the graph of $y = 3 \sin x$ finishes one complete cycle as x grows from 0 to 2π. Note that the graph of $y = 3 \sin x$, shown in Figure 3-14, looks like a puffed-up version of $y = \sin x$. The graph of $y = \sin x$ has been drawn in the figure for reference.

$y = 3 \sin x$

x	y
0	0
$\frac{\pi}{2}$	3
π	0
$\frac{3\pi}{2}$	-3
2π	0

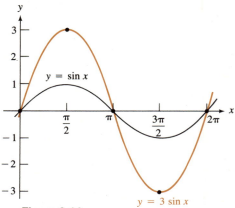

Figure 3-14

Example 2   Graph $y = 2 \cos x$ and $y = -2 \cos x$ on the same set of coordinate axes.

Solution   As x increases from 0 to π, the values of the cosine function decrease from 1 to -1. As x increases from π to 2π, the values of the cosine function increase from -1 back to their starting value of 1. Thus, as x increases from 0 to 2π, the values of $2 \cos x$ must begin at 2, decrease to -2 at $x = \pi$, and then increase back to 2 at $x = 2\pi$. The graph of $y = 2 \cos x$ appears in Figure 3-15.

As x increases from 0 to 2π, the values of $-2 \cos x$ must begin at -2, increase to $+2$ at $x = \pi$, and then decrease back to -2 at $x = 2\pi$. The graph of $y = -2 \cos x$ also appears in Figure 3-15. Note that the two graphs are reflections of each other, with the x-axis as the axis of symmetry.

$y = 2 \cos x$

x	y
0	2
$\frac{\pi}{2}$	0
π	-2
$\frac{3\pi}{2}$	0
2π	2

$y = -2 \cos x$

x	y
0	-2
$\frac{\pi}{2}$	0
π	2
$\frac{3\pi}{2}$	0
2π	-2

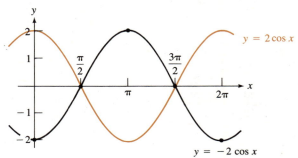

Figure 3-15

Example 3   Graph $y = \sin 2x$.

Solution   The graph of $y = \sin x$ completes one cycle as x increases from 0 to 2π. Thus, the graph of $y = \sin 2x$ must complete one cycle as $2x$ increases from 0 to 2π, or as x increases from 0 to π. Because the graph of $y = \sin 2x$ completes one cycle as x increases from 0 to π, the period of $y = \sin 2x$ must be π. Note that the graph of $y = \sin 2x$ oscillates twice as fast as the graph of $y = \sin x$. That is, it squeezes two cycles into the same space required for one cycle of the graph of $y = \sin x$. The graph of $y = \sin 2x$, along with a table of values, is shown in Figure 3-16. Note that the coefficient of x is a number greater than 1 and that the curve is compressed.

$y = \sin 2x$

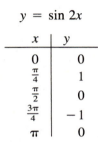

x	y
0	0
$\frac{\pi}{4}$	1
$\frac{\pi}{2}$	0
$\frac{3\pi}{4}$	-1
π	0

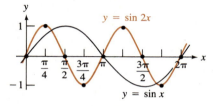

Figure 3-16

Example 4   Graph $y = \cos \frac{1}{3}x$.

Solution   As x increases from 0 to 2π, the graph of $y = \cos x$ completes one cycle. As $\frac{1}{3}x$ increases from 0 to 2π, (that is, as x increases from 0 to 6π), the graph of $y = \cos \frac{1}{3}x$ completes one cycle. Thus, the period of the graph of $y = \cos \frac{1}{3}x$ is 6π. The graph, along with a table of values, is shown in Figure 3-17. Note that the coefficient of x is a fraction less than 1 and that the curve is stretched.

$y = \cos \frac{1}{3}x$

x	y
0	1
$\frac{3\pi}{2}$	0
3π	-1
$\frac{9\pi}{2}$	0
6π	1

Figure 3-17

Example 5   Graph $y = \cos \pi x$.

Solution   As x increases from 0 to 2π, the graph of $y = \cos x$ completes one cycle. As πx increases from 0 to 2π (that is, as x increases from 0 to 2), the graph of $y = \cos \pi x$ completes one cycle. Thus, the period of the graph of $y = \cos \pi x$ is 2. The graph of $y = \cos \pi x$ appears in Figure 3-18.

$y = \cos \pi x$

x	y
0	1
$\frac{1}{2}$	0
1	-1
$\frac{3}{2}$	0
2	1
$\frac{5}{2}$	0
3	-1
$\frac{7}{2}$	0
4	1
$\frac{9}{2}$	0
5	-1
$\frac{11}{2}$	0
6	1

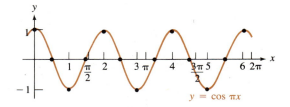

Figure 3-18

One cycle of the graph of $y = a \sin bx$ is completed as bx increases from 0 to 2π. When $bx = 0$, then x must be 0. When $bx = 2\pi$, then $x = \frac{2\pi}{b}$. It follows that one cycle of the curve $y = a \sin bx$ is completed as x itself increases from 0 to $\left|\frac{2\pi}{b}\right|$. Thus, the period of the graph of $y = a \sin bx$ is $\left|\frac{2\pi}{b}\right|$.

The largest value attained by $\sin bx$ is 1. Therefore, the largest value that can be attained by $a \sin bx$ is $|a| \cdot 1 = |a|$. This value is called the **amplitude** of the graph of $y = a \sin bx$.

A similar argument applies to determine the period and the amplitude of the graph of $y = a \cos bx$. These facts are summarized in the following statement.

(3.10) | **Period and Amplitude.**   The **period** of the graph of $\begin{Bmatrix} y = a \sin bx \\ y = a \cos bx \end{Bmatrix}$ is $\left|\dfrac{2\pi}{b}\right|$, and the **amplitude** is $|a|$.

Example 6   Graph $y = 5 \sin 7x$.

Solution   The amplitude is 5, and the period is $\frac{2\pi}{7}$. One cycle of $y = 5 \sin 7x$ appears in Figure 3-19.

$y = 5 \sin 7x$

x	y
0	0
$\frac{\pi}{14}$	5
$\frac{\pi}{7}$	0
$\frac{3\pi}{14}$	-5
$\frac{2\pi}{7}$	0

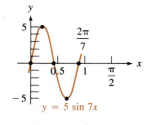

Figure 3-19

Example 7   Graph $y = 2 \cos \frac{1}{2}x$.

Solution   The amplitude is 2, and the period is

$$\frac{2\pi}{\frac{1}{2}} = 4\pi$$

Because $2 \cos \frac{1}{2}x$ is zero when $x = \ldots, -\pi, \pi, 3\pi, \ldots$, the curve intersects the x-axis at these points. One cycle of $y = 2 \cos \frac{1}{2}x$ appears in Figure 3-20. Note that the coefficient of x (which is a fraction less than 1) stretches the curve.

$$y = 2 \cos \frac{1}{2}x$$

x	y
0	2
π	0
2π	-2
3π	0
4π	2

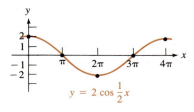

$y = 2 \cos \frac{1}{2}x$

Figure 3-20

Exercise 3.4

In Exercises 1–16, find the amplitude and period of each function. **Do not construct a graph.**

sin function only

1. $y = 2 \sin x$ **2.** $y = 3 \cos x$

3. $y = \cos 9x$ **4.** $y = -\sin 11x$

5. $y = \sin \frac{1}{3}x$ **6.** $y = \cos \frac{1}{4}x$

7. $y = -\cos 0.2x$ **8.** $y = \sin 0.25x$

9. $y = 3 \sin \frac{1}{2}x$ **10.** $y = \frac{1}{2} \cos \frac{1}{3}x$

11. $y = -\frac{1}{2} \cos \pi x$ **12.** $y = 17 \sin 2\pi x$

13. $y = 3 \sin 2\pi x$ *period = 1*   **14.** $y = 8 \cos \pi x$

15. $y = -\frac{1}{3} \sin \frac{3x}{\pi}$ **16.** $y = -\frac{5}{3} \cos \frac{x}{\pi}$

In Exercises 17–26, graph each pair of functions over the indicated interval.

17. $y = \sin x$ and $y = 2 \sin x$, $0 \leq x \leq 2\pi$

18. $y = \cos x$ and $y = -3 \cos x$, $0 \leq x \leq 2\pi$

19. $y = \cos x$ and $y = -\frac{1}{3} \cos x$, $0 \leq x \leq 2\pi$

20. $y = \sin x$ and $y = \dfrac{1}{2} \sin x$, $0 \leqslant x \leqslant 2\pi$

21. $y = \sin x$ and $y = \sin 2x$, $0 \leqslant x \leqslant 2\pi$

22. $y = \cos x$ and $y = \cos 3x$, $0 \leqslant x \leqslant 2\pi$

23. $y = \cos x$ and $y = \cos \dfrac{1}{3} x$, $0 \leqslant x \leqslant 6\pi$

24. $y = \sin x$ and $y = \sin \dfrac{1}{2} x$, $0 \leqslant x \leqslant 4\pi$

25. $y = \sin x$ and $y = \sin \pi x$, $0 \leqslant x \leqslant 2\pi$

26. $y = \cos x$ and $y = \cos \pi x$, $0 \leqslant x \leqslant 2\pi$

In Exercises 27–40, graph each given function over an interval that is at least one period long.

27. $y = 3 \cos x$

28. $y = 4 \sin x$

29. $y = -\sin x$

30. $y = -\cos x$

31. $y = \cos 2x$

32. $y = \sin 3x$

33. $y = -\sin \dfrac{x}{4}$

34. $y = \cos \dfrac{x}{4}$

35. $y = 3 \sin \pi x$

36. $y = -2 \cos \pi x$

37. $y = \dfrac{1}{2} \cos 4x$

38. $y = -4 \sin 3x$

39. $y = -4 \sin \dfrac{x}{2}$

40. $y = \dfrac{1}{3} \cos \dfrac{x}{2}$

3.5 GRAPHS OF FUNCTIONS INVOLVING TAN X, COT X, CSC X, AND SEC X

Because the reference angles of x and $x + \pi$ are equal (see Figure 3-21) and because the tangents of angles in nonadjacent quadrants agree in sign, we have

$$\tan x = \tan(x + \pi)$$

for all x. Furthermore, the number π is the smallest positive number p for which $\tan x = \tan(x + p)$. Therefore, the period of the tangent function is π.

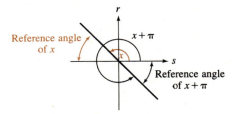

Figure 3-21

Because $\tan 0 = 0$, the graph of $y = \tan x$ passes through the origin. See Figure 3-22. As x gets very close to $\frac{\pi}{2}$, the value of $|\tan x|$ becomes very large. Since a value for $\tan \frac{\pi}{2}$ does not exist, the graph of $y = \tan x$ cannot intersect the line $x = \frac{\pi}{2}$. However, the graph of $y = \tan x$ does approach the vertical line $x = \frac{\pi}{2}$ as x approaches $\frac{\pi}{2}$. The vertical line $x = \frac{\pi}{2}$ is called an **asymptote.** Other vertical asymptotes of the tangent function are the vertical lines $x = \frac{3\pi}{2}$, $x = \frac{5\pi}{2}$, and so on. The graph of $y = \tan x$ approaches these lines, but it never touches them.

For all x between $\frac{\pi}{2}$ and π, the values of $\tan x$ are negative, returning to 0 when $x = \pi$. The graph of $y = \tan x$ is shown in Figure 3-22. In the exercises, you will be asked to explain why the tangent function has no defined amplitude.

$y = \tan x$

x	y
0	0
$\frac{\pi}{4}$	1
$\frac{3\pi}{4}$	-1
π	0
$\frac{5\pi}{4}$	1
$\frac{7\pi}{4}$	-1

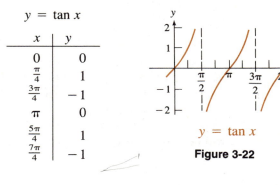

$y = \tan x$

Figure 3-22

Similarly, the cotangent function has a period of π, with vertical asymptotes at multiples of π. The graph of $y = \cot x$ is shown in Figure 3-23. In the exercises, you will be asked to explain why the cotangent function has no defined amplitude.

$y = \cot x$

x	y
$\frac{\pi}{4}$	1
$\frac{\pi}{2}$	0
$\frac{3\pi}{4}$	-1
$\frac{5\pi}{4}$	1
$\frac{3\pi}{2}$	0
$\frac{7\pi}{4}$	-1

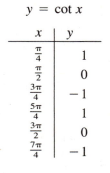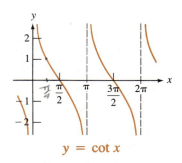

$y = \cot x$

Figure 3-23

Because $\csc x$ is the reciprocal of $\sin x$, it has the same period as $\sin x$ and is undefined whenever $\sin x$ is 0—at $x = 0$, at $x = \pi$, at $x = 2\pi$, and so on. These values determine the vertical asymptotes for the graph of $y = \csc x$. The graph of $y = \csc x$ appears in Figure 3-24. In the exercises, you will be asked to explain why the cosecant function has no defined amplitude.

$y = \csc x$

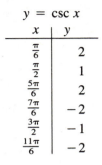

x	y
$\frac{\pi}{6}$	2
$\frac{\pi}{2}$	1
$\frac{5\pi}{6}$	2
$\frac{7\pi}{6}$	-2
$\frac{3\pi}{2}$	-1
$\frac{11\pi}{6}$	-2

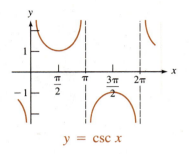

$y = \csc x$

Figure 3-24

Similarly, the graph of $y = \sec x$ has a period of 2π, because $\sec x$ is the reciprocal of $\cos x$ and $\cos x$ has a period of 2π. There is no value for the secant function at $x = \frac{\pi}{2}$, at $x = \frac{3\pi}{2}$, and so on because the cosine of these values is 0. Thus, the vertical lines $x = \frac{\pi}{2}$, $x = \frac{3\pi}{2}$, . . . are the vertical asymptotes for the graph of $y = \sec x$. The graph is shown in Figure 3-25. In the exercises, you will be asked to explain why the secant function has no defined amplitude.

$y = \sec x$

x	y
0	1
$\frac{\pi}{3}$	2
$\frac{2\pi}{3}$	-2
π	-1
$\frac{4\pi}{3}$	-2
$\frac{5\pi}{3}$	2
2π	1

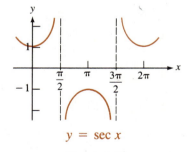

$y = \sec x$

Figure 3-25

Example 1   Graph $y = \tan 3x$.

Solution   The graph of $y = \tan 3x$ intersects the x-axis when $\tan 3x = 0$. This is true when $3x = 0$, when $3x = \pi$, when $3x = 2\pi$, and so on. Hence, the graph intersects the x-axis at $x = 0$,   at $x = \frac{\pi}{3}$,   at $x = \frac{2\pi}{3}$,   and so on. The graph of $y = \tan 3x$ completes one cycle as $3x$ increases from 0 to π—that is, as x itself increases from 0 to $\frac{\pi}{3}$. A value for $\tan 3x$ is undefined at $3x = \frac{\pi}{2}$ or at $x = \frac{\pi}{6}$. The period is $\frac{\pi}{3}$, and the vertical asymptotes are $x = \frac{\pi}{6}$, $x = \frac{\pi}{2}$, $x = \frac{5\pi}{6}$,   and so on. The graph is shown in Figure 3-26.

$y = \tan 3x$

x	y
0	0
$\frac{\pi}{3}$	0
$\frac{2\pi}{3}$	0

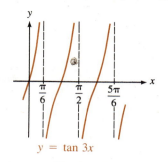

$y = \tan 3x$

Figure 3-26

Example 2   Graph $y = 2 \csc x$ on the interval from 0 to 2π.

Solution   Because $\csc x$ is the reciprocal of $\sin x$, it has the same period as $\sin x$ and is undefined whenever $\sin x$ is 0. Thus, in the interval from 0 to 2π, a value for $2 \csc x$ is undefined at $x = 0$,   at $x = \pi$,   and at $x = 2\pi$. These values determine vertical asymptotes for the graph of $y = 2 \csc x$. A table of values and the graph of $y = 2 \csc x$ appear in Figure 3-27.

$y = 2 \csc x$

x	y
$\frac{\pi}{6}$	4
$\frac{\pi}{2}$	2
$\frac{5\pi}{6}$	4
$\frac{7\pi}{6}$	-4
$\frac{3\pi}{2}$	-2
$\frac{11\pi}{6}$	-4

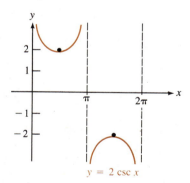

$y = 2 \csc x$

Figure 3-27

Example 3   Graph $y = -2 \sec 2x$ on the interval from 0 to 2π.

Solution   Because $\sec 2x$ is the reciprocal of $\cos 2x$, it has the same period as $\cos 2x$. Thus, the period of $y = -2 \sec 2x$ is $\frac{2\pi}{2}$ or π. Furthermore, a value for $\sec 2x$ is undefined whenever $\cos 2x$ is 0. In the interval from 0 to 2π, this occurs when $x = \frac{\pi}{4}$,   when $x = \frac{3\pi}{4}$,   when $x = \frac{5\pi}{4}$,   and when $x = \frac{7\pi}{4}$. These values determine the vertical asymptotes for the graph $y = -2 \sec 2x$. The graph of $y = -2 \sec 2x$ appears in Figure 3-28.

$$y = -2 \sec 2x$$

x	y
0	-2
$\frac{\pi}{2}$	2
π	-2
$\frac{3\pi}{2}$	2
2π	-2

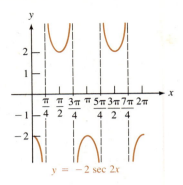

Figure 3-28

We summarize the results of this section.

(3.11)

Period and Amplitude. The **period** of

$$\left.\begin{cases} y = a \tan bx \\ y = a \cot bx \end{cases}\right\} \quad \text{is} \quad \left|\frac{\pi}{b}\right|.$$

The **period** of

$$\left.\begin{cases} y = a \csc bx \\ y = a \sec bx \end{cases}\right\} \quad \text{is} \quad \left|\frac{2\pi}{b}\right|.$$

The tangent, cotangent, cosecant, and secant functions have no defined amplitude.

Exercise 3.5

In Exercises 1–18, give the period of each of the given functions. **Do not construct a graph.**

1. $y = 3 \tan x$ **2.** $y = 2 \csc x$

3. $y = \frac{1}{2} \sec x$ **4.** $y = \frac{1}{3} \cot x$

5. $y = \frac{1}{3} \tan 3x$ **6.** $y = \frac{1}{2} \sec 2x$

7. $y = -2 \csc \pi x$ **8.** $y = -3 \tan \pi x$

9. $y = 3 \sec \frac{x}{3}$ **10.** $y = -2 \tan \frac{\pi x}{3}$

11. $y = \frac{7}{2} \cot \frac{2\pi x}{3}$ **12.** $y = -\frac{2}{3} \csc \frac{\pi x}{3}$

13. $y = 3 \csc \dfrac{\pi x}{2}$

14. $y = -4 \sec \dfrac{2\pi x}{5}$

15. $y = -\cot \dfrac{x}{2\pi}$

16. $y = 7 \csc \dfrac{x}{4\pi}$

17. $y = -\dfrac{2}{5} \sec \dfrac{3x}{\pi}$

18. $y = \dfrac{7}{9} \cot \dfrac{2x}{\pi}$

In Exercises 19–30, graph each given function over the indicated interval.

19. $y = 2 \tan x, \quad \dfrac{-\pi}{2} < x < \dfrac{3\pi}{2}$

20. $y = 2 \csc x, \quad 0 < x < 2\pi$

21. $y = -3 \sec x, \quad 0 \leqslant x \leqslant 2\pi$

22. $y = -\csc 2x, \quad 0 < x < \dfrac{3\pi}{2}$

23. $y = \cot 2x, \quad 0 < x < \pi$

24. $y = \sec 3x, \quad 0 \leqslant x \leqslant \dfrac{2\pi}{3}$

25. $y = -2 \tan \dfrac{x}{2}, \quad 0 < x < 2\pi$

26. $y = -3 \csc \dfrac{x}{2}, \quad 0 < x < 4\pi$

27. $y = 2 \sec 2x, \quad 0 \leqslant x \leqslant 2\pi$

28. $y = 2 \csc 2x, \quad 0 < x < 2\pi$

29. $y = -2 \cot \dfrac{\pi}{4}x, \quad 0 < x < 4$

30. $y = -2 \sec \dfrac{\pi}{4}x, \quad 0 \leqslant x \leqslant 8$

31. Explain why amplitude has no meaning when discussing the tangent, cotangent, secant, and cosecant functions.

32. For what values of x, if any, are the values of $\sin x$ and $\csc x$ equal?

33. For what values of x, if any, are the values of $\cos x$ and $\sec x$ equal?

34. For what values of x, if any, are the values of $\tan x$ and $\cot x$ equal?

3.6 VERTICAL AND HORIZONTAL TRANSLATIONS OF THE TRIGONOMETRIC FUNCTIONS

In applied work we often encounter graphs of trigonometric functions that have been shifted in either a vertical or a horizontal direction. Such shifts are called **translations.** The graph of a function can be shifted vertically (a vertical translation) by adding a constant value to the function. The graph of a function can be shifted horizontally (a horizontal translation) by adding a constant value to the argument of the function. The first two examples illustrate vertical translations, and the rest involve horizontal translations.

Example 1 Graph $y = 2 + \cos x$.

Solution You already know that $y = \cos x$ has a period of 2π, has values that lie between -1 and 1, and has x-intercepts at $x = \frac{\pi}{2}$, at $x = \frac{3\pi}{2}$, at $x = \frac{5\pi}{2}$, and so on. The values of $2 + \cos x$ will be similar to the values of $\cos x$, except that each will be increased by 2. Thus, the graph of $y = 2 + \cos x$ has a period of 2π and has values that lie between $2 + (-1)$ and $2 + 1$, or between 1 and 3. The curve will intersect the line $y = 2$ at $x = \frac{\pi}{2}$, at $x = \frac{3\pi}{2}$, at $x = \frac{5\pi}{2}$, and so on. A table of values and the graph of $y = 2 + \cos x$ are shown in Figure 3-29.

$y = 2 + \cos x$

x	y
0	3
$\frac{\pi}{2}$	2
π	1
$\frac{3\pi}{2}$	2
2π	3

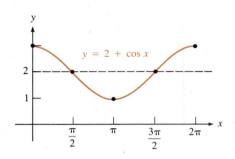

Figure 3-29

Example 2   Graph $y = -3 + \tan \frac{x}{2}$

Solution   The function $y = \tan \frac{x}{2}$ has a period of $\frac{\pi}{1/2}$, or 2π, and has vertical asymptotes at $x = \pi$, at $x = 3\pi$, at $x = 5\pi$, and so on. The values of $-3 + \tan \frac{x}{2}$ will be similar to the values of $\tan \frac{x}{2}$, except that each value of $\tan \frac{x}{2}$ will be decreased by 3. The graph of $y = -3 + \tan \frac{x}{2}$ appears in Figure 3-30.

$y = -3 + \tan \frac{x}{2}$

x	y
0	-3
2π	-3
4π	-3

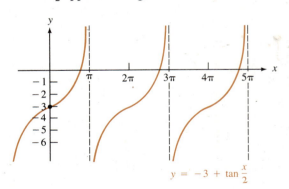

$y = -3 + \tan \frac{x}{2}$

Figure 3-30

The preceding two examples suggest the following facts.

(3.12)

> **Vertical Shift.**   The graph of $y = k + a \sin bx$ is identical to the graph of $y = a \sin bx$, except that it is shifted $|k|$ units
>
> $$\begin{Bmatrix} \text{up} \\ \text{down} \end{Bmatrix} \quad \text{if } k \text{ is} \quad \begin{Bmatrix} \text{positive} \\ \text{negative} \end{Bmatrix}$$
>
> A similar statement is true for each of the other trigonometric functions.

Example 3   Graph $y = \sin\left(x + \dfrac{\pi}{6}\right)$ and $y = \sin x$ on the same set of coordinate axes.

Solution   One complete cycle of $y = \sin x$ is described as x increases from 0 to 2π. Similarly, one cycle of $y = \sin(x + \frac{\pi}{6})$ is completed when $x + \frac{\pi}{6}$ increases from 0 to 2π—that is, when x itself increases from $-\frac{\pi}{6}$ to $(2\pi - \frac{\pi}{6})$. At $x = -\frac{\pi}{6}$, $\sin(x + \frac{\pi}{6})$ is zero. The graph of $y = \sin(x + \frac{\pi}{6})$ looks just like the graph of $y = \sin x$, except that it is offset to the left by a distance of $\frac{\pi}{6}$. The graph appears in Figure 3-31.

$$y = \sin\left(x + \tfrac{\pi}{6}\right)$$

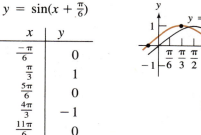

x	y
$-\frac{\pi}{6}$	0
$\frac{\pi}{3}$	1
$\frac{5\pi}{6}$	0
$\frac{4\pi}{3}$	-1
$\frac{11\pi}{6}$	0

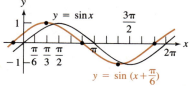

Figure 3-31

The distance that a graph is shifted to the left or to the right is called the **phase shift** of the graph.

(3.13)
Phase Shift.   The graph of $y = a \sin b(x + c)$ is identical to the graph of $y = a \sin bx$, except that it is shifted $|c|$ units to the

$$\begin{Bmatrix} \text{left} \\ \text{right} \end{Bmatrix} \quad \text{if } c \text{ is} \quad \begin{Bmatrix} \text{positive} \\ \text{negative} \end{Bmatrix}.$$

The number $|c|$ is called the **phase shift** of the graph. A similar statement is true for $y = a \cos b(x + c)$.

Example 4   Graph $y = 3 \cos\left(2x - \dfrac{\pi}{3}\right)$.

Solution   Rewrite

$$y = 3 \cos\left(2x - \frac{\pi}{3}\right)$$

in the form $y = a \cos b(x + c)$ by factoring out a 2 from the binomial $2x - \frac{\pi}{3}$. Thus, you have

$$y = 3 \cos\left(2x - \frac{\pi}{3}\right)$$

$$= 3 \cos 2\left(x + \frac{-\pi}{6}\right)$$

From the equation $y = 3 \cos 2(x + \frac{-\pi}{6})$, you can read that the amplitude is 3, the period is $\frac{2\pi}{2} = \pi$, and the graph looks like that of $y = 3 \cos 2x$. However, it is shifted $\frac{\pi}{6}$ units to the right. With $y = 3 \cos 2x$ included for reference, the graph appears in Figure 3-32.

$$y = 3 \cos(2x - \tfrac{\pi}{3})$$

x	y
$\frac{\pi}{6}$	3
$\frac{5\pi}{12}$	0
$\frac{2\pi}{3}$	-3
$\frac{11\pi}{12}$	0
$\frac{7\pi}{6}$	3

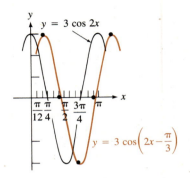

Figure 3-32

Example 5   Graph $y = \tan\left(x - \dfrac{\pi}{4}\right)$.

Solution   One complete cycle of $y = \tan x$ is described as x increases from 0 to π. Thus, one cycle of $y = \tan(x - \frac{\pi}{4})$ is completed when $x - \frac{\pi}{4}$ increases from 0 to π—that is, when x itself increases from $\frac{\pi}{4}$ to $\frac{5\pi}{4}$. At $x = \frac{\pi}{4}$, the value of $\tan(x - \frac{\pi}{4})$ is zero. The graph of $y = \tan(x - \frac{\pi}{4})$ looks just like the graph of $y = \tan x$, except that it is shifted to the right by a distance of $\frac{\pi}{4}$. The graph appears in Figure 3-33.

$$y = \tan(x - \tfrac{\pi}{4})$$

x	y
0	-1
$\frac{\pi}{4}$	0
$\frac{\pi}{2}$	1
$\frac{3\pi}{4}$	undefined
π	-1
$\frac{5\pi}{4}$	0
$\frac{3\pi}{2}$	1
$\frac{7\pi}{4}$	undefined

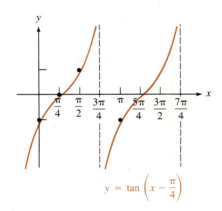

Figure 3-33

The following statement summarizes the facts involving phase shifts of the tangent and the cotangent functions.

(3.14)

Phase Shift.   The graph of

$$\left.\begin{cases} y = a \tan b(x + c) \\ y = a \cot b(x + c) \end{cases}\right\}$$

is identical to the graph of $\left.\begin{cases} y = a \tan bx \\ y = a \cot bx \end{cases}\right\}$

except that it is shifted $|c|$ units to the $\begin{cases} \text{left} \\ \text{right} \end{cases}$ if c is $\begin{cases} \text{positive} \\ \text{negative} \end{cases}$.

The number $|c|$ is called the **phase shift** of the graph.

Example 6   Graph $y = \csc\left(3x - \dfrac{\pi}{2}\right)$.

Solution   Factor 3 out of the binomial $3x - \dfrac{\pi}{2}$ to write the equation in the form $y = \csc b(x + c)$.

$$y = \csc\left(3x - \frac{\pi}{2}\right) = \csc 3\left[x + \left(-\frac{\pi}{6}\right)\right]$$

As with the sine function, the period of the cosecant function is $\dfrac{2\pi}{b}$ or in this case $\dfrac{2\pi}{3}$. The graph looks like that of $y = \csc 3x$, but it is shifted $\dfrac{\pi}{6}$ units to the right. The graph appears in Figure 3-34.

$$y = \csc(3x - \tfrac{\pi}{2})$$

x	y
0	-1
$\dfrac{\pi}{3}$	1
$\dfrac{2\pi}{3}$	-1

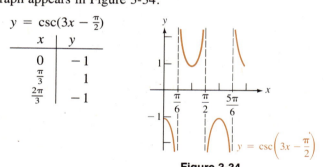

Figure 3-34

The following statement summarizes the facts involving phase shifts of the cosecant and secant functions.

(3.15)

Phase Shift.   The graph of

$$\left.\begin{cases} y = a \csc b(x + c) \\ y = a \sec b(x + c) \end{cases}\right\}$$

is identical to the graph of $\left.\begin{cases} y = a \csc bx \\ y = a \sec bx \end{cases}\right\}$

except that it is shifted $|c|$ units to the $\begin{cases} \text{left} \\ \text{right} \end{cases}$ if c is $\begin{cases} \text{positive} \\ \text{negative} \end{cases}$.

The number $|c|$ is called the **phase shift** of the graph.

Exercise 3.6

In Exercises 1–12, give the number of units and the direction (either up or down) that each trigonometric function has been shifted. Also give the period. **Do not draw the graph.**

1. $y = 2 + \sin x$

2. $y = -4 + \cos x$

3. $y = \tan x - 1$

4. $y = \csc x + 3$

5. $y = 7 + 9 \sec 5x$

6. $y = 7 + 9 \cot 5x$

7. $y = 3 - \sin x$

8. $y = -2 - \cos x$

9. $y + 5 = \csc 2x$

10. $y + 5 = \cot 2x$

11. $y = 2(3 + \tan \pi x)$

12. $y = -4(1 - \sec \pi x)$

In Exercises 13–30, give the period and the phase shift (including direction), if any, of each given function. **Do not draw the graph.**

13. $y = \sin\left(x - \dfrac{\pi}{3}\right)$

14. $y = \cos\left(x + \dfrac{\pi}{4}\right)$

15. $y = \cos\left(x + \dfrac{\pi}{6}\right)$

16. $y = -\sin\left(x - \dfrac{\pi}{2}\right)$

17. $y + 2 = 3 \cos 2\pi x$

18. $y = 3 \sin \dfrac{2x}{\pi}$

19. $y = \tan(x - \pi)$

20. $y = \csc\left(x + \dfrac{\pi}{6}\right)$

21. $y = -\sec\left(x + \dfrac{\pi}{4}\right)$

22. $y = -2 \sec\left(x - \dfrac{\pi}{3}\right)$

23. $y = \sin(2x + \pi)$

24. $y = \cos(2x - \pi)$

25. $y = \tan\left(\dfrac{\pi x}{2} + \dfrac{\pi}{4}\right)$

26. $y = \csc\left(\dfrac{2\pi x}{3} + \dfrac{\pi}{9}\right)$

27. $y = 2 \sec\left(\dfrac{1}{3}x - 6\pi\right)$

28. $y = 2 \cot\left(\dfrac{\pi}{10} + \dfrac{x}{5}\right)$

29. $2y = 3 \cot\left(7x - \dfrac{21}{2}\pi\right)$

30. $17y = \sec\left(\dfrac{x}{5} + \dfrac{\pi}{4}\right)$

In Exercises 31–46, graph each function through at least one period.

31. $y = -4 + \sin x$

32. $y + 2 = \tan \dfrac{x}{2}$

33. $y = 3 - \sec x$

34. $y = 1 + \csc x$

35. $y = \cot \dfrac{x}{2} - 2$

36. $y = 1 - 2 \cos x$

37. $y = \sin\left(x + \dfrac{\pi}{2}\right)$

38. $y = -\cos\left(x - \dfrac{\pi}{2}\right)$

39. $y = \tan\left(x - \dfrac{\pi}{2}\right)$ **40.** $y = \csc\left(x + \dfrac{\pi}{4}\right)$

41. $y = \cos(2x + \pi)$ **42.** $y = \sin(3x - \pi)$

43. $y = \sec\left(3x + \dfrac{\pi}{2}\right)$ **44.** $y = \tan\left(\dfrac{x}{2} - \dfrac{\pi}{2}\right)$

45. $y - 1 = \csc\left(2x - \dfrac{\pi}{6}\right)$ **46.** $y + 2 = \cot\left(\dfrac{x}{3} - \dfrac{\pi}{2}\right)$

3.7   GRAPHS OF OTHER TRIGONOMETRIC FUNCTIONS

There are other graphs of trigonometric functions that are of interest.

Example 1   Graph $y = \sin^4 x$ for all x from 0 to 2π.

Solution   The function can be graphed by taking the fourth power of each y value of $y = \sin x$. Because the exponent is the even number 4, the graph of $y = \sin^4 x$ will never go below the x-axis. A table of values and the graph are shown in Figure 3-35. Note that the shape of this curve differs from the shape of the sine curve.

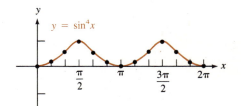

x	$\sin x$	$y = \sin^4 x$
0	0	0
$\frac{\pi}{6}$	.5	.06
$\frac{\pi}{3}$	.87	.56
$\frac{\pi}{2}$	1	1
$\frac{2\pi}{3}$	.87	.56
$\frac{5\pi}{6}$	.5	.06
π	0	0
$\frac{7\pi}{6}$	$-.5$	.06
$\frac{4\pi}{3}$	$-.87$	.56
$\frac{3\pi}{2}$	-1	1
$\frac{5\pi}{3}$	$-.87$	.56
$\frac{11\pi}{6}$	$-.5$	.06
2π	0	0

Figure 3-35

Example 2   Graph $y = x + \cos x$.

Solution   It would be possible to make an extensive table of values, plot the points, and draw the curve. However, there is an easier way. Use your knowledge of the functions $y = x$ and $y = \cos x$, and sketch each one separately as in Figure 3-36. Then pick several numbers x such as x_1, and add the values of x_1 and $\cos x_1$ together to obtain the y value of a point on the graph of $y = x + \cos x$.

For example, if $x_1 = 0$, the value of $\cos x_1 = 1$. Thus, the point $(0, 0 + 1)$ or $(0, 1)$ is on the graph of $y = x + \cos x$. As another example, if $x_1 = 1$, the value of $\cos x_1 \approx 0.54$. Thus, the point $(1, 1 + 0.54)$ or $(1, 1.54)$ is on the graph of $y = x + \cos x$. It is easy to use a compass to add these values of y. Set your compass so that it spans segment BA and transfer that length to form segment CD, locating point D on the desired graph. However, if $x_1 = \pi$, the value of $\cos x_1$ is negative. In this case, you must subtract the magnitudes of the y values. Do this by using a compass to transfer segment MN to position PQ to locate point Q on the desired graph. Repeat this process of adding y values to determine several points. Then join them with a smooth curve to obtain the graph of $y = x + \cos x$.

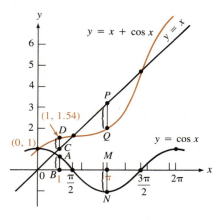

Figure 3-36

Example 3   Graph $y = \cos x + 2 \sin 2x$.

Solution   Proceed as in Example 2. For values of x between 0 and $\frac{\pi}{2}$ such as $\frac{\pi}{4}$, the values of $\cos x$ and $2 \sin 2x$ are both positive. In this case, you must add the y values. Do this by transferring segments such as MN to new positions such as PQ. This locates point Q, which is on the graph of $y = \cos x + 2 \sin 2x$. For values of x that are between π and $\frac{3\pi}{2}$, such as $\frac{5\pi}{4}$, the y value of $2 \sin 2x$ is positive while the y value of $\cos x$ is negative. In this case, subtract the magnitudes of the y values of the two functions. Do this by transferring segments such as AB to new positions such as CD to locate point D on the graph of $y = \cos x + 2 \sin 2x$. See Figure 3-37.

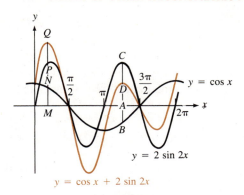

Figure 3-37

The process illustrated in Examples 2 and 3 is called **addition of ordinates.**

Exercise 3.7

Graph each of the following functions. A calculator will be helpful.

1. $y = \cos^4 x$

2. $y = -\tan^4 x$

3. $y = -\sin^3 x$

4. $y = \cos^3 x$

5. $y = \sin^{10} x$

6. $y = \cot^{10} x$

7. $y = \sin^{100} x$

8. $y = \cos^{101} x$

9. $y^3 = \sin x$

10. $y^3 = \tan x$

11. $y = \cos^2 3x$

12. $y = -\sin^2 \dfrac{x}{2}$

13. $y = \sin x + \cos x$

14. $y = 2 \sin x + \cos x$

15. $y = \sin \dfrac{x}{2} + \sin x$

16. $y = x + \sin x$

17. $y = -x + \cos x$

18. $y = x - \cos x$

19. $y = 2 \sin x + 2 \cos \dfrac{x}{2}$

20. $y = \sin^2 x + \cos^2 x$

21. $y = 2 \sin x \cos x$

22. $y = \cos^2 x - \sin^2 x$

23. $y = |\sin x|$

24. $y = |\cos x|$

25. $y = \sin x^2$

26. $y = \sin \dfrac{1}{x}, \quad x \neq 0$

27. $y = \sin \sqrt{x}, \quad x \geq 0$

28. $y = \cos \sqrt{x}, \quad x \geq 0$

3.8 APPLICATIONS OF RADIAN MEASURE

When the domains of the trigonometric functions are considered as real numbers, many applications are made possible. We will mention three of these that come from the fields of electronics and mechanics.

1. Alternating current

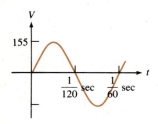

Figure 3-38

The humming sound that you hear in a fluorescent light fixture or a toy train transformer is caused by electric current reversing itself 120 times each second. The voltage available at a wall outlet is described by the formula

$$V = V_0 \sin (2\pi f t)$$

where V_0 is approximately 155 volts, f is the frequency, and t is time. In the United States, the frequency is 60 cycles per second (or 60 hertz), and time is measured in seconds. The graph of $V = V_0 \sin (2\pi f t) = 155 \sin (120 \pi t)$ is given in Figure 3-38. When $t = \frac{1}{60}$, $V = 155 \sin\left(120\pi \cdot \frac{1}{60}\right) = 155 \sin 2\pi = 0$. One cycle is completed in $\frac{1}{60}$ of a second, so there are 60 cycles a second.

2. Simple harmonic motion

If an object is suspended from a ceiling by a spring, and then pushed up and released, it will start to bounce. See Figure 3-39. Its position above (positive) or below (negative) the equilibrium position is given by a formula involving another circular function:

$$y = A \cos \left(\sqrt{\frac{k}{m}}\, t \right)$$

The coefficient A, called the **amplitude,** represents the distance the object is pushed above the equilibrium position. The constant k, called the **spring constant,** depends on the stiffness of the spring, and m is the mass of the object. The variable t represents the time in seconds since the object was released.

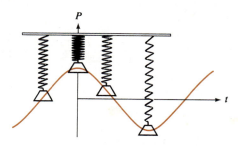

Figure 3-39

The first cycle is completed when

$$\sqrt{\frac{k}{m}}\ t = 2\pi$$

or when

$$t = \sqrt{\frac{m}{k}}\ 2\pi$$

The value

$$\sqrt{\frac{m}{k}}\ 2\pi$$

is called the **period of oscillation** and indicates the number of seconds per complete cycle. Its reciprocal

$$\frac{1}{2\pi}\sqrt{\frac{k}{m}}$$

is the **frequency of the oscillation,** indicating the number of cycles per second.

Example 1 A mass of 4 g, attached to a spring with a spring constant of 9 g/cm, is raised above the equilibrium position and released. Find the period and the frequency of the oscillation.

Solution The period of the oscillation is given by the value

$$\sqrt{\frac{m}{k}}\ 2\pi$$

Substituting **4** for *m* and **9** for *k* gives

$$\begin{aligned}
\text{period} &= \sqrt{\frac{m}{k}}\ 2\pi \\
&= \sqrt{\frac{4}{9}}\ 2\pi \\
&= \frac{2}{3}\cdot 2\pi \\
&= \frac{4\pi}{3}
\end{aligned}$$

It takes the mass $\frac{4\pi}{3}$ seconds, or about 4.2 seconds, to return to its starting point. The frequency of the oscillation is the reciprocal of the period. Thus, you have

$$\text{frequency} = \frac{1}{\text{period}}$$

$$= \frac{3}{4\pi}$$

The frequency is $\frac{3}{4\pi}$, or approximately 0.24 hertz (cycles per second).

Any motion described by an equation of the form

$$y = A \cos kt$$

is called **simple harmonic motion.** We give two more examples of such motion.

A pendulum swinging through a small arc makes an angle θ with the vertical where θ is given by the formula

$$\theta = A \cos \sqrt{\frac{g}{l}}\; t$$

The coefficient A is the maximum amplitude of the swing, l is the length of the pendulum, and g is a constant related to the force of gravity. The motion described by the pendulum is simple harmonic motion.

If a floating object is pushed under water and then released, it will oscillate as it bobs up and down. For certain shapes, the position y relative to the equilibrium position is given by the equation

$$y = A \cos \sqrt{\frac{k}{w}}\; t$$

where A is the maximum displacement from equilibrium, w is the object's weight, and k depends on the density of the water and the force of gravity. The bobbing action of such a floating object also describes simple harmonic motion.

3. Sound waves

When a tuning fork is struck, its tongs vibrate at a rate that depends upon its dimensions and the material of which it is made. See Figure 3-40. The vibrations of the fork set the surrounding air into vibration, and it is these tiny periodic variations in pressure that our ears perceive as a musical tone.

The vibrations of a tuning fork and the surrounding air are described by the equation

$$y = A \sin (2\pi ft)$$

Figure 3-40

where f is the frequency, t is the time, and A is the amplitude of the oscillations.

The first cycle is completed when $2\pi ft = 2\pi$, or when $t = \frac{1}{f}$. The reciprocal of the frequency is the period—the time it takes to complete one cycle.

Sound travels through air at approximately 1100 feet per second. The distance that sound can travel in the time of one cycle is $1100 \cdot \frac{1}{f}$ feet. This distance is called the **wavelength** of that sound in air.

The rate at which the pressure varies determines the frequency, or pitch, of the tone. In the United States, 440 pressure variations per second is the pitch recognized as musical "A." Middle "C" on the piano is 261.6 variations per second. Physicists use a different scale of pitch. For them, middle "C" is 256 variations per second. If the frequency of a tone is doubled, the pitch of the tone is one octave higher. Human ears are sensitive to frequencies between about 30 and 20,000 vibrations per second, although the top end of that range usually decreases with age.

A vibrating tuning fork is very quiet if you hold it in your hand. However, if you touch it to a surface such as a table top, the surface becomes a sounding board, and makes the weak vibrations of the tuning fork much louder. This is because the larger surface is capable of transferring the vibrations to a larger volume of air.

A thorough knowledge of the nature of sound and of vibrations in both solid materials and air is needed for the proper design of the sounding board of such musical instruments as the violin, piano, and guitar. This is because the vibrations of the sounding board also add distinctive colorations to the tone. This is one reason why a grand piano's tone is fuller than that of a spinet.

The analysis of these more complex tones still involves the sine function, but in the form of

$$y = A \sin(2\pi f t) + A_1 \sin(2\pi f_1 t) + A_2 \sin(2\pi f_2 t) + \cdots$$

where $f_1, f_2, \ldots$ and $A_1, A_2 \ldots$ are the frequencies and amplitudes of the overtones present in the original sound.

Exercise 3.8

1. In some European countries, the power distribution network operates at a frequency of 50 hertz. If V_0 is 310 volts in Europe, express V as a function of t.

2. A radio station operates on a federally assigned frequency of 1400 kilohertz (1 kilohertz is 1000 hertz). If radio waves travel at 300 million m/sec, what is the wavelength of the radio station's signal?

3. A spring with a spring constant of 6 Newtons per meter hangs from a ceiling and supports a mass of 24 kg. The mass is pulled to a starting point a few centimeters below the equilibrium position and released. How long will it take to return to the starting point?

4. What is the frequency of the oscillation in Exercise 3?

5. A spring supports a mass of 12 g. The frequency of its oscillation is $\frac{1}{\pi}$ hertz (cycles per second). What is the spring constant? The units of your answer will be dynes per cm.

6. For a given spring, what is the effect on the period of its oscillation if a mass suspended by the spring is doubled?

7. A pendulum 1 m long is set to swinging through a small amplitude. Let $g = 9.8$ m/sec^2. Find the period and the frequency of the oscillation.

8. A cubical box floats on its side and oscillates with a period of $\frac{1}{3}$ second. What is its weight? Let $k = 200,000$ lb/sec^2.

9. One tine of a tuning fork takes 0.0035 seconds to complete one vibration. What is the frequency of the tone?

10. A tuning fork vibrates at a frequency of 256 cycles per second. How long does it take to complete one cycle?

11. Construct the graph of $y = A \sin(2\pi f t)$.

12. Construct the graph of $y = A \cos kt$.

CHAPTER SUMMARY

Key Words

addition of ordinates   (3.7)
amplitude   (3.4)
angular velocity   (3.2)
arc length   (3.1)
argument of
   a function   (3.4)
asymptote   (3.5)
circular functions   (3.3)

frequency of an oscillation   (3.8)
linear velocity   (3.2)
period   (3.4)
periodic function   (3.4)
phase shift   (3.4)
radian   (3.1)
sector of a circle   (3.1)
simple harmonic motion   (3.8)

Key Ideas

(3.1) If a central angle θ of a circle with radius of r intercepts an arc of length s, then the radian measure of angle θ is s/r.

π radians $= 180°$

If a central angle θ in a circle with radius r is in radians, the length s of the intercepted arc is given by the formula $s = r\theta$.

The area of a sector of a circle with radius r and a central angle θ (measured in radians) is given by the formula $A = \frac{1}{2}r^2\theta$.

(3.2) Linear and angular velocity of a point on a rotating wheel are related by the formula $v = r\omega$, where $v = s/t$, $\omega = \theta/t$, and r is the radius of the wheel.

(3.3) It is possible to think of trigonometric functions as functions with domains and ranges that are real numbers.

(3.4) The period of $y = a \sin bx$ and $y = a \cos bx$ is $\left|\dfrac{2\pi}{b}\right|$. The amplitude of each of these functions is $|a|$.

(3.5)   The period of $y = a \tan bx$ and $y = a \cot bx$ is $\left|\dfrac{\pi}{b}\right|$.

The period of $y = a \csc bx$ and $y = a \sec bx$ is $\left|\dfrac{2\pi}{b}\right|$.

(3.6)   The graph of $y = k + a \sin bx$ is identical to the graph of $y = a \sin bx$, except that it is shifted $|k|$ units

$\begin{Bmatrix} \text{up} \\ \text{down} \end{Bmatrix}$ if k is $\begin{Bmatrix} \text{positive} \\ \text{negative} \end{Bmatrix}$

A similar statement is true for each of the other trigonometric functions.

The graph of $y = a \sin b(x + c)$ is identical to the graph of $y = a \sin bx$, except that it is shifted $|c|$ units to the

$\begin{Bmatrix} \text{left} \\ \text{right} \end{Bmatrix}$ if c is $\begin{Bmatrix} \text{positive} \\ \text{negative} \end{Bmatrix}$

A similar statement is true for each of the other trigonometric functions.

(3.7)   Functions that are sums of simpler functions can be graphed by using the method of addition of ordinates.

REVIEW EXERCISES

In Review Exercises 1–4, change each angle to radians.

1. 105°   **2.** 325°   **3.** 318°   **4.** −105°

In Review Exercises 5–8, each angle is expressed in radians. Change each angle to degrees.

5. $\dfrac{19}{6}\pi$ **6.** $-\dfrac{5}{6}\pi$ **7.** 7π **8.** 8

In Review Exercises 9–12, find the values of the given function without using a calculator. Then, check your work with a calculator.

9. $\sin \dfrac{5\pi}{6}$ **10.** $\cos\left(-\dfrac{13}{6}\pi\right)$ **11.** $\tan\left(-\dfrac{\pi}{3}\right)$ **12.** $\csc \dfrac{\pi}{6}$

13. The latitude of Springfield, Illinois, is 39.8° N. How far is Springfield from the equator? Use 3960 mi to estimate the radius of the earth.

14. Des Moines, Iowa, is approximately 2870 mi north of the equator. If the radius of the earth is about 3960 mi, find the latitude of Des Moines.

15. Find the area of a sector of a circle if the sector has a central angle of 15° and the circle has a radius of 12 cm.

16. Find the angular velocity of the earth in rad/sec.

17. A truck is traveling 55 mph. How fast are the 32-in. diameter tires spinning in rpm?

18. A vehicle has 40-in. diameter tires that are making 100 rpm. How fast is the vehicle going in mph?

*In Review Exercises 19–20, find the coordinates of the point P on the unit circle that corresponds to each given real number. **Do not use a calculator.***

19. $\dfrac{7\pi}{6}$

20. $\dfrac{13\pi}{4}$

In Review Exercises 21–28, use a calculator to find the values of the given functions. Give answers to the nearest ten-thousandth.

21. $\sin(5\pi)$

22. $\cos 7$

23. $\tan(2 + \pi)$

24. $\csc 3\pi$

25. $\sec(-1)$

26. $\cot\left(\dfrac{5\pi}{3}\right)$

27. $\sin(\pi^3)$

28. $\cos \sqrt{\pi}$

In Review Exercises 29–32, find the amplitude and the period of the given function.

29. $y = 4 \sin 3x$

30. $y = \dfrac{\cos 4x}{8}$

31. $y = -\dfrac{1}{3} \cos \dfrac{x}{3}$

32. $y = 0.875 \sin \dfrac{1}{4} x$

In Review Exercises 33–36, find the vertical shift and phase shift, if any, of each of the given functions.

33. $y = 2 + \tan x$

34. $y = \csc\left(2x + \dfrac{\pi}{3}\right)$

35. $y - 4 = 3 \sin\left(\dfrac{x}{7} + \dfrac{3}{2}\right)$

36. $y = -\cos\left(\dfrac{1}{5}x - \dfrac{1}{2}\pi\right) - 1$

In Review Exercises 37–50, graph the given function.

37. $y = 4 \sin x$

38. $y = 0.5 \cos x$

39. $y = \cos \dfrac{x}{4}$

40. $y = \tan \dfrac{\pi x}{3}$

41. $y = 3 + \sin x$

42. $y = -2 + \tan x$

43. $y = 2 \sin\left(x - \dfrac{5\pi}{6}\right)$

44. $y = \tan\left(x - \dfrac{2\pi}{3}\right)$

45. $y = \dfrac{1}{2} \tan\left(x + \dfrac{\pi}{4}\right)$

46. $y = 2 \sin 3x$

47. $y = 2 \sin^2 x$

48. $y = \dfrac{1}{2} \cos^2 x$

49. $y = 2 \cos x + \sin \dfrac{x}{2}$

50. $y = \dfrac{x}{2} + \cos \dfrac{x}{2}$

CHAPTER FOUR

TRIGONOMETRIC IDENTITIES

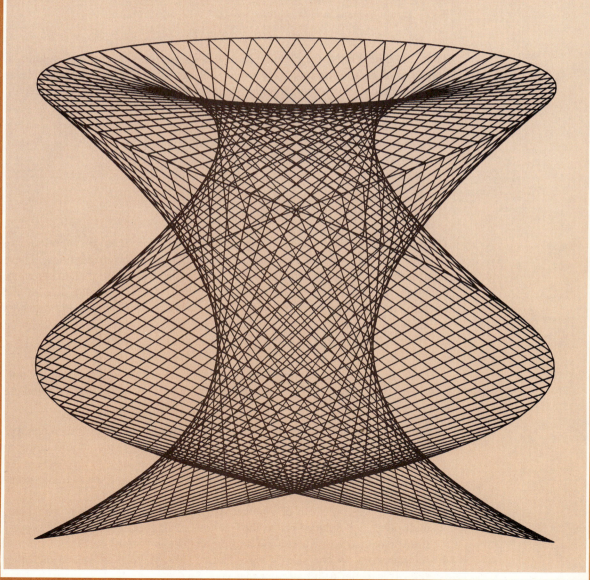

We have used trigonometry to solve problems such as finding heights of flagpoles and distances across rivers. However, trigonometry is useful in another, more theoretical way. Because the trigonometric functions are related to each other and to certain algebraic expressions, they are often used to transform difficult and awkward mathematical expressions into more manageable ones. To explore some of these relationships and to gain facility in manipulating trigonometric expressions, we turn our attention to trigonometric identities.

4.1   TRIGONOMETRIC IDENTITIES

There are three kinds of equations: **impossible equations** that have no solutions, **conditional equations** that have some, but not all, values of the variable as solutions, and **identities** that have all values of the variable as solutions. Because this chapter deals with identities, we formalize a definition.

(4.1)

> **Definition.**   An **identity** is an equation that is true for every value of its variable(s) for which each side of the equation is defined.

The equations

$$3(x + 2) = 3x + 6 \quad \text{and} \quad x^2 - y^2 = (x + y)(x - y)$$

for example, are true for all values of x and y. Thus, they are identities. In Section 2.2, you studied several fundamental trigonometric identities. These relationships and the rules of algebra will be used to show that many more equations are identities. In doing so, we usually work with one side of the equation only, manipulating it until it is transformed into the other side. Success at verifying identities comes only with practice. Sometimes the work may be frustrating, but it is a skill that is indispensable in later work.

Hints on proving identities

Here are some suggestions that will make the work easier. They do not, however, guarantee success, because each identity presents its own unique challenge.

1. You must memorize the fundamental relationships on page 30. Whenever you see one side of one of these identities, the other side should come to mind immediately.
2. Start with the more complicated side of the equation, and try to transform it into the other side.
3. Do not forget the rules of algebra. Trigonometric fractions may be added, multiplied, and reduced just as the fractions of algebra. Trigonometric expressions may also be factored and similar terms combined.
4. As you work on one side of the equation, always keep an eye on the other side. It is easier to hit a target that you can see.

5. If one side of the equation contains only a single trigonometric function, eliminate all other functions, if possible, from the other side.
6. It is sometimes helpful to change all functions into sine and cosine functions before proceeding.
7. If the numerator or denominator of a fraction contains a factor of $1 + \sin x$ (or $1 - \sin x$), consider multiplying both the numerator and the denominator by the conjugate $1 - \sin x$ (or $1 + \sin x$). This operation creates a factor of $1 - \sin^2 x$, which may be replaced with $\cos^2 x$. The same idea applies to factors of $1 \pm \cos x$, $\sec x \pm 1$, and $\sec x \pm \tan x$.

Example 1   Verify that $\dfrac{\tan x}{\sin x} = \sec x$ is an identity.

Solution   Work on the left side because it is the more complicated.

$$\begin{array}{l|l}
\dfrac{\tan x}{\sin x} & \sec x \\[2em]
= \dfrac{\dfrac{\sin x}{\cos x}}{\dfrac{\sin x}{1}} & \text{Replace } \tan x \text{ with } \dfrac{\sin x}{\cos x}. \\[2em]
= \dfrac{\sin x}{\cos x} \cdot \dfrac{1}{\sin x} & \text{Invert and multiply.} \\[1.5em]
= \dfrac{1}{\cos x} & \text{Divide out the } \sin x. \\[1.5em]
= \sec x & \text{Replace } \dfrac{1}{\cos x} \text{ with } \sec x.
\end{array}$$

Because the left side has been transformed into the right side, the identity is verified. ■

Example 2   Verify the identity $\tan x + \cot x = \csc x \sec x$.

Solution   Work with the left side of this equality and change $\tan x$ and $\cot x$ into expressions containing $\sin x$ and $\cos x$. Then proceed as follows.

$$\begin{array}{l|l}
\tan x + \cot x & \csc x \sec x \\[1.5em]
= \dfrac{\sin x}{\cos x} + \dfrac{\cos x}{\sin x} & \\[1.5em]
= \dfrac{\sin x \sin x}{\sin x \cos x} + \dfrac{\cos x \cos x}{\sin x \cos x} & \begin{array}{l}\text{Change each fraction to a}\\\text{fraction with a common}\\\text{denominator.}\end{array} \\[1.5em]
= \dfrac{\sin^2 x + \cos^2 x}{\sin x \cos x} & \text{Add the fractions.}
\end{array}$$

$$= \frac{1}{\sin x \cos x}$$ Replace $\sin^2 x + \cos^2 x$ with 1.

$$= \frac{1}{\sin x} \cdot \frac{1}{\cos x}$$ Rewrite as two fractions.

$$= \csc x \sec x$$

Because the left side has been transformed into the right side, the identity is verified. ■

Example 3   Verify that $1 + \tan x = \sec x(\cos x + \sin x)$.

Solution   Applying the distributive property to the right side and remembering that $\sec x \cos x = 1$ is the strategy:

$$1 + \tan x \quad \bigg| \quad \sec x(\cos x + \sin x)$$

$$= \sec x \cos x + \sec x \sin x \qquad \text{Remove parentheses.}$$

$$= 1 + \sec x \sin x \qquad \text{Recall that } \sec x \cos x = 1.$$

$$= 1 + \frac{\sin x}{\cos x} \qquad \text{Recall that } \sec x = \frac{1}{\cos x}.$$

$$= 1 + \tan x \qquad \text{Recall that } \frac{\sin x}{\cos x} = \tan x.$$

The identity is verified. ■

Example 4   Verify that $(1 + \tan^2 x) \cos^2 x = 1$.

Solution   Change the terms of the left side into sines and cosines and remove the parentheses.

$$(1 + \tan^2 x)\cos^2 x \quad \bigg| \quad 1$$

$$= \left(1 + \frac{\sin^2 x}{\cos^2 x}\right) \cos^2 x \qquad \text{Recall that } \tan^2 x = \frac{\sin^2 x}{\cos^2 x}.$$

$$= \cos^2 x + \sin^2 x \qquad \text{Remove parentheses.}$$

$$= 1 \qquad \text{Recall that } \cos^2 x + \sin^2 x = 1.$$

The identity is verified. ■

Example 5   Verify that $\cos^4 x - \sin^4 x = 1 - 2 \sin^2 x$.

Solution   The left side is the difference of two squares and can be factored. Because your target—the expression on the right side—involves only the sine function, eliminate the cosine function from the left side.

$$\cos^4 x - \sin^4 x \quad \bigg| \quad 1 - 2 \sin^2 x$$

$$= (\cos^2 x + \sin^2 x)(\cos^2 x - \sin^2 x)$$

$$= 1 \cdot (\cos^2 x - \sin^2 x)$$

$$= \cos^2 x - \sin^2 x$$

$$= 1 - \sin^2 x - \sin^2 x$$
$$= 1 - 2 \sin^2 x$$

Remember that $\cos^2 x = 1 - \sin^2 x$.

The identity is verified.   ■

Example 6   Verify that $\dfrac{1}{\sec x - \tan x} - \dfrac{1}{\sec x + \tan x} = 2 \tan x$.

Solution   Find the common denominator for the fractions on the left and add them.

$$\frac{1}{\sec x - \tan x} - \frac{1}{\sec x + \tan x}$$

$$= \frac{(\sec x + \tan x)}{(\sec x - \tan x)(\sec x + \tan x)} - \frac{(\sec x - \tan x)}{(\sec x + \tan x)(\sec x - \tan x)}$$

$$= \frac{\sec x + \tan x - \sec x + \tan x}{\sec^2 x - \tan^2 x}$$

$$= \frac{2 \tan x}{1}$$

$$= 2 \tan x$$

$2 \tan x$

The identity is verified.   ■

In Examples 1–6, all of the work done in verifying each identity was performed on one side of the equation only. Occasionally, if the equation seems to warrant it, mathematicians work on both sides of the identity independently until each side is transformed into a common third expression. When following this strategy, it is important that each step in the process be reversible so that each side of the identity can be derived from the common third expression. It is also important to note that each side must be worked on independently. It is incorrect to multiply or divide both sides of an equation whose truth you are trying to establish by an expression containing a variable, because the resulting equation might not be equivalent to the given equation.

Example 7   Verify that $\dfrac{1 - \cos x}{1 + \cos x} = (\csc x - \cot x)^2$.

Solution   In this example, change the left side and the right side of the equation to a common third expression. The left side can be changed as follows:

$$\frac{1 - \cos x}{1 + \cos x} = \frac{(1 - \cos x)(1 - \cos x)}{(1 + \cos x)(1 - \cos x)}$$

$$= \frac{(1 - \cos x)^2}{1 - \cos^2 x}$$

$$= \frac{(1 - \cos x)^2}{\sin^2 x}$$

$$= \left(\frac{1 - \cos x}{\sin x}\right)^2$$

The right side can be changed as follows:

$$(\csc x - \cot x)^2 = \left(\frac{1}{\sin x} - \frac{\cos x}{\sin x}\right)^2$$

$$= \left(\frac{1 - \cos x}{\sin x}\right)^2$$

Each side has been transformed independently into the expression

$$\left(\frac{1 - \cos x}{\sin x}\right)^2$$

Because each step is reversible, it follows that the given equation is an identity. ◼

Example 8   For what values of x is $\sqrt{1 - \cos^2 x} = \sin x$?

Solution   Take the square root of both sides of the identity $1 - \cos^2 x = \sin^2 x$.

$$\sqrt{1 - \cos^2 x} = \sqrt{\sin^2 x}$$
$$\sqrt{1 - \cos^2 x} = |\sin x|$$

The equation $\sqrt{1 - \cos^2 x} = |\sin x|$ is an identity, but because $\sin x$ is sometimes negative, $\sqrt{1 - \cos^2 x} = \sin x$ is *not* an identity. If you think of x as a real number from 0 to 2π, $\sin x$ is nonnegative for $0 \leqslant x \leqslant \pi$. Hence, $\sqrt{1 - \cos^2 x} = \sin x$ only if x is a number from 0 to π, from 2π to 3π, from 4π to 5π, and so on. ◼

When verifying identities, always remember to write the variable associated with a trigonometric function. The notation "$\cos x$" represents a numerical value, but the notation "cos" is meaningless.

Exercise 4.1

SET A   *In Exercises 1–30, indicate whether the statement is an identity. If so, verify it. If not, explain why. Remember that, if an equation is false for one value of its variable, it cannot be an identity.*

1. $\sin x + \sin x = 2 \sin x$

2. $\sin x + \sin(\pi - x) = 0$

3. $\sin \dfrac{1}{x} = \csc x$

4. $\sec = \dfrac{1}{\cos}$

5. $\sin x + \cos x = 1$

6. $\sec^2 x - \tan^2 x = 1$

7. $(\sin x + \cos x)^2 = 1 + 2 \sin x \cos x$ **8.** $(\sec x - \tan x)^2 = 1$

9. $\cos^2 x + \sin^4 x = 1$ **10.** $\sqrt{\sin^2 x + \cos^2 x} = \sin x + \cos x$

11. $\dfrac{1}{\sin} = \csc$ **12.** $\dfrac{1}{\tan x} = \dfrac{\cos x}{\sin x}$

13. $\dfrac{1}{\sin x} = \cos x$ **14.** $\tan \alpha \cos \alpha \csc \alpha = 1$

15. $\cot \beta \sec \beta \sin \beta = 1$ **16.** $\sin^2 \alpha + \cos^2 \beta = 1$

17. $1 + \tan^2 x = \sec^2 y$ **18.** $\sqrt{1 - \sin^2 x} = |\cos x|$

19. $\sqrt{1 - \sin^2 x} = \cos x$ **20.** $\sqrt{1 - \cos^2 x} = \sin x$

21. $\sin^4 x - \cos^4 x = \sin^2 x - \cos^2 x$

22. $\sin^2 x + 2 \sin x + 1 = (\sin x + 1)^2$

23. $(1 + \sin x)(1 - \sin x) = \cos^2 x$ **24.** $(1 - \cos x)(1 + \cos x) = \sin^2 x$

25. $\sec^2 z + 1 = \tan^2 z$ **26.** $\sin(60° - x) = \sin 60° - \sin x$

27. $\cos(\pi + x) + \cos x = \pi$ **28.** $\tan(\pi + x) - \tan x = 0$

29. $\sin 2x = 2 \sin x$ **30.** $\cos 2x = 2 \cos x$

In Exercises 31–70, verify the given identity.

31. $\dfrac{1 - \cos^2 x}{\sin x} = \sin x$ **32.** $\dfrac{\cot x}{\csc x} = \cos x$

33. $\dfrac{1 + \tan^2 x}{\sec^2 x} = 1$ **34.** $\tan x \csc x = \sec x$

35. $\dfrac{\sin x}{\cos^2 x - 1} = -\csc x$ **36.** $\dfrac{1}{1 - \sin^2 x} = 1 + \tan^2 x$

37. $\dfrac{1 - \cos^2 x}{1 - \sin^2 x} = \tan^2 x$ **38.** $\sin^2 x + \cos^2 x = \cos^2 x \sec^2 x$

39. $\dfrac{1 - \sin^2 x}{1 + \tan^2 x} = \cos^4 x$ **40.** $\tan x \sin x = \dfrac{\csc x}{\cot x + \cot^3 x}$

41. $\dfrac{\cos x(\cos x + 1)}{\sin x} = \cos x \cot x + \cot x$

42. $\cos^2 x \csc x - \csc x = -\sin x$

43. $\sin^2 x \sec x - \sec x = -\cos x$

44. $(\sin x - \cos x)(1 + \sin x \cos x) = \sin^3 x - \cos^3 x$

45. $\sin^2 x - \tan^2 x = -\sin^2 x \tan^2 x$ **46.** $\dfrac{1 + \cot x}{\csc x} = \sin x + \cos x$

47. $\dfrac{1 - \csc x}{\cot x} = \tan x - \sec x$ **48.** $\dfrac{\cos^2 x - \tan^2 x}{\sin^2 x} = \cot^2 x - \sec^2 x$

49. $\dfrac{1}{\sec x - \tan x} = \tan x + \sec x$ **50.** $\csc^2 x + \sec^2 x = \csc^2 x \sec^2 x$

51. $\dfrac{\cos x}{\cot x} + \dfrac{\sin x}{\tan x} = \sin x + \cos x$ **52.** $\dfrac{\cos x}{1 + \sin x} = \sec x - \tan x$

53. $\dfrac{\cos x}{1 - \sin x} = \dfrac{1 + \sin x}{\cos x}$

54. $\cos^2 x + \sin x \cos x = \dfrac{\cos x(\cot x + 1)}{\csc x}$

55. $(\sin x + \cos x)^2 + (\sin x - \cos x)^2 = 2$

56. $\dfrac{\sec x + 1}{\tan x} = \dfrac{\tan x}{\sec x - 1}$

57. $\dfrac{\cos x - \cot x}{\cos x \cot x} = \dfrac{\sin x - 1}{\cos x}$ **58.** $\cos^4 x - \sin^4 x = 2 \cos^2 x - 1$

59. $\dfrac{1}{1 + \sin x} + \dfrac{1}{1 - \sin x} = 2 \sec^2 x$

60. $\sqrt{\dfrac{1 - \sin x}{1 + \sin x}} = \sec x - \tan x$ with $0 < x < \dfrac{\pi}{2}$

61. $\sqrt{\dfrac{1 - \sin x}{1 + \sin x}} = \dfrac{1 - \sin x}{\cos x}$ with $0 < x < \dfrac{\pi}{2}$

62. $-\sqrt{\dfrac{\sec x - 1}{\sec x + 1}} = \dfrac{1 - \sec x}{\tan x}$ with $0 < x < \dfrac{\pi}{2}$

63. $\sqrt{\dfrac{\csc x - \cot x}{\csc x + \cot x}} = \dfrac{\sin x}{1 + \cos x}$ with $0 < x < \dfrac{\pi}{2}$

64. $\dfrac{\cos x}{1 - \sin x} - \dfrac{1}{\cos x} = \tan x$

65. $\dfrac{1}{\sec x(1 + \sin x)} = \sec x(1 - \sin x)$ **66.** $\dfrac{\cot x - \cos x}{\cot x \cos x} = \dfrac{1 - \sin x}{\cos x}$

67. $\dfrac{1}{\tan x(\csc x + 1)} = \tan x(\csc x - 1)$

68. $\dfrac{(\cos x + 1)^2}{\sin^2 x} = 2 \csc^2 x + 2 \csc x \cot x - 1$

69. $\dfrac{\csc x}{\sec x - \csc x} = \dfrac{\cot^2 x + \csc x \sec x + 1}{\tan^2 x - \cot^2 x}$

70. $(\sin x + \cos x)^2 - (\sin x - \cos x)^2 = 4 \sin x \cos x$

SET B *In Exercises 71–86, verify the given identity.*

71. $\dfrac{\sin^2 x \tan x + \sin^2 x}{\sec x - 2 \sin x \tan x} = \dfrac{\sin^2 x}{\cos x - \sin x}$

72. $\dfrac{\sin x + \cos x + 1}{\sin x + \cos x - 1} = \csc x \sec x + \csc x + \sec x + 1$

73. $\dfrac{\sec x + \tan x}{\csc x + \cot x} = \dfrac{\cot x - \csc x}{\tan x - \sec x}$

74. $\dfrac{\cos x + \sin x}{\cos x - \sin x} = \dfrac{\csc x + 2\cos x}{\csc x - 2\sin x}$

75. $\dfrac{3\cos^2 x + 11\sin x - 11}{\cos^2 x} = \dfrac{3\sin x - 8}{1 + \sin x}$

76. $\dfrac{3\sin^2 x + 5\cos x - 5}{\sin^2 x} = \dfrac{3\cos x - 2}{1 + \cos x}$

77. $\dfrac{\cos x + \sin x + 1}{\cos x - \sin x - 1} = \dfrac{1 + \cos x}{-\sin x}$

78. $\dfrac{1 + \sin x + \cos x}{1 + \sin x - \cos x} = \dfrac{\sin x}{1 - \cos x}$

79. $\dfrac{1 + \sin x}{\cos x} = \dfrac{\sin x - \cos x + 1}{\sin x + \cos x - 1}$

80. $\dfrac{\tan x + \sec x + 1}{\tan x + \sec x - 1} = \dfrac{1 + \cos x}{\sin x}$

81. $\dfrac{\csc x + 1 + \cot x}{\csc x + 1 - \cot x} = \dfrac{1 + \cos x}{\sin x}$

82. $\dfrac{3\cot x \csc x - 2\csc^2 x}{\csc^2 x + \cot x \csc x} - 3 = -5(\csc^2 x - \csc x \cot x)$

83. $\dfrac{\sin^3 x \cos x + \cos x - \sin^2 x \cos x - \cos x \sin x}{\cos^4 x} = \sec x - \tan x$

84. $\dfrac{\sin x + \sin x \cos x - \sin x \cos^2 x - \sin x \cos^3 x}{\cos^4 x - 2\cos^2 x + 1} = \csc x + \cot x$

85. $\dfrac{1}{\sec x + \csc x - \sec x \csc x} = \dfrac{\sin x + \cos x + 1}{2}$

86. $\dfrac{2(\sin x + 1)}{1 + \cot x + \csc x} = \sin x + 1 - \cos x$

4.2 IDENTITIES INVOLVING SUMS AND DIFFERENCES OF TWO ANGLES

We will often encounter expressions that contain a trigonometric function of either a sum or a difference of two angles. It is tempting to believe that an expression such as $\cos(A + B)$, for example, is equal to $\cos A + \cos B$. However, this is *not* true, as the following work shows.

$$\cos(30° + 60°) = \cos 90° = 0$$
$$\cos 30° + \cos 60° = \frac{\sqrt{3}}{2} + \frac{1}{2} = \frac{\sqrt{3} + 1}{2}$$

Hence, we have

$$\cos(30° + 60°) \neq \cos 30° + \cos 60°$$

In general, a trigonometric function of the sum (or difference) of two angles is not equal to the sum (or difference) of the trigonometric functions of each angle. It is possible, however, to develop formulas for finding a trigonometric function of the sum (or difference) of two angles. We will use the distance formula to derive a formula to evaluate the expression $\cos(A + B)$.

We draw angles A and $-B$ in standard position on the unit circle as in Figure 4-1. We locate point R on the circle so that angle POR is equal to angle B. We form triangles POQ and ROS. Because these two isosceles triangles have equal vertex angles, they are congruent. Hence, $RS = PQ$. The coordinates of points P, Q, R, and S are:

P: $(\cos A, \sin A)$
Q: $(\cos [-B], \sin [-B]) = (\cos B, -\sin B)$
R: $(\cos [A + B], \sin [A + B])$
S: $(1, 0)$

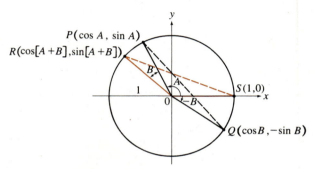

Figure 4-1

Because $RS = PQ$, their squares are also equal. Hence, $RS^2 = PQ^2$. By the distance formula, we have

$$[\cos(A + B) - 1]^2 + [\sin(A + B) - 0]^2$$
$$= (\cos A - \cos B)^2 + (\sin A + \sin B)^2$$

or

$$\cos^2(A + B) - 2\cos(A + B) + 1 + \sin^2(A + B)$$
$$= \cos^2 A - 2\cos A \cos B + \cos^2 B + \sin^2 A + 2\sin A \sin B + \sin^2 B$$

Because $\cos^2 \theta + \sin^2 \theta = 1$ for any angle θ, the previous equation can be written as

$$1 - 2\cos(A + B) + 1 = 1 - 2\cos A \cos B + 1 + 2\sin A \sin B$$
$$2 - 2\cos(A + B) = 2 - 2\cos A \cos B + 2\sin A \sin B$$
$$-2\cos(A + B) = -2\cos A \cos B + 2\sin A \sin B$$

Dividing both sides of the equation above by -2 gives the identity*

(4.2) $\cos(A + B) = \cos A \cos B - \sin A \sin B$

---

*Keep in mind that this formula, and all others developed in this chapter, are also true if A and B represent real numbers.

Example 1 Find cos 75° without using tables or a calculator.

Solution Note that cos 75° = cos(45° + 30°) and use the formula for the cosine of the sum of two angles.

$$\cos(A + B) = \cos A \cos B - \sin A \sin B$$
$$\cos(45° + 30°) = \cos 45° \cos 30° - \sin 45° \sin 30°$$
$$= \frac{\sqrt{2}}{2} \cdot \frac{\sqrt{3}}{2} - \frac{\sqrt{2}}{2} \cdot \frac{1}{2}$$
$$\cos 75° = \frac{\sqrt{6} - \sqrt{2}}{4}$$

■

The identity for the cosine of the difference of two angles follows from the fact that $A - B = A + (-B)$.

$$\cos(A - B) = \cos[A + (-B)]$$
$$= \cos A \cos(-B) - \sin A \sin(-B)$$

Because $\cos(-B) = \cos B$ and $\sin(-B) = -\sin B$, this result can be rewritten as

(4.3) $$\cos(A - B) = \cos A \cos B + \sin A \sin B$$

Example 2 Find the value of $\cos \frac{\pi}{12}$ without using tables or a calculator.

Solution Because $\frac{\pi}{12} = \frac{\pi}{4} - \frac{\pi}{6}$, it follows that

$$\cos \frac{\pi}{12} = \cos\left(\frac{\pi}{4} - \frac{\pi}{6}\right)$$
$$= \cos\frac{\pi}{4} \cos\frac{\pi}{6} + \sin\frac{\pi}{4} \sin\frac{\pi}{6}$$
$$= \frac{\sqrt{2}}{2} \cdot \frac{\sqrt{3}}{2} + \frac{\sqrt{2}}{2} \cdot \frac{1}{2}$$
$$= \frac{\sqrt{6} + \sqrt{2}}{4}$$

■

In Section 2.4, you learned that if θ is an acute angle, any trigonometric function of θ is equal to the cofunction of the complement of θ. This property may be extended to all angles. To this end, we find a value for $\cos(90° - \theta)$.

$$\cos(90° - \theta) = \cos 90° \cos \theta + \sin 90° \sin \theta$$
$$= 0 \cdot \cos \theta + 1 \cdot \sin \theta$$
$$= \sin \theta$$

Thus, for any angle θ, we have

(4.4) $\sin \theta = \cos(90° - \theta)$

Now note that $\theta = 90° - (90° - \theta)$ and find a value for $\cos \theta$.

$$\begin{aligned}
\cos \theta &= \cos[90° - (90° - \theta)] \\
&= \cos 90° \cos(90° - \theta) + \sin 90° \sin(90° - \theta) \\
&= \sin(90° - \theta)
\end{aligned}$$

Thus, for any angle θ, we have

(4.5) $\cos \theta = \sin(90° - \theta)$

Equations (4.4) and (4.5) show that the sine of any angle θ is equal to the cosine of $(90° - \theta)$ and that the cosine of any angle θ is equal to the sine of $(90° - \theta)$. These facts can be used to develop a formula for $\sin(A + B)$.

If $A + B$ is substituted for θ in Equation (4.4), we have

$$\begin{aligned}
\sin(A + B) &= \cos[90° - (A + B)] \\
&= \cos[(90° - A) - B] \\
&= \cos(90° - A)\cos B + \sin(90° - A)\sin B
\end{aligned}$$

Because $\cos(90° - A) = \sin A$ and $\sin(90° - A) = \cos A$, it follows that

(4.6) $\sin(A + B) = \sin A \cos B + \cos A \sin B$

To find a formula for $\sin(A - B)$, proceed as follows.

$$\begin{aligned}
\sin(A - B) &= \sin[A + (-B)] \\
&= \sin A \cos(-B) + \cos A \sin(-B) \\
&= \sin A \cos B - \cos A \sin B
\end{aligned}$$

This gives the identity

(4.7) $\sin(A - B) = \sin A \cos B - \cos A \sin B$

Example 3   Derive a formula for $\sin(A + B + C)$.

Solution
$$\begin{aligned}
\sin(A + B + C) &= \sin[(A + B) + C] \\
&= \sin(A + B)\cos C + \cos(A + B)\sin C \\
&= (\sin A \cos B + \cos A \sin B)\cos C \\
&\quad + (\cos A \cos B - \sin A \sin B)\sin C \\
&= \sin A \cos B \cos C + \cos A \sin B \cos C \\
&\quad + \cos A \cos B \sin C - \sin A \sin B \sin C
\end{aligned}$$

To find identities for the tangent of the sum and difference of two angles, we make use of the identity

$$\tan x = \frac{\sin x}{\cos x}$$

and substitute $A + B$ for x.

$$\tan(A + B) = \frac{\sin(A + B)}{\cos(A + B)}$$

$$= \frac{\sin A \cos B + \cos A \sin B}{\cos A \cos B - \sin A \sin B}$$

To simplify this result, we divide both the numerator and denominator of the fraction by $\cos A \cos B$.

$$\tan(A + B) = \frac{\dfrac{\sin A \cos B}{\cos A \cos B} + \dfrac{\cos A \sin B}{\cos A \cos B}}{\dfrac{\cos A \cos B}{\cos A \cos B} - \dfrac{\sin A \sin B}{\cos A \cos B}}$$

$$= \frac{\tan A + \tan B}{1 - \tan A \tan B}$$

This gives the identity

(4.8) $$\tan(A + B) = \frac{\tan A + \tan B}{1 - \tan A \tan B}$$

If we substitute $-B$ for B in Equation (4.8) and simplify, we obtain an identity for the tangent of the difference of two angles.

(4.9) $$\tan(A - B) = \frac{\tan A - \tan B}{1 + \tan A \tan B}$$

Example 4   Given that $\sin \alpha = \dfrac{12}{13}$, α in QI, $\cos \beta = -\dfrac{4}{5}$, and β in QII, find $\sin(\beta - \alpha)$.

Solution   Because $\sin \alpha = \frac{12}{13}$ and α is in QI, you can draw Figure 4-2(a) and determine that $\cos \alpha = \frac{5}{13}$. Because $\cos \beta = -\frac{4}{5}$ and β is in QII, you can draw Figure 4-2(b) and determine that $\sin \beta = \frac{3}{5}$. You can substitute $\frac{3}{5}$ for $\sin \beta$, $\frac{5}{13}$ for $\cos \alpha$, $-\frac{4}{5}$ for $\cos \beta$,   and $\frac{12}{13}$ for $\sin \alpha$ in the following identity and simplify.

$$\sin(\beta - \alpha) = \sin \beta \cos \alpha - \cos \beta \sin \alpha$$

$$= \frac{3}{5}\left(\frac{5}{13}\right) - \left(-\frac{4}{5}\right)\left(\frac{12}{13}\right)$$

$$= \frac{15}{65} + \frac{48}{65}$$

$$= \frac{63}{65}$$

Thus, you have

$$\sin(\beta - \alpha) = \frac{63}{65}$$

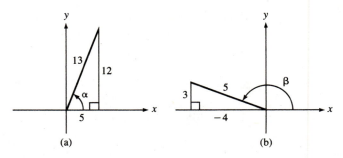

Figure 4-2

Example 5   Verify the identity $\tan(\theta + 45°) = \dfrac{1 + \tan \theta}{1 - \tan \theta}$.

Solution   Use the formula for the tangent of the sum of two angles.

$$
\begin{array}{c|c}
\tan(\theta + 45°) & \dfrac{1 + \tan \theta}{1 - \tan \theta} \\[2ex]
= \dfrac{\tan \theta + \tan 45°}{1 - \tan \theta \tan 45°} & \\[2ex]
= \dfrac{\tan \theta + 1}{1 - \tan \theta} & \\[2ex]
= \dfrac{1 + \tan \theta}{1 - \tan \theta} &
\end{array}
$$

The identity is verified.

Exercise 4.2

1. Find $\sin 195°$ from the trigonometric functions of $45°$ and $150°$.

2. Find $\cos 195°$ from the trigonometric functions of $45°$ and $150°$.

3. Find $\tan 195°$ from the trigonometric functions of $225°$ and $30°$.

4. Find $\tan 165°$ from the trigonometric functions of $210°$ and $45°$.

5. Find $\cos \dfrac{11\pi}{12}$ from the trigonometric functions of $\dfrac{\pi}{6}$ and $\dfrac{3\pi}{4}$.

6. Find $\sin \dfrac{11\pi}{12}$ from the trigonometric functions of $\dfrac{\pi}{6}$ and $\dfrac{3\pi}{4}$.

7. Find $\cos \dfrac{19\pi}{12}$ from the trigonometric functions of $\dfrac{11\pi}{6}$ and $\dfrac{\pi}{4}$.

8. Find $\tan \dfrac{19\pi}{12}$ from the trigonometric functions of $\dfrac{11\pi}{6}$ and $\dfrac{\pi}{4}$.

In Exercises 9–16, choose two appropriate angles and use an identity of the section to evaluate each expression. **Do not use a calculator.**

9. $\sin 255°$

10. $\cos 285°$

11. $\tan 105°$

12. $\cot 255°$

13. $\cos \dfrac{\pi}{12}$

14. $\sin \dfrac{7\pi}{12}$

15. $\sin \dfrac{5\pi}{12}$

16. $\cos \dfrac{13\pi}{12}$

17. Show that $\sin(60° + \theta) = \dfrac{\sqrt{3}}{2}\cos\theta + \dfrac{1}{2}\sin\theta$.

18. Show that $\cos\left(\dfrac{\pi}{2} + x\right) = -\sin x$.

19. Show that $\tan(\pi + x) = \tan x$.

20. Show that $\sin\left(\dfrac{3\pi}{2} - x\right) = -\cos x$.

21. Show that $\cos(\pi - x) = -\cos x$.

22. Show that $\tan\left(\dfrac{\pi}{4} - x\right) = \dfrac{1 - \tan x}{1 + \tan x}$.

In Exercises 23–30, express each quantity as a single function of one angle.

23. $\sin 10° \cos 30° + \cos 10° \sin 30°$

24. $\cos 20° \cos 30° - \sin 20° \sin 30°$

25. $\dfrac{\tan 75° + \tan 40°}{1 - \tan 75° \tan 40°}$

26. $\sin 100° \cos 80° - \cos 100° \sin 80°$

27. $\cos 120° \cos 40° + \sin 120° \sin 40°$

28. $\dfrac{\tan 37° - \tan 125°}{1 + \tan 37° \tan 125°}$

29. $\sin x \cos 2x + \sin 2x \cos x$

30. $\cos 2y \cos 3y - \sin 2y \sin 3y$

31. Given that $\sin\alpha = \frac{3}{5}$, α in QII, and $\cos\beta = -\frac{12}{13}$, and β in QII, find $\sin(\alpha + \beta)$ and $\cos(\alpha - \beta)$.

32. Given that $\sin\alpha = \frac{7}{25}$ with α in QI, and $\sin\beta = \frac{15}{17}$ with β in QI, find $\sin(\alpha - \beta)$ and $\cos(\alpha + \beta)$.

33. Given that $\tan\alpha = -\frac{5}{12}$ with α in QII, and $\tan\beta = \frac{15}{8}$ with β in QI, find $\tan(\alpha + \beta)$ and $\tan(\alpha - \beta)$.

34. Given that $\sin\alpha = \frac{12}{13}$ with α in QI, and $\sin(\alpha + \beta) = \frac{24}{25}$ with $\alpha + \beta$ in QII, find $\sin\beta$ and $\cos\beta$.

35. Given that $\cos\beta = -\frac{15}{17}$ with β in QII, and $\sin(\alpha - \beta) = -\frac{24}{25}$ with $\alpha - \beta$ in QIV, find $\sin\alpha$ and $\cos\alpha$.

36. Given that $\sin(\alpha + \beta) = \frac{3}{5}$ with $\alpha + \beta$ in QI, and $\cos(\alpha - \beta) = \frac{12}{13}$ with $\alpha - \beta$ in QIV, find $\sin 2\alpha$ and $\cos 2\beta$.

In Exercises 37–51, verify each identity.

37. $\sin(30° + \theta) - \cos(60° + \theta) = \sqrt{3}\sin\theta$

38. $\sin(60° + \theta) - \cos(30° + \theta) = \sin\theta$

39. $\sin(30° + \theta) + \cos(60° + \theta) = \cos\theta$

40. $\sin(A + B) - \sin(A - B) = 2 \cos A \sin B$

41. $\cos(A + B) + \cos(A - B) = 2 \cos A \cos B$

42. $\sin(A + B) \sin(A - B) = \sin^2 A - \sin^2 B$

43. $\cos(A + B) \cos(A - B) = \cos^2 A + \cos^2 B - 1$

44. $\cot(A + B) = \dfrac{\cot A \cot B - 1}{\cot A + \cot B}$

45. $\cot(A - B) = \dfrac{\cot A \cot B + 1}{\cot B - \cot A}$

46. $\cos(A + B) - \cos(A - B) = -2 \sin A \sin B$

47. $\sin(A + B) + \sin(A - B) = 2 \sin A \cos B$

48. $-\tan A = \cot\left(A + \dfrac{\pi}{2}\right)$

49. $\dfrac{\tan A + \tan B}{1 + \tan A \tan B} = \dfrac{\sin(A + B)}{\cos(A - B)}$ **50.** $\dfrac{\cos(A + B)}{\sin(A - B)} = \dfrac{\cot A - \tan B}{1 - \cot A \tan B}$

51. $\dfrac{\cos(A - B)}{\sin(A + B)} = \dfrac{1 + \tan A \tan B}{\tan A + \tan B}$

52. Derive a formula for $\cos(A + B + C)$.

53. Derive a formula for $\sin(A - B - C)$.

54. Derive a formula for $\cos(A - B - C)$.

55. Derive a formula for $\tan(A + B + C)$.

4.3   THE DOUBLE-ANGLE IDENTITIES

Trigonometric identities are like olives in a jar—once you get one out, the rest come easily. There are several more identities that follow directly from the results in the previous section.

If angles A and B are equal, the formulas for the sine of the sum of two angles and the cosine of the sum of two angles may be transformed into formulas called the **double-angle identities.**

Let $A = B$ in the identity $\sin(A + B) = \sin A \cos B + \cos A \sin B$, and simplify.

$$\sin(A + A) = \sin A \cos A + \cos A \sin A$$

(4.10)   **$\sin 2A = 2 \sin A \cos A$**

Similarly, let $A = B$ in the identity $\cos(A + B) = \cos A \cos B - \sin A \sin B$, and simplify.

$$\cos(A + A) = \cos A \cos A - \sin A \sin A$$

(4.11)   **$\cos 2A = \cos^2 A - \sin^2 A$**

There are two variations of this identity for $\cos 2A$. To obtain one of them, we substitute $1 - \cos^2 A$ for $\sin^2 A$ in Identity (4.11), and simplify.

$$\cos 2A = \cos^2 A - (1 - \cos^2 A)$$

(4.12) $\cos 2A = 2\cos^2 A - 1$

To obtain the other, we substitute $1 - \sin^2 A$ for $\cos^2 A$ in Identity (4.11), and simplify.

$$\cos 2A = 1 - \sin^2 A - \sin^2 A$$

(4.13) $\cos 2A = 1 - 2\sin^2 A$

To find the double-angle identity for the tangent function, we let $A = B$ in the identity

$$\tan(A + B) = \frac{\tan A + \tan B}{1 - \tan A \tan B}$$

and simplify.

$$\tan(A + A) = \frac{\tan A + \tan A}{1 - \tan A \tan A}$$

(4.14) $\tan 2A = \dfrac{2\tan A}{1 - \tan^2 A}$

Example 1   Simplify the expression $\sin 5A \cos 5A$.

Solution   Because the given expression is one-half of $2\sin 5A \cos 5A$, it is one-half of $\sin 2(5A)$.

$$\sin 5A \cos 5A = \frac{1}{2}(2\sin 5A \cos 5A)$$

$$= \frac{1}{2}\sin 2(5A)$$

$$= \frac{1}{2}\sin 10A$$

■

Example 2   Use a double-angle identity to evaluate $\tan 2(60°)$.

Solution   Use the identity $\tan 2A = \dfrac{2\tan A}{1 - \tan^2 A}$ and substitute $60°$ for A.

$$\tan 2A = \frac{2\tan A}{1 - \tan^2 A}$$

$$\tan 2(60°) = \frac{2\tan(60°)}{1 - \tan^2(60°)}$$

$$= \frac{2\sqrt{3}}{1 - 3}$$

$$= \frac{2\sqrt{3}}{-2}$$

$$= -\sqrt{3}$$

Example 3   Find the **a.** sine, **b.** cosine, and **c.** tangent of 2θ if $\cos \theta = \dfrac{12}{13}$ and θ is in QIV.

Solution   Use the information that $\cos \theta = \frac{12}{13}$ and that θ is in QIV to sketch the angle in standard position as in Figure 4-3. From the figure, it follows that

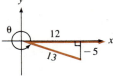

$$\sin \theta = \frac{-5}{13} = -\frac{5}{13}$$

$$\tan \theta = \frac{-5}{12} = -\frac{5}{12}$$

Figure 4-3   Use the double-angle identities to find the functions of 2θ.

a. $\sin 2\theta = 2 \sin \theta \cos \theta = 2\left(-\dfrac{5}{13}\right)\left(\dfrac{12}{13}\right) = -\dfrac{120}{169}$

b. $\cos 2\theta = \cos^2 \theta - \sin^2 \theta = \left(\dfrac{12}{13}\right)^2 - \left(-\dfrac{5}{13}\right)^2 = \dfrac{119}{169}$

c. $\tan 2\theta = \dfrac{2\left(-\dfrac{5}{12}\right)}{1 - \left(-\dfrac{5}{12}\right)^2} = \dfrac{-\dfrac{10}{12}}{1 - \dfrac{25}{144}} = \dfrac{-\dfrac{5}{6}}{\dfrac{119}{144}} = -\dfrac{5}{\cancel{6}} \cdot \dfrac{\overset{24}{\cancel{144}}}{119} = -\dfrac{120}{119}$

Example 4   Verify the identity $\cos 2A = \cos^4 A - \sin^4 A$.

Solution   Because the right side factors, work on that side.

$$
\begin{array}{c|l}
\cos 2A & \cos^4 A - \sin^4 A \\
& = (\cos^2 A + \sin^2 A)(\cos^2 A - \sin^2 A) \\
& = 1 \cdot (\cos^2 A - \sin^2 A) \\
& = \cos 2A
\end{array}
$$

Example 5   Verify the identity $\cos A \sin 2A = 2 \sin A - 2 \sin^3 A$.

Solution   Although the right side looks more complicated, work with the left side because it involves a double angle.

$$
\begin{array}{l|l}
\cos A \sin 2A & 2 \sin A - 2 \sin^3 A \\
= \cos A (2 \sin A \cos A) & \\
= 2 \cos^2 A \sin A & \\
= 2(1 - \sin^2 A) \sin A & \\
= 2 \sin A - 2 \sin^3 A &
\end{array}
$$

Example 6 Verify the identity $\cos 2A \sec^2 A = 2 - \sec^2 A$.

Solution Work on the left side to eliminate the double angle.

$$
\begin{array}{l|l}
\cos 2A \sec^2 A & 2 - \sec^2 A \\
= (2\cos^2 A - 1)\sec^2 A & \\
= 2\cos^2 A \sec^2 A - \sec^2 A & \\
= 2 - \sec^2 A &
\end{array}
$$
∎

Example 7 Verify that $\sin(A + B) + \sin(A - B) = 2\sin A \cos B$.

Solution

$$
\begin{array}{l|l}
\sin(A + B) + \sin(A - B) & 2\sin A \cos B \\
= \sin A \cos B + \cos A \sin B + \sin A \cos B - \cos A \sin B & \\
= \sin A \cos B + \sin A \cos B & \\
= 2\sin A \cos B &
\end{array}
$$
∎

Example 8 Verify the identity $2\cot 2A = \cot A - \tan A$.

Solution Work with the left side of the equation.

$$
\begin{array}{l|l}
2\cot 2A & \cot A - \tan A \\
= 2\left(\dfrac{1}{\tan 2A}\right) & \\
= 2\left(\dfrac{1 - \tan^2 A}{2\tan A}\right) & \\
= \dfrac{1 - \tan^2 A}{\tan A} & \\
= \dfrac{1}{\tan A} - \dfrac{\tan^2 A}{\tan A} & \\
= \cot A - \tan A &
\end{array}
$$
∎

Exercise 4.3

In Exercises 1–26, write the given expression in terms of a single trigonometric function of twice the given angle.

1. $2\sin\alpha\cos\alpha$

2. $2\cos^2\alpha - 1$

3. $2\sin 3\theta \cos 3\theta$

4. $2\cos^2 2A - 1$

5. $\cos^2\beta - \sin^2\beta$

6. $1 - 2\sin^2\beta$

7. $2\cos^2\dfrac{B}{2} - 1$

8. $\sin 5\theta \cos 5\theta$

9. $4\sin\theta\cos\theta$

10. $\cos^2\dfrac{\theta}{2} - \sin^2\dfrac{\theta}{2}$

11. $4 \sin^2 2\theta \cos^2 2\theta$

12. $2 - 4 \sin^2 6B$

13. $\cos^2 \alpha - \dfrac{1}{2}$

14. $2 - 4 \sin^2 \dfrac{\alpha}{4}$

15. $\cos^2 9\theta - \sin^2 9\theta$

16. $4 \sin 4B \cos 4B$

17. $4 \sin^2 5\theta \cos^2 5\theta$

18. $\dfrac{2 \tan A}{1 - \tan^2 A}$

19. $\dfrac{2 \tan 4C}{1 - \tan^2 4C}$

20. $3 - 6 \cos^2 6x$

21. $\dfrac{\tan \dfrac{A}{2}}{\dfrac{1}{2} - \dfrac{1}{2} \tan^2 \dfrac{A}{2}}$

22. $\dfrac{2 \tan \alpha}{2 - \sec^2 \alpha}$

23. $\cos^4 4x - \sin^4 4x$

24. $\dfrac{1}{2} \sec \theta \csc \theta$

25. $1 - 2 \cos^2 5x$

26. $\dfrac{1 - \tan^2 3x}{2 \tan 3x}$

In Exercises 27–38, use a double-angle identity to find the value of the given expression. Then, check your work by evaluating the expression directly.

27. $\sin 2(30°)$ **28.** $\cos 2(30°)$ **29.** $\tan 2(45°)$ **30.** $\cot 2(135°)$

31. $\cos 2(240°)$ **32.** $\sin 2(150°)$ **33.** $\cot 2(225°)$ **34.** $\tan 2(315°)$

35. $\sin 2\left(\dfrac{\pi}{3}\right)$ **36.** $\cos 2\left(\dfrac{7\pi}{6}\right)$ **37.** $\cos 2\left(\dfrac{11\pi}{6}\right)$ **38.** $\sin 2\left(\dfrac{5\pi}{3}\right)$

In Exercises 39–50, find the exact value of the sine, cosine, and tangent of 2θ using the given information.

39. $\sin \theta = \dfrac{12}{13}$ and θ is in QI

40. $\cos \theta = \dfrac{5}{13}$ and θ is in QIV

41. $\tan \theta = \dfrac{12}{5}$ and θ is in QIII

42. $\sin \theta = \dfrac{3}{5}$ and θ is in QII

43. $\cos \theta = -\dfrac{4}{5}$ and θ is in QIII

44. $\cot \theta = -\dfrac{15}{8}$ and θ is in QIV

45. $\sin \theta = -\dfrac{24}{25}$ and θ is in QIV

46. $\cos \theta = -\dfrac{7}{25}$ and θ is in QIII

47. $\sin \theta = \dfrac{40}{41}$ and $\cos \theta$ is positive

48. $\sin \theta = -\dfrac{40}{41}$ and $\tan \theta$ is negative

49. $\cos \theta = \dfrac{40}{41}$ and $\tan \theta$ is negative

50. $\cos \theta = 0$

In Exercises 51–77, verify the given identity.

51. $\dfrac{\tan 2x}{\sin 2x} = \sec 2x$

52. $\dfrac{\cot 2x}{\cos 2x} = \csc 2x$

53. $2 \csc 2A = \sec A \csc A$

54. $\sin 2A + 2 \sin A = \dfrac{-2 \sin^3 A}{\cos A - 1}$

55. $\cos 2x - \dfrac{\sin 2x}{\cos x} + 2 \sin^2 x = 1 - 2 \sin x$

56. $2 \cos^4 \theta + 2 \sin^2 \theta \cos^2 \theta - 1 = \cos 2\theta$

57. $\sin 2A - 2 \sin A = -\dfrac{2 \sin^3 A}{\cos A + 1}$

58. $2 \cos A - 1 = \dfrac{\cos 2A + \cos A}{\cos A + 1}$

59. $2 \sin^4 \theta + 2 \sin^2 \theta \cos^2 \theta - 1 = -\cos 2\theta$

60. $\sin 2\theta - \tan \theta = \cos 2\theta \tan \theta$

61. $2 \cos A + 1 = \dfrac{\cos 2A - \cos A}{\cos A - 1}$

62. $(\sin A + \cos A)^2 = 1 + \sin 2A$

63. $\sec 2\theta = \dfrac{\tan \theta + \cot \theta}{\cot \theta - \tan \theta}$

64. $\sin 4x = 4 \cos x \sin x - 8 \cos x \sin^3 x$

65. $4 \csc^2 2A = \sec^2 A + \csc^2 A$

66. $1 - \dfrac{1}{2} \sin 2x = \dfrac{\sin^3 x + \cos^3 x}{\sin x + \cos x}$

67. $\cos 4x = 8 \cos^4 x - 8 \cos^2 x + 1$

68. $-\tan 2x = \dfrac{2}{\tan x - \cot x}$

69. $2 \tan 2x = \dfrac{\sin x - \cos x}{\sin x + \cos x} - \dfrac{\sin x + \cos x}{\sin x - \cos x}$

70. $\sin^4 \theta = \dfrac{\cos 4\theta}{8} - \dfrac{\cos 2\theta}{2} + \dfrac{3}{8}$

71. $\dfrac{\sin x + \sin 2x}{\cos x - \cos 2x} = \dfrac{\sec x + 2}{\csc x - \cot x + \tan x}$

72. $\dfrac{\cos x - \sin x}{\cos 2x} = \dfrac{\cos x + \sin x}{1 + \sin 2x}$

73. $\dfrac{1 - \tan^2 x}{1 + \tan^2 x} = \cos 2x$

74. $\dfrac{\sec^2 x}{\sec 2x} = 2 - \sec^2 x$

75. $\dfrac{1 + \tan x}{1 - \tan x} = \dfrac{\cos 2x}{1 - \sin 2x}$

76. $\cos 2x = \dfrac{1}{\tan 2x \tan x + 1}$

77. $\tan A \sin 2A + \cos 2A = 1$

4.4   THE HALF-ANGLE IDENTITIES

If the trigonometric functions of angle A are known, the formulas of the last section allow us to find the functions of $2A$. The identities in this section will enable us to find the trigonometric functions of $\frac{1}{2}A$.

We begin by solving the identity $\cos 2\theta = 2 \cos^2 \theta - 1$ for $\cos \theta$.

$$\cos 2\theta = 2 \cos^2 \theta - 1$$
$$2 \cos^2 \theta = 1 + \cos 2\theta$$

$$\cos^2 \theta = \frac{1 + \cos 2\theta}{2}$$

$$\cos \theta = \pm \sqrt{\frac{1 + \cos 2\theta}{2}}$$

With an appropriate choice of the sign preceding the radical, this identity is true for all values of θ. To derive the first **half-angle identity,** we let $\theta = \frac{A}{2}$.

(4.15) $$\cos \frac{A}{2} = \pm \sqrt{\frac{1 + \cos A}{2}}$$

The sign preceding the radical in Equation (4.15) is determined by the quadrant in which $\frac{A}{2}$ lies.

   In a similar fashion, we solve $\cos 2\theta = 1 - 2 \sin^2 \theta$ for $\sin \theta$ and let $\theta = \frac{A}{2}$ to obtain the half-angle identity for the sine function.

$$\cos 2\theta = 1 - 2 \sin^2 \theta$$

$$2 \sin^2 \theta = 1 - \cos 2\theta$$

$$\sin^2 \theta = \frac{1 - \cos 2\theta}{2}$$

$$\sin \theta = \pm \sqrt{\frac{1 - \cos 2\theta}{2}}$$

(4.16) $$\sin \frac{A}{2} = \pm \sqrt{\frac{1 - \cos A}{2}}$$

Again, the " $+$ " or " $-$ " sign is chosen by the quadrant in which $\frac{A}{2}$ lies.

Example 1   Use a half-angle identity to find $\sin 15°$.

Solution   Because $15°$ is $\frac{1}{2}(30°)$, it follows that

$$\sin 15° = \sin \frac{30°}{2} = + \sqrt{\frac{1 - \cos 30°}{2}}$$

$$= + \sqrt{\frac{1 - \dfrac{\sqrt{3}}{2}}{2}}$$

$$= + \sqrt{\frac{2 - \sqrt{3}}{4}}$$

$$= + \frac{\sqrt{2 - \sqrt{3}}}{2}$$

The " + " sign is chosen because 15° is a first-quadrant angle and the sine of a first-quadrant angle is positive.   ■

Example 2   Use a half-angle identity to find $\cos \dfrac{7\pi}{12}$.

Solution   Because $\dfrac{7\pi}{12}$ is $\dfrac{1}{2} \cdot \dfrac{7\pi}{6}$, it follows that

$$\cos \frac{7\pi}{12} = \cos \frac{1}{2} \cdot \frac{7\pi}{6} = -\sqrt{\frac{1 + \cos \dfrac{7\pi}{6}}{2}}$$

$$= -\sqrt{\frac{1 - \dfrac{\sqrt{3}}{2}}{2}}$$

$$= -\frac{\sqrt{2 - \sqrt{3}}}{2}$$

Here, the " − " sign is chosen because $\frac{7\pi}{12}$ is a second-quadrant angle and the cosine of a second-quadrant angle is negative.   ■

The half-angle tangent identities are derived from the half-angle sine and cosine identities. To develop the identity involving $\tan \frac{A}{2}$, we substitute $\frac{A}{2}$ for θ in the relationship

$$\tan \theta = \frac{\sin \theta}{\cos \theta}$$

and proceed as follows.

$$\tan \frac{A}{2} = \frac{\sin \dfrac{A}{2}}{\cos \dfrac{A}{2}} = \frac{\pm \sqrt{\dfrac{1 - \cos A}{2}}}{\pm \sqrt{\dfrac{1 + \cos A}{2}}}$$

$$= \pm \sqrt{\frac{1 - \cos A}{1 + \cos A}}$$

If the fraction within the radical is multiplied by 1 written in the form

$$\frac{1 - \cos A}{1 - \cos A}$$

further simplification results.

$$\tan \frac{A}{2} = \pm \sqrt{\frac{(1 - \cos A)(1 - \cos A)}{(1 + \cos A)(1 - \cos A)}}$$

$$= \pm \sqrt{\frac{(1 - \cos A)^2}{1 - \cos^2 A}}$$

$$= \pm \sqrt{\frac{(1 - \cos A)^2}{\sin^2 A}}$$

(4.17) $$\tan \frac{A}{2} = \frac{1 - \cos A}{\sin A}$$

Identity (4.17) has an advantage over the previous half-angle identity because it is not necessary to choose the appropriate " + " or " − " sign. The selection is automatic because $1 - \cos A$ is never negative and $\sin A$ and $\tan \frac{A}{2}$ always agree in sign. You will be asked to prove this in an exercise.

Another form is also possible. Because

$$\tan \frac{A}{2} = \frac{1 - \cos A}{\sin A}$$

it follows that

$$\tan \frac{A}{2} = \frac{(1 - \cos A)(1 + \cos A)}{\sin A (1 + \cos A)}$$

$$= \frac{1 - \cos^2 A}{\sin A (1 + \cos A)}$$

$$= \frac{\sin^2 A}{\sin A (1 + \cos A)}$$

(4.18) $$\tan \frac{A}{2} = \frac{\sin A}{1 + \cos A}$$

Example 3   Use the identity $\tan \dfrac{A}{2} = \dfrac{1 - \cos A}{\sin A}$ to find $\tan \dfrac{\pi}{8}$.

Solution   Because $\dfrac{\pi}{8}$ is $\dfrac{1}{2}\left(\dfrac{\pi}{4}\right)$, it follows that

$$\tan \frac{\pi}{8} = \tan \frac{\frac{\pi}{4}}{2} = \frac{1 - \cos \frac{\pi}{4}}{\sin \frac{\pi}{4}}$$

$$= \frac{1 - \dfrac{\sqrt{2}}{2}}{\dfrac{\sqrt{2}}{2}}$$

$$= \frac{2 - \sqrt{2}}{\sqrt{2}} \qquad \text{Multiply numerator and denominator by 2.}$$

$$= \sqrt{2} - 1$$

Example 4   Use the identity $\tan \dfrac{A}{2} = \dfrac{\sin A}{1 + \cos A}$ to find $\tan 157.5°$.

Solution   Because $157.5°$ is $\dfrac{315°}{2}$, then

$$\tan 157.5° = \tan \frac{315°}{2} = \frac{\sin 315°}{1 + \cos 315°}$$

$$= \frac{-\dfrac{\sqrt{2}}{2}}{1 + \dfrac{\sqrt{2}}{2}}$$

$$= \frac{-\dfrac{\sqrt{2}}{2}}{\dfrac{2 + \sqrt{2}}{2}}$$

$$= \frac{-\sqrt{2}}{2 + \sqrt{2}}$$

$$= \frac{-\sqrt{2}}{(2 + \sqrt{2})} \cdot \frac{(2 - \sqrt{2})}{(2 - \sqrt{2})}$$

$$= \frac{-\sqrt{2}(2 - \sqrt{2})}{2}$$

$$= 1 - \sqrt{2}$$

This number is negative, as it must be for the tangent of a second-quadrant angle.

Example 5 Find the **a.** sine, **b.** cosine, and **c.** tangent of $\dfrac{\theta}{2}$, if $\sin \theta = \dfrac{3}{5}$ and $\dfrac{\pi}{2} < \theta < \pi$.

Solution Use the information that $\sin \theta = \frac{3}{5}$ and that θ is in QII to sketch the angle in standard position as in Figure 4-4. The value of $\cos \theta$ can be read from the figure.

$$\cos \theta = -\frac{4}{5}$$

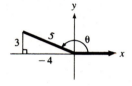

Figure 4-4

Use the half-angle identities to find the functions of $\dfrac{\theta}{2}$.

a. $\sin \dfrac{\theta}{2} = \sqrt{\dfrac{1 - \left(\dfrac{-4}{5}\right)}{2}} = \sqrt{\dfrac{9}{10}} = \dfrac{3\sqrt{10}}{10}$

b. $\cos \dfrac{\theta}{2} = \sqrt{\dfrac{1 + \left(\dfrac{-4}{5}\right)}{2}} = \sqrt{\dfrac{1}{10}} = \dfrac{\sqrt{10}}{10}$

c. $\tan \dfrac{\theta}{2} = \dfrac{\sin \dfrac{\theta}{2}}{\cos \dfrac{\theta}{2}} = \dfrac{\dfrac{3\sqrt{10}}{10}}{\dfrac{\sqrt{10}}{10}} = 3$

Choose the radicals to be positive, because if $\dfrac{\pi}{2} < \theta < \pi$, then $\dfrac{\theta}{2}$ is in QI. ∎

Example 6 Write $\dfrac{\sin(-20A)}{-\cos 20A - 1}$ as a trigonometric function of $10A$.

Solution Because $\sin(-20A) = -\sin 20A$, it follows that

$$\frac{\sin(-20A)}{-\cos 20A - 1} = \frac{-\sin 20A}{-(1 + \cos 20A)} = \frac{\sin 20A}{1 + \cos 20A}$$

One of the identities for $\tan \dfrac{\theta}{2}$ is

$$\tan \frac{\theta}{2} = \frac{\sin \theta}{1 + \cos \theta}$$

Hence, you have

$$\frac{\sin 20A}{1 + \cos 20A} = \tan \frac{20A}{2} = \tan 10A$$

∎

Example 7   Verify the identity $\tan \dfrac{\theta}{2} = \csc \theta - \cot \theta$.

Solution   Work on the left side to get rid of the half-angle.

$$\tan \dfrac{\theta}{2} \quad \Big| \quad \csc \theta - \cot \theta$$

$$= \dfrac{1 - \cos \theta}{\sin \theta}$$

$$= \dfrac{1}{\sin \theta} - \dfrac{\cos \theta}{\sin \theta}$$

$$= \csc \theta - \cot \theta$$

Example 8   Verify the identity $2 \sin^2 \dfrac{x}{2} \tan x = \tan x - \sin x$.

Solution   Work on the left side.

$$2 \sin^2 \dfrac{x}{2} \tan x \quad \Big| \quad \tan x - \sin x$$

$$= 2 \cdot \dfrac{1 - \cos x}{2} \cdot \dfrac{\sin x}{\cos x}$$

$$= \dfrac{\sin x - \sin x \cos x}{\cos x}$$

$$= \dfrac{\sin x}{\cos x} - \sin x$$

$$= \tan x - \sin x$$

Example 9   Verify the identity $\dfrac{2 \tan \dfrac{x}{2}}{\sin x} = \sec^2 \dfrac{x}{2}$.

Solution   Again, work on the left side.

$$\dfrac{2 \tan \dfrac{x}{2}}{\sin x} \quad \Big| \quad \sec^2 \dfrac{x}{2}$$

$$= \dfrac{2 \cdot \dfrac{\sin x}{1 + \cos x}}{\sin x}$$

$$= \dfrac{2}{1 + \cos x}$$

$$= \dfrac{1}{\dfrac{1 + \cos x}{2}}$$

$$= \frac{1}{\cos^2 \frac{x}{2}}$$

$$= \sec^2 \frac{x}{2}$$

Exercise 4.4

In Exercises 1–12, use half-angle identities to find the required values. **Do not use a calculator.**

1. cos 15°   **2.** tan 15°   **3.** tan 105°   **4.** sin 105°

5. $\sin \frac{\pi}{8}$ **6.** $\cos \frac{\pi}{8}$ **7.** $\cos \frac{\pi}{12}$ **8.** $\sin \frac{\pi}{12}$

9. tan 165°   **10.** cos 165°   **11.** $\cot \frac{5\pi}{4}$ **12.** $\tan \frac{7\pi}{4}$

In Exercises 13–24, use the given information to find the exact value of the sine, cosine, and tangent of θ/2. Assume that 0° ≤ θ < 360°.

13. $\sin \theta = \frac{3}{5}$;   θ in QI

14. $\cos \theta = \frac{12}{13}$;   θ in QI

15. $\tan \theta = \frac{4}{3}$;   θ in QIII

16. $\tan \theta = \frac{3}{4}$;   θ in QIII

17. $\cos \theta = \frac{8}{17}$;   θ in QIV

18. $\sin \theta = -\frac{7}{25}$;   θ in QIII

19. $\cot \theta = \frac{40}{9}$;   θ in QI

20. $\sec \theta = -\frac{41}{40}$;   θ in QII

21. $\csc \theta = \frac{17}{8}$;   θ in QII

22. $\csc \theta = -\frac{5}{3}$;   θ in QIV

23. $\sec \theta = \frac{3}{2}$;   θ in QIV

24. $\cos \theta = -0.1$;   θ in QIII

In Exercises 25–36, write the given expression as a single trigonometric function of half the given angle.

25. $\sqrt{\dfrac{1 + \cos 30°}{2}}$

26. $\sqrt{\dfrac{1 - \cos 30°}{2}}$

27. $\dfrac{1 - \cos 200°}{\sin 200°}$

28. $\dfrac{\sin 50°}{1 + \cos 50°}$

29. $\csc 80° - \cot 80°$

30. $\dfrac{2 \tan 140°}{\sin 280°}$ (280° is the given angle)

31. $\sqrt{\dfrac{1 - \cos 2\pi}{1 + \cos 2\pi}}$

32. $\dfrac{1 - \cos 4\theta}{\sin 4\theta}$

33. $\dfrac{1 - \cos \dfrac{x}{2}}{\sin \dfrac{x}{2}}$

34. $\dfrac{1 - \cos 2x}{1 + \cos 2x}$

35. $\dfrac{\sin 10A}{1 + \cos 10A}$

36. $\dfrac{1 + \cos 4x}{1 - \cos 4x}$

In Exercises 37–50, verify the given identity.

37. $\sin^2 \dfrac{\theta}{2} = \dfrac{1}{2} (1 - \cos \theta)$

38. $\cos^2 \dfrac{\theta}{2} = \dfrac{1}{2} (1 + \cos \theta)$

39. $\sec^2 \dfrac{\theta}{2} = \dfrac{2}{1 + \cos \theta}$

40. $\csc^2 \dfrac{\theta}{2} = \dfrac{2}{1 - \cos \theta}$

41. $\cot \dfrac{\theta}{2} = \dfrac{1 + \cos \theta}{\sin \theta}$

42. $\cot \dfrac{\theta}{2} = \dfrac{\sin \theta}{1 - \cos \theta}$

43. $\csc^2 \dfrac{\theta}{2} = 2 \csc^2 \theta + 2 \cot \theta \csc \theta$

44. $\sin^2 \dfrac{\theta}{2} = \dfrac{\sec \theta - 1}{2 \sec \theta}$

45. $- \sec B = \dfrac{\sec^2 \dfrac{B}{2}}{\sec^2 \dfrac{B}{2} - 2}$

46. $\tan \left(\dfrac{\pi}{4} + \dfrac{\theta}{2} \right) = \dfrac{1 + \cos \theta + \sin \theta}{1 + \cos \theta - \sin \theta}$

47. $\csc \theta = \dfrac{1}{2} \csc \dfrac{\theta}{2} \sec \dfrac{\theta}{2}$

48. $\dfrac{1}{2} \sin x \tan \dfrac{x}{2} \csc^2 \dfrac{x}{2} = 1$

49. $\tan \dfrac{B}{2} \cos B + \tan \dfrac{B}{2} = \sin B$

50. $\left(\cos \dfrac{\alpha}{2} - \sin \dfrac{\alpha}{2} \right)^2 = 1 - \sin \alpha$

In Exercises 51–54, assume that A + B + C = 180°. Verify each identity.

51. $\sin A + \sin B + \sin C = 4 \cos \dfrac{A}{2} \cos \dfrac{B}{2} \cos \dfrac{C}{2}$

52. $\cos A + \cos B - \cos C = 4 \cos \dfrac{A}{2} \cos \dfrac{B}{2} \sin \dfrac{C}{2} - 1$

53. $\tan A + \tan B + \tan C = \tan A \tan B \tan C$

54. $\cos A + \cos B + \cos C = 1 + 4 \sin \dfrac{A}{2} \sin \dfrac{B}{2} \sin \dfrac{C}{2}$

55. Show that $\sin A$ and $\tan \dfrac{A}{2}$ always agree in sign.

4.5   SUM-TO-PRODUCT AND PRODUCT-TO-SUM IDENTITIES

The craftsmanship of a skilled cabinetmaker is dependent on the quality of tools he has and on the skill with which he uses them. The trigonometric identities developed so far are important tools in the mathematician's tool box. You have verified many other identities to gain practice with these tools. Just as some woodworking tools are used more often than others, some identities find greater use than others. The identities discussed in this section are used less often than the ones previously developed, but they are indispensible when the need does arise. They are useful whenever it is necessary to convert a product of two trigonometric functions into a sum, or a sum into a product.

If the identities for $\sin(x + y)$ and $\sin(x - y)$ are added, some of these new identities result.

1. $\sin(x + y) = \sin x \cos y + \cos x \sin y$
2. $\sin(x - y) = \sin x \cos y - \cos x \sin y$

Adding Equations 1 and 2 causes the $\cos x \sin y$ term to drop out. The result is shown in Equation 3.

3. $\sin(x + y) + \sin(x - y) = 2 \sin x \cos y$

We divide both sides of Equation 3 by 2 to get

(4.19) $\sin x \cos y = \dfrac{1}{2} [\sin(x + y) + \sin(x - y)]$

Identity (4.19) is used to convert a product of the sine and cosine functions into a sum.

Example 1   Calculate the value of $\sin 67.5° \cos 22.5°$.

Solution   Let $x = 67.5°$, $y = 22.5°$, and use Identity (4.19).

$$\sin x \cos y = \frac{1}{2} [\sin(x + y) + \sin(x - y)]$$

$$\sin 67.5° \cos 22.5° = \frac{1}{2} [\sin(67.5° + 22.5°) + \sin(67.5° - 22.5°)]$$

$$= \frac{1}{2} (\sin 90° + \sin 45°)$$

$$= \frac{1}{2} \left(1 + \frac{\sqrt{2}}{2} \right)$$

$$= \frac{1}{2} + \frac{\sqrt{2}}{4}$$

We now develop a formula to convert the sum of two sines into a product. To do so, we let $A = x + y$ and $B = x - y$, and solve for x and y. We first solve

$$\begin{cases} A = x + y \\ B = x - y \end{cases}$$

for x by adding the equations.

$$A + B = 2x$$

$$x = \frac{1}{2}(A + B)$$

To find y, we subtract one equation from the other.

$$A - B = 2y$$

$$y = \frac{1}{2}(A - B)$$

We substitute

$$A \text{ for } x + y, \quad B \text{ for } x - y, \quad \frac{A + B}{2} \text{ for } x, \quad \text{and} \quad \frac{A - B}{2} \text{ for } y$$

in Equation 3 to get

(4.20) $$\sin A + \sin B = 2 \sin \frac{A + B}{2} \cos \frac{A - B}{2}$$

Example 2   Verify the identity $\sin 3\theta + \sin \theta = 2 \sin 2\theta \cos \theta$.

Solution   Use Identity (4.20) and let $A = 3\theta$ and $B = \theta$. Work on the left side.

$$\sin 3\theta + \sin \theta \qquad \qquad \Big| \quad 2 \sin 2\theta \cos \theta$$

$$= 2 \sin \frac{3\theta + \theta}{2} \cos \frac{3\theta - \theta}{2}$$

$$= 2 \sin \frac{4\theta}{2} \cos \frac{2\theta}{2}$$

$$= 2 \sin 2\theta \cos \theta$$

Example 3   Find the value of $\sin 75° + \sin 15°$.

Solution   Again, use Identity (4.20), and let $A = 75°$ and $B = 15°$.

$$\sin A + \sin B = 2 \sin \frac{A + B}{2} \cos \frac{A - B}{2}$$

$$\sin 75° + \sin 15° = 2 \sin \frac{75° + 15°}{2} \cos \frac{75° - 15°}{2}$$

$$= 2 \sin \frac{90°}{2} \cos \frac{60°}{2}$$

$$= 2 \sin 45° \cos 30°$$

$$= 2 \cdot \frac{\sqrt{2}}{2} \cdot \frac{\sqrt{3}}{2}$$

$$= \frac{\sqrt{6}}{2}$$

If Identities 1 and 2 are subtracted, more identities result.

1. $$\sin(x + y) = \sin x \cos y + \cos x \sin y$$
2. $$\sin(x - y) = \sin x \cos y - \cos x \sin y$$
3. $\sin(x + y) - \sin(x - y) = 2 \cos x \sin y$

Dividing both sides of Equation 3 by 2 gives a new identity.

(4.21) $\cos x \sin y = \dfrac{1}{2}[\sin(x + y) - \sin(x - y)]$

This identity is used to convert a product into a difference. If we let $x + y = A$ and $x - y = B$, then

$$x = \frac{A + B}{2} \quad \text{and} \quad y = \frac{A - B}{2}$$

We can substitute the values for $x + y$, $x - y$, x, and y into Equation 3 to obtain the next identity.

(4.22) $\sin A - \sin B = 2 \cos \dfrac{A + B}{2} \sin \dfrac{A - B}{2}$

This formula is used to convert the difference of the sines of two angles into a product.

If the formulas for the cosines of the sum and the difference of two angles are added, still more identities result.

4. $\begin{cases} \cos(x + y) = \cos x \cos y - \sin x \sin y \\ \cos(x - y) = \cos x \cos y + \sin x \sin y \end{cases}$

5.

After adding, the $\sin x \sin y$ terms drop out, as shown in Equation 6.

6. $\cos(x + y) + \cos(x - y) = 2 \cos x \cos y$

We divide both sides of Equation 6 by 2 to obtain a new identity.

(4.23) $\cos x \cos y = \dfrac{1}{2}[\cos(x + y) + \cos(x - y)]$

This identity is used when a product of cosines needs to be changed to a sum. To derive an identity to convert from sums to products, we let $x + y = A$ and $x - y = B$. Then

$$x = \frac{A + B}{2} \quad \text{and} \quad y = \frac{A - B}{2}$$

We use Equation 6 and substitute to get

(4.24) $\cos A + \cos B = 2 \cos \dfrac{A + B}{2} \cos \dfrac{A - B}{2}$

We obtain the final identities when we subtract Equation 4 from Equation 5 and make the usual substitutions.

(4.25) $\sin x \sin y = \dfrac{1}{2}[\cos(x - y) - \cos(x + y)]$

(4.26) $\cos A - \cos B = -2 \sin \dfrac{A + B}{2} \sin \dfrac{A - B}{2}$

We list all of the previous formulas for easy reference.

(4.27)

The product-to-sum formulas:

$$\sin A \cos B = \frac{1}{2}[\sin(A + B) + \sin(A - B)]$$

$$\cos A \sin B = \frac{1}{2}[\sin(A + B) - \sin(A - B)]$$

$$\sin A \sin B = \frac{1}{2}[\cos(A - B) - \cos(A + B)]$$

$$\cos A \cos B = \frac{1}{2}[\cos(A + B) + \cos(A - B)]$$

(4.28)

The sum-to-product formulas:

$$\sin A + \sin B = 2 \sin \frac{A + B}{2} \cos \frac{A - B}{2}$$

$$\sin A - \sin B = 2 \cos \frac{A + B}{2} \sin \frac{A - B}{2}$$

$$\cos A + \cos B = 2 \cos \frac{A + B}{2} \cos \frac{A - B}{2}$$

$$\cos A - \cos B = -2 \sin \frac{A + B}{2} \sin \frac{A - B}{2}$$

Example 4   Write $\cos 2\theta + \cos 6\theta$ as a product of two functions.

Solution   Substitute 2θ for A and 6θ for B in the identity for $\cos A + \cos B$, and simplify.

$$\cos 2\theta + \cos 6\theta = 2 \cos \frac{2\theta + 6\theta}{2} \cos \frac{2\theta - 6\theta}{2}$$

$$= 2 \cos 4\theta \cos(-2\theta)$$

$$= 2 \cos 4\theta \cos 2\theta \qquad \text{Remember that}$$
$$\cos(-2\theta) = \cos 2\theta. \qquad ■$$

Example 5   Verify the identity $\tan 3A = \dfrac{\sin 4A + \sin 2A}{\cos 4A + \cos 2A}$.

Solution   Work on the right side of the equation.

$$\tan 3A \quad \left| \quad \frac{\sin 4A + \sin 2A}{\cos 4A + \cos 2A} \right.$$

$$= \frac{2 \sin \dfrac{4A + 2A}{2} \cos \dfrac{4A - 2A}{2}}{2 \cos \dfrac{4A + 2A}{2} \cos \dfrac{4A - 2A}{2}}$$

$$= \frac{2 \sin 3A \cos A}{2 \cos 3A \cos A}$$

$$= \frac{\sin 3A}{\cos 3A}$$

$$= \tan 3A$$

Exercise 4.5

In Exercises 1–10, evaluate each of the products. **Do not use a calculator or tables.**

1. $\cos 75° \cos 15°$

2. $\sin 15° \cos 75°$

3. $\sin 165° \sin 105°$

4. $\sin 15° \cos 15°$

5. $\cos 22.5° \cos 67.5°$

6. $\cos 105° \sin 15°$

7. $\sin \dfrac{\pi}{12} \sin \dfrac{5\pi}{12}$

8. $\sin \dfrac{5\pi}{12} \cos \dfrac{13\pi}{12}$

9. $\cos \dfrac{7\pi}{12} \cos \dfrac{5\pi}{12}$

10. $\cos \dfrac{7\pi}{12} \sin \dfrac{13\pi}{12}$

In Exercises 11–20, express each quantity as a product.

11. $\sin 30° + \sin 40°$

12. $\sin 75° - \sin 70°$

13. $\cos 110° + \cos 220°$

14. $\cos 305° - \cos 15°$

15. $\sin 100° - \sin 50°$

16. $\sin 250° + \sin 200°$

17. $\cos 280° - \cos 105°$

18. $\cos 305° + \cos 300°$

19. $\sin 5\theta + \sin \theta$

20. $\cos 7\theta - \cos 5\theta$

In Exercises 21–30, express each quantity as a product and find its value. **Do not use a calculator or tables.**

21. $\cos 75° + \cos 15°$

22. $\sin 15° + \sin 75°$

23. $\sin 165° - \sin 105°$

24. $\sin 15° - \sin 75°$

25. $\cos 165° - \cos 105°$

26. $\cos 105° - \cos 15°$

27. $\sin \dfrac{\pi}{12} + \sin \dfrac{5\pi}{12}$

28. $\sin \dfrac{5\pi}{12} - \sin \dfrac{13\pi}{12}$

29. $\cos \dfrac{7\pi}{12} + \cos \dfrac{5\pi}{12}$

30. $\sin \dfrac{7\pi}{12} + \sin \dfrac{13\pi}{12}$

In Exercises 31–40, express each quantity as a sum or difference and find its value. **Do not use a calculator or tables.**

31. $\sin 75° \cos 15°$

32. $\cos 75° \sin 15°$

33. $\sin 75° \sin 15°$

34. $\cos 75° \cos 15°$

35. $\cos \dfrac{\pi}{12} \sin \dfrac{5\pi}{12}$

36. $\sin \dfrac{\pi}{12} \cos \dfrac{5\pi}{12}$

37. $\cos \dfrac{7\pi}{12} \cos \dfrac{5\pi}{12}$

38. $\sin \dfrac{5\pi}{12} \sin \dfrac{7\pi}{12}$

39. $\sin 105° \sin 15°$

40. $\cos 105° \cos 15°$

In Exercises 41–50, verify each identity.

41. $\dfrac{\sin A + \sin B}{\sin A - \sin B} = \tan \dfrac{1}{2}(A + B) \cot \dfrac{1}{2}(A - B)$

42. $\dfrac{\sin A + \sin B}{\cos A + \cos B} = \tan \dfrac{1}{2}(A + B)$

43. $\dfrac{\sin A + \sin B}{\cos A - \cos B} = -\cot \dfrac{1}{2}(A - B)$ **44.** $\dfrac{\cos A + \cos B}{\sin A - \sin B} = \cot \dfrac{1}{2}(A - B)$

45. $\dfrac{\cos A + \cos 5A}{\cos A - \cos 5A} = \cot 3A \cot 2A$

46. $\sin^2 A - \sin^2 B = \sin(A + B)\sin(A - B)$

47. $\cos^2 A - \cos^2 B = \sin(B + A)\sin(B - A)$

48. $\cos 2A(1 + 2\cos A) = \cos A + \cos 2A + \cos 3A$

49. $2\cos 5A \sin 2A = \sin 7A - \sin 3A$

50. $\cot 7A \cot 5A = \dfrac{\cos 12A + \cos 2A}{\cos 2A - \cos 12A}$

4.6 SUMS OF THE FORM A SIN X + B COS X

It is an interesting and useful fact that the graph of $y = A \sin x + B \cos x$ is a sine curve of the form $y = k \sin(x + \phi)$. The amplitude k and the phase shift angle ϕ are determined by the values of A and B. Some algebraic manipulations will establish this result and provide the proper values for k and ϕ.

From the terms of the expression $A \sin x + B \cos x$, factor out a common factor of $\sqrt{A^2 + B^2}$. This is not difficult if you realize that the product

$$\sqrt{A^2 + B^2} \cdot \frac{A}{\sqrt{A^2 + B^2}}$$

is just A, and that the product

$$\sqrt{A^2 + B^2} \cdot \frac{B}{\sqrt{A^2 + B^2}}$$

is just B. Thus, we have

(4.29) $A \sin x + B \cos x = \sqrt{A^2 + B^2} \left(\dfrac{A}{\sqrt{A^2 + B^2}} \sin x + \dfrac{B}{\sqrt{A^2 + B^2}} \cos x \right)$

Because the sum of the squares of the coefficients

$$\frac{A}{\sqrt{A^2 + B^2}} \quad \text{and} \quad \frac{B}{\sqrt{A^2 + B^2}}$$

is 1, one of these coefficients is $\sin \phi$ and one is $\cos \phi$, for some angle ϕ. Let ϕ be an angle such that

$$\sin \phi = \frac{B}{\sqrt{A^2 + B^2}} \quad \text{and} \quad \cos \phi = \frac{A}{\sqrt{A^2 + B^2}}$$

Then Equation (4.29) becomes

$$A \sin x + B \cos x = \sqrt{A^2 + B^2} \, (\cos \phi \sin x + \sin \phi \cos x)$$
$$= \sqrt{A^2 + B^2} \sin(x + \phi)$$

In summary, we have the following theorem.

(4.30)

> **Theorem.** $A \sin x + B \cos x = k \sin(x + \phi)$ where $k = \sqrt{A^2 + B^2}$ and ϕ is any angle for which
>
> $$\sin \phi = \frac{B}{\sqrt{A^2 + B^2}} \quad \text{and} \quad \cos \phi = \frac{A}{\sqrt{A^2 + B^2}}$$

Example 1   Express $3 \sin x + 4 \cos x$ in the form $k \sin(x + \phi)$.

Solution   Begin by evaluating k.

$$k = \sqrt{A^2 + B^2} = \sqrt{3^2 + 4^2} = \sqrt{25} = 5$$

ϕ is an angle such that $\sin \phi = \frac{4}{5} = 0.8$ and $\cos \phi = \frac{3}{5} = 0.6$. Use a calculator to determine that $\phi \approx 53.1°$. Thus, you have

$$3 \sin x + 4 \cos x \approx 5 \sin(x + 53.1°)$$

■

Example 2   Express $5 \sin 3x - 12 \cos 3x$ as a single expression involving the sine function only.

Solution   The expression $5 \sin 3x - 12 \cos 3x$ can be written as $k \sin(3x + \phi)$, where $k = \sqrt{5^2 + (-12)^2} = \sqrt{25 + 144} = \sqrt{169} = 13$. The angle ϕ is such that $\sin \phi = -\frac{12}{13} \approx -0.9231$ and $\cos \phi = \frac{5}{13} \approx 0.3846$. Because $\sin \phi$ is negative and $\cos \phi$ is positive, ϕ must lie in the fourth quadrant; $\phi \approx -67.4°$. Hence, you have

$$5 \sin 3x - 12 \cos 3x \approx 13 \sin(3x - 67.4°)$$

■

Example 3   Express $3 \sin \theta - \sqrt{3} \cos \theta$ in the form $k \cos(\theta + \phi)$.

Solution   First, express the quantity in the form $k \sin(\theta + \phi)$.

$$k = \sqrt{3^2 + (-\sqrt{3})^2} = \sqrt{9 + 3} = \sqrt{12} = 2\sqrt{3}$$

$$\sin \phi = \frac{-\sqrt{3}}{2\sqrt{3}} = -\frac{1}{2}$$

$$\cos \phi = \frac{3}{2\sqrt{3}} = \frac{\sqrt{3}}{2}$$

Because sin ϕ is negative and cos ϕ is positive, ϕ is a fourth-quadrant angle. A fourth-quadrant angle with a sine of $-\frac{1}{2}$ is $-30°$. Hence, you have

$$3 \sin \theta - \sqrt{3} \cos \theta = 2\sqrt{3} \sin(\theta - 30°)$$

Because $\sin x = \cos(90° - x)$, it follows that

$$3 \sin \theta - \sqrt{3} \cos \theta = 2\sqrt{3} \cos[90° - (\theta - 30°)]$$
$$= 2\sqrt{3} \cos(120° - \theta)$$

Because $\cos(120° - \theta) = \cos[-(120° - \theta)] = \cos(\theta - 120°)$, this can be written as

$$3 \sin \theta - \sqrt{3} \cos \theta = 2\sqrt{3} \cos(\theta - 120°) \qquad ■$$

Example 4   Graph the function $y = \cos x - \sin x$.

Solution   Note that $\cos x - \sin x$ or $-\sin x + \cos x$ can be changed to $k \sin(x + \phi)$ with

$$k = \sqrt{(-1)^2 + 1^2} = \sqrt{2}, \quad \sin \phi = \frac{1}{\sqrt{2}} = \frac{\sqrt{2}}{2}, \quad \text{and} \quad \cos \phi = \frac{-1}{\sqrt{2}} = -\frac{\sqrt{2}}{2}$$

Because sin ϕ is positive and cos ϕ is negative, angle ϕ is a second-quadrant angle. The second-quadrant angle with a sine of $\frac{\sqrt{2}}{2}$ is $\frac{3\pi}{4}$ radians. Hence, $y = \cos x - \sin x$ is equivalent to

$$y = \sqrt{2} \sin \left(x + \frac{3\pi}{4} \right)$$

The graph of this function is a simple sine curve with amplitude of $\sqrt{2}$ and a phase shift of $\frac{3\pi}{4}$ to the left. The graph appears in Figure 4-5. Note that the function $y = \cos x - \sin x$ could be graphed using the method of addition of ordinates, which was discussed in Section 3.6.

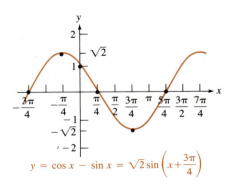

$$y = \cos x - \sin x = \sqrt{2} \sin \left(x + \frac{3\pi}{4} \right)$$

Figure 4-5

Exercise 4.6

In Exercises 1–12, write the given expression in the form k sin(x + φ).

1. $6 \sin x + 8 \cos x$

2. $12 \sin x + 5 \cos x$

3. $6 \sin x - 8 \cos x$

4. $12 \sin x - 5 \cos x$

5. $2 \sin x + \cos x$

6. $\sin x - \cos x$

7. $\sin x + \cos x$

8. $2 \sin x - \cos x$

9. $-\sin x + 5 \cos x$

10. $\sqrt{3} \sin x + 3 \cos x$

11. $\sqrt{3} \sin x - 3 \cos x$

12. $-\sin x - 5 \cos x$

13. Verify that the sum of the squares of the coefficients of sin *x* and cos *x* in Equation (4.29) is equal to 1.

14. If, in the development of this section, the assignments of sin φ and cos φ had been interchanged, what would the results have been?

In Exercises 15–20, use the method of Example 4 to graph each function.

15. $y = \sin x + \cos x$

16. $y = \sin x - \cos x$

17. $y = \sin x - \sqrt{3} \cos x$

18. $y = \sqrt{3} \sin x + \cos x$

19. $y = \sin 2x + \sqrt{3} \cos 2x$

20. $y = \sin 2x - \sqrt{3} \cos 2x$

$$2 \sin \left(2x + \frac{\pi}{3}\right)$$

4.7   APPLICATIONS OF THE TRIGONOMETRIC IDENTITIES

The results of this chapter have made it clear that the six basic circular functions are interrelated in many ways. So far, your work has been theoretical, proving given identities and developing new ones. We now turn our attention to four applications that use trigonometric identities.

1.   Finding the angle between two lines

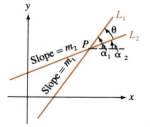

Figure 4-6

If Lines L_1 and L_2, with angles of inclination α_1 and α_2 and slopes m_1 and m_2, intersect at point *P* (see Figure 4-6), then the angle θ from L_2 to L_1 is the difference of the angles of inclination $\alpha_1 - \alpha_2$.

$$\theta = \alpha_1 - \alpha_2$$

This angle, θ, can be found by first finding tan θ. By an identity of Section 4.2, it follows that

$$\tan \theta = \tan(\alpha_1 - \alpha_2)$$

$$= \frac{\tan \alpha_1 - \tan \alpha_2}{1 + \tan \alpha_1 \tan \alpha_2}$$

Recall that the tangent of the angle of inclination of a line is its rise divided by its run—that is, the slope of that line. Hence, we have

$$\tan \alpha_1 = m_1$$

and

$$\tan \alpha_2 = m_2$$

and

$$\tan \theta = \frac{m_1 - m_2}{1 + m_1 m_2}$$ Remember that θ is the angle from L_2 to L_1.

Example 1   Find the angle from the line $y = 3x + 1$ to the line $y = -2x + 3$.

Solution   Let L_1 be the line $y = -2x + 3$ and L_2 be the line $y = 3x + 1$. Because the slope of $y = mx + b$ is m, the slope of $L_1 = -2$ and the slope of $L_2 = 3$. See Figure 4-7. Because $m_1 = -2$ and $m_2 = 3$, you have

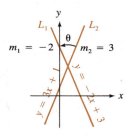

Figure 4-7

$$\begin{aligned}
\tan \theta &= \frac{m_1 - m_2}{1 + m_1 m_2} \\
&= \frac{-2 - 3}{1 + (-2)(3)} \\
&= \frac{-5}{-5} \\
&= 1
\end{aligned}$$

The only angle less than $180°$ whose tangent is 1 is $45°$. Hence, the angle from $y = 3x + 1$ to $y = -2x + 3$ is $45°$.   ■

2. Simple harmonic motion

If a weight, hanging by a spring from the ceiling, is pushed upward a small distance and released, it starts to bounce. Its position at any time t can be found by the formula

$$y = A \cos (2\pi f t)$$

This formula was discussed in Section 3.8.

   If, as the weight is released, it is given a push upward, it will still continue to bounce. However, its position depends on time in a more complicated way.

$$y = A \sin (2\pi f t) + B \cos (2\pi f t)$$

where A, B, and f are dependent on the characteristics of the spring, the upward displacement, and the initial push.

Example 2   A weight is attached to a certain spring and is pushed to start it bouncing. Suppose the position y depends on t according to the formula

$$y = 3 \sin 6t + 4 \cos 6t$$

What is the amplitude of the oscillation?

Solution   By the identity of Section 4.6, it follows that

$$A \sin x + B \cos x = \sqrt{A^2 + B^2} \sin(x + \phi)$$
$$y = 3 \sin 6t + 4 \cos 6t = \sqrt{3^2 + 4^2} \sin(t + \phi)$$
$$y = 5 \sin(t + \phi)$$

The oscillation may still be described by the sine function. Because the amplitude is 5, the weight is at most 5 units above (or below) the equilibrium position. ■

3. Parametric equations

Suppose that two variables (such as x and y) each depend on a third variable (such as t) by the equations

$$\begin{cases} x = \sin t \\ y = \cos t \end{cases}$$

These equations are called **parametric equations** with parameter t. Values of various pairs (x, y) may be computed by letting t assume values from 0 to 2π.

Example 3   What is the graph of the system of equations

$$\begin{cases} x = \sin t \\ y = \cos t \end{cases}$$

Solution   Eliminate the parameter t to obtain an equation in the variables x and y. This can be done by squaring each equation and then adding them.

$$x^2 = \sin^2 t$$
$$y^2 = \cos^2 t$$
$$x^2 + y^2 = \sin^2 t + \cos^2 t$$

Use the identity $\sin^2 t + \cos^2 t = 1$ to obtain

$$x^2 + y^2 = 1$$

This is the equation of a circle, centered at the origin, with a radius of 1. ■

4. Beat frequencies

It is difficult to recognize that two tones are distinct if they are very close in pitch and sounded in succession. If they are sounded together, however, the two tones beat against each other. The combined sound has a noticeable warble, slow or rapid, depending on how far apart the initial frequencies are. A piano tuner listens for these lower beat frequencies and tries to eliminate them by tightening or loosening the individual piano strings.

If two equally loud tones of frequencies f_1 and f_2 are sounded together, the combined amplitude A is

$$A = k \sin(2\pi f_1 t) + k \sin(2\pi f_2 t)$$
$$= k[\sin(2\pi f_1 t) + \sin(2\pi f_2 t)]$$

where k is the amplitude of each of the individual tones. By the identity

$$\sin x + \sin y = 2 \cos \frac{x - y}{2} \sin \frac{x + y}{2}$$

the combined amplitude can be written in a different form.

$$A = k(\sin 2\pi f_1 t + \sin 2\pi f_2 t)$$

$$= k\left(2 \cos \frac{2\pi f_1 t - 2\pi f_2 t}{2} \sin \frac{2\pi f_1 t + 2\pi f_2 t}{2} \right)$$

$$= 2k \cos \left[2\pi \left(\frac{f_1 - f_2}{2} \right) t \right] \sin \left[2\pi \left(\frac{f_1 + f_2}{2} \right) t \right]$$

The factor

$$\sin 2\pi \left(\frac{f_1 + f_2}{2} \right) t$$

represents a sound of frequency $(f_1 + f_2)/2$, which is the average of the frequencies of the two original tones. The other factor

$$\cos 2\pi \left(\frac{f_1 - f_2}{2} \right) t$$

represents a time-varying amplitude, changing at the frequency $(f_1 - f_2)/2$, or half of the difference of the original frequencies. It is this factor that describes the slow warble of volume of the combined sound, because its frequency $(f_1 - f_2)/2$ is quite low when f_1 and f_2 are close to each other. The graph of the sound formed by two tones that are close in pitch appears in Figure 4-8.

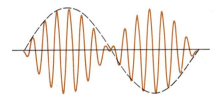

Figure 4-8

Example 4 If the C below middle C on a piano is depressed, a felt-covered hammer strikes two strings. If they are not carefully tuned, the tones beat with each other and the note sounds sour. A piano tuner counts 20 pulsations of sound in a 10-second time interval. How far apart are the frequencies of the two piano strings?

Solution   Figure 4-8 represents one cycle of the time-varying amplitude but two pulsations of sound. Twenty pulsations in 10 seconds is two pulsations per second, or a beat frequency of 1 cycle per second.

$$\frac{f_1 - f_2}{2} = 1$$

$$f_1 - f_2 = 2$$

The two strings are tuned at 2 cycles per second apart.   ∎

Exercise 4.7

1. Find the angle from the line $y = 2x - 5$ to the line $y = 3x + 1$.

2. Find the angle from $y = 3x - 8$ to $y = 5x$.

3. Show that if lines L_1 and L_2 are perpendicular and each has slope, then

$$m_1 = -\frac{1}{m_2}$$

4. If in Example 1 in the text, L_1 had been $y = 3x + 1$ and L_2 had been $y = -2x + 3$, what would have been the angle between the lines?

5. What is the amplitude of the oscillation described by $y = 5 \sin 3t + 12 \cos 3t$?

6. What is the amplitude of the oscillation described by $y = 6 \sin \pi t - 8 \cos \pi t$?

7. A piano tuner counts 50 pulsations of sound in 20 seconds. How far apart are the frequencies of two piano strings that the tuner has struck?

8. Two tuning forks, cut for frequencies of 2000 hertz (cycles per second) and 2003 hertz, are struck and simultaneously touched to a sounding board. How many pulsations of the resulting sound will be heard each second?

In Exercises 9–12, eliminate the parameter t from the equations to obtain an equation in x and y.

9. $\begin{cases} x = a \sin t \\ y = b \cos t \end{cases}$ 10. $\begin{cases} x = \sec t \\ y = \tan t \end{cases}$

11. $\begin{cases} x = 2 \tan t \\ y = 6 \cot t \end{cases}$ 12. $\begin{cases} x = \sin t \\ y = \csc t \end{cases}$

CHAPTER SUMMARY

Key Words

beat frequency   (4.7)
conditional equations   (4.1)
identities   (4.1)

impossible equations   (4.1)
parametric equations   (4.7)

Key Ideas

(4.1) Identities are verified by manipulating one side of an identity algebraically until it is transformed so that it is identical to the other side. An alternative method is to work on both sides of an identity independently until each side is transformed into some common third expression.

(4.2) Identities involving sums and differences of two angles.

$$\cos(A + B) = \cos A \cos B - \sin A \sin B$$
$$\cos(A - B) = \cos A \cos B + \sin A \sin B$$
$$\sin \theta = \cos(90° - \theta)$$
$$\cos \theta = \sin(90° - \theta)$$
$$\sin(A + B) = \sin A \cos B + \cos A \sin B$$
$$\sin(A - B) = \sin A \cos B - \cos A \sin B$$
$$\tan(A + B) = \frac{\tan A + \tan B}{1 - \tan A \tan B}$$
$$\tan(A - B) = \frac{\tan A - \tan B}{1 + \tan A \tan B}$$

(4.3) The double-angle identities.

$$\sin 2A = 2 \sin A \cos A$$
$$\cos 2A = \cos^2 A - \sin^2 A$$
$$\cos 2A = 2 \cos^2 A - 1$$
$$\cos 2A = 1 - 2 \sin^2 A$$
$$\tan 2A = \frac{2 \tan A}{1 - \tan^2 A}$$

(4.4) The half-angle identities.

$$\cos \frac{A}{2} = \pm \sqrt{\frac{1 + \cos A}{2}}$$
$$\sin \frac{A}{2} = \pm \sqrt{\frac{1 - \cos A}{2}}$$
$$\tan \frac{A}{2} = \pm \sqrt{\frac{1 - \cos A}{1 + \cos A}}$$
$$\tan \frac{A}{2} = \frac{1 - \cos A}{\sin A}$$
$$\tan \frac{A}{2} = \frac{\sin A}{1 + \cos A}$$

(4.5) Product-to-sum and sum-to-product identities.

$$\sin A \cos B = \frac{1}{2}[\sin(A + B) + \sin(A - B)]$$

$$\cos A \sin B = \frac{1}{2}[\sin(A + B) - \sin(A - B)]$$

$$\sin A \sin B = \frac{1}{2}[\cos(A - B) - \cos(A + B)]$$

$$\cos A \cos B = \frac{1}{2}[\cos(A + B) + \cos(A - B)]$$

$$\sin A + \sin B = 2 \sin \frac{A + B}{2} \cos \frac{A - B}{2}$$

$$\sin A - \sin B = 2 \cos \frac{A + B}{2} \sin \frac{A - B}{2}$$

$$\cos A + \cos B = 2 \cos \frac{A + B}{2} \cos \frac{A - B}{2}$$

$$\cos A - \cos B = -2 \sin \frac{A + B}{2} \sin \frac{A - B}{2}$$

(4.6) Sums of the form $A \sin x + B \cos x$.

$A \sin x + B \cos x = k \sin(x + \phi)$, where $k = \sqrt{A^2 + B^2}$ and ϕ is any angle for which

$$\sin \phi = \frac{B}{\sqrt{A^2 + B^2}} \quad \text{and} \quad \cos \phi = \frac{A}{\sqrt{A^2 + B^2}}$$

REVIEW EXERCISES

In Review Exercises 1–10, find the required value by using trigonometric functions of 300° and 45°.

1. $\sin 345°$ **2.** $\cos 345°$ **3.** $\tan 345°$ **4.** $\sin 255°$

5. $\cos 255°$ **6.** $\tan 255°$ **7.** $\sin 600°$ **8.** $\cos 600°$

9. $\tan 600°$ **10.** $\sin 22.5°$

11. Evaluate $\cos 22.5°$.   **12.** Evaluate $\tan 22.5°$.

In Review Exercises 13–24, express each quantity as a single function of one angle, and simplify if possible.

13. $\sin 20° \cos 51° + \cos 20° \sin 51°$ **14.** $\dfrac{\tan 20° - \tan 51°}{1 + \tan 20° \tan 51°}$

15. $\cos 35° \cos 15° + \sin 35° \sin 15°$ **16.** $\cos \dfrac{3\pi}{11} \cos \dfrac{\pi}{11} - \sin \dfrac{3\pi}{11} \sin \dfrac{\pi}{11}$

17. $\sin \dfrac{2\pi}{7} \cos \dfrac{\pi}{7} - \cos \dfrac{2\pi}{7} \sin \dfrac{\pi}{7}$

18. $2 \cos 17° \sin 17°$

19. $2 \cos^2 17° - 1$

20. $\dfrac{2 \tan 217°}{1 - \tan^2 217°}$

21. $4 \sin \dfrac{3\pi}{8} \cos \dfrac{3\pi}{8}$

22. $1 - 2 \sin^2 \dfrac{\pi}{3}$

23. $\cos^4 \dfrac{\pi}{13} - \sin^4 \dfrac{\pi}{13}$

24. $8 \sin^2 \dfrac{\theta}{2} \cos^2 \dfrac{\theta}{2}$

In Review Exercises 25–36, write the right-hand side of the indicated formula.

25. $\sin(\theta + \alpha) =$

26. $\cos(\theta + \alpha) =$

27. $\tan(\theta + \alpha) =$

28. $\sin(\theta - \alpha) =$

29. $\cos(\theta - \alpha) =$

30. $\tan(\theta - \alpha) =$

31. $\sin 2\theta =$

32. $\cos 2\theta =$

33. $\tan 2\theta =$

34. $\sin \dfrac{\theta}{2} =$

35. $\cos \dfrac{\theta}{2} =$

36. $\tan \dfrac{\theta}{2} =$

37. Show that $\cos(60° + \theta) = \dfrac{1}{2} (\cos \theta - \sqrt{3} \sin \theta)$.

38. Show that $\sin \left(\dfrac{3\pi}{2} - \theta \right) = -\cos \theta$.

39. Show that $\tan(180° - \theta) = -\tan \theta$.

40. Show that $\sin(120° + \theta) = \dfrac{1}{2}(\sqrt{3} \cos \theta - \sin \theta)$.

41. Show that $\cos(300° - \theta) = \dfrac{1}{2} (\cos \theta - \sqrt{3} \sin \theta)$.

42. Show that $\tan \left(\dfrac{\pi}{4} + \theta \right) = \dfrac{1 + \tan \theta}{1 - \tan \theta}$.

43. Show that $\sin x = \dfrac{\sin 2x}{2 \cos x}$.

44. Show that $\cos \theta = \pm \sqrt{\sin^2 \theta + \cos 2\theta}$.

In Review Exercises 45–50, use the given information to find the sine, cosine, and tangent of 2θ.

45. $\cos \theta = -\dfrac{12}{13}$; θ in QII

46. $\sin \theta = -\dfrac{5}{13}$; θ in QIII

47. $\tan \theta = -\dfrac{20}{21}$; θ in QIV

48. $\cos \theta = \dfrac{21}{29}$; θ in QI

49. $\sin \theta = -\dfrac{4}{5}$; θ in QIII

50. $\cot \theta = -\dfrac{3}{4}$; θ in QII

In Review Exercises 51–56, use the given information to find the sine, cosine, and tangent of θ/2. Assume that $0° \leq \theta < 360°$.

51. $\cos \theta = \dfrac{12}{13}$;   θ in QI

52. $\sin \theta = -\dfrac{5}{13}$;   θ in QIV

53. $\tan \theta = -\dfrac{5}{12}$;   θ in QIV

54. $\cos \theta = -\dfrac{3}{5}$;   θ in QII

55. $\sin \theta = \dfrac{4}{5}$;   θ in QII

56. $\cot \theta = \dfrac{3}{4}$;   θ in QIII

*In Review Exercises 57–60, evaluate each of the products. **Do not use a calculator or tables.***

57. $\sin 285° \cos 15°$

58. $\cos 285° \sin 15°$

59. $\sin \dfrac{5\pi}{4} \sin \dfrac{\pi}{12}$

60. $\cos \dfrac{\pi}{12} \cos \dfrac{5\pi}{4}$

In Review Exercises 61–64, express each quantity as a product.

61. $\sin 5° + \sin 7°$

62. $\sin 312° - \sin 140°$

63. $\cos \dfrac{3\pi}{5} - \cos \dfrac{\pi}{5}$

64. $\cos \dfrac{2\pi}{7} + \cos \dfrac{3\pi}{7}$

In Review Exercises 65–68, express each quantity as a product and find its value without using a calculator or tables.

65. $\sin 285° + \sin 15°$

66. $\sin 15° - \sin 285°$

67. $\cos \dfrac{5\pi}{12} - \cos \dfrac{\pi}{12}$

68. $\cos \dfrac{\pi}{12} + \cos \dfrac{5\pi}{12}$

In Review Exercises 69–72, verify each identity.

69. $\cot^2 x \sec x = \dfrac{\csc x}{\tan x}$

70. $\cos x(\sec x + \cos x) = 2 - \sin^2 x$

71. $\sec^2 x(\csc x - \sin x) = \csc x$

72. $\dfrac{1}{\sec \theta - \tan \theta} + \dfrac{1}{\sec \theta + \tan \theta} = 2 \sec \theta$

CHAPTER FIVE

TRIGONOMETRIC EQUATIONS AND THE INVERSE TRIGONOMETRIC FUNCTIONS

We have discussed equations that were identities. We now consider trigonometric equations that are true for some, but not all, values of the variable.

5.1   TRIGONOMETRIC EQUATIONS

The equation $\sin 2x = 2 \sin x$ is not an identity because it is not true for all values of x. It is true, however, for *some* values of x. For example, the equation is true if $x = 0$, as the following check shows.

$$\sin 2x = 2 \sin x$$
$$\sin 2(0) \overset{?}{=} 2 \sin 0$$
$$\sin 0 \overset{?}{=} 2 \cdot 0$$
$$0 = 0$$

Other values of x, such as π, 2π, 3π, and so on also satisfy this equation. The process of finding *all* such values is called **solving** the equation. Solving a trigonometric equation is similar to solving an equation in algebra. We must apply such familiar techniques as combining terms, adding terms to both sides of an equation, factoring, and so on. However, we must also use the trigonometric formulas and identities previously discussed. We begin by providing a formal solution of the equation $\sin 2x = 2 \sin x$.

Example 1   Solve $\sin 2x = 2 \sin x$ for all values of x, where x is a real number.

Solution   Use the identity for $\sin 2x$ to rewrite the left side of the equation.

$$\sin 2x = 2 \sin x$$
$$2 \sin x \cos x = 2 \sin x$$

Divide both sides of the previous equation by 2, subtract $\sin x$ from both sides, and proceed as follows. Do not begin by dividing both sides of the equation by $\sin x$. Dividing by an expression that might be zero could cause you to lose a solution of the equation.

$\sin x \cos x = \sin x$	Divide both sides by 2.
$\sin x \cos x - \sin x = 0$	Subtract $\sin x$ from both sides.
$\sin x(\cos x - 1) = 0$	Factor out $\sin x$.
$\sin x = 0 \quad$ or $\quad \cos x - 1 = 0$	Set each factor equal to 0.
$\cos x = 1$	

The solutions to the equation $\sin x \cos x = \sin x$ are all real numbers whose sine is 0 or whose cosine is 1. Numbers with a sine of 0 are

$$\ldots, \, -4\pi, \, -3\pi, \, -2\pi, \, -\pi, \, 0, \, \pi, \, 2\pi, \, 3\pi, \, 4\pi, \, \ldots$$

Numbers with a cosine of 1 are

$$\ldots, -4\pi, -2\pi, 0, 2\pi, 4\pi, \ldots$$

Thus, the solutions of the given equation are the real numbers

$$\ldots, -4\pi, -3\pi, -2\pi, -\pi, 0, \pi, 2\pi, 3\pi, 4\pi, \ldots$$

Verify that each of these solutions satisfies the given equation. ■

Example 2 Solve $2 \sin^2 \theta + \sin \theta = 1$ for all values of θ, where $0° \le \theta < 360°$.

Solution After subtracting 1 from both sides of the equation, factor the left-hand side. Then, set each factor equal to zero, and solve each equation for θ.

$$2 \sin^2 \theta + \sin \theta = 1$$
$$2 \sin^2 \theta + \sin \theta - 1 = 0$$
$$(2 \sin \theta - 1)(\sin \theta + 1) = 0$$

$2 \sin \theta - 1 = 0$	or	$\sin \theta + 1 = 0$
$2 \sin \theta = 1$		$\sin \theta = -1$
$\sin \theta = \dfrac{1}{2}$		
$\theta = 30°, 150°$		$\theta = 270°$

These are the three solutions that lie in the designated interval from $0°$ to $360°$. Verify that all three values satisfy the given equation. ■

Example 3 Solve $\sin 3x = \dfrac{1}{2}$ for all x, where $0 \le x < 2\pi$.

Solution There are infinitely many values of $3x$ that will satisfy the equation $\sin 3x = \frac{1}{2}$. Because you are given that $0 \le x < 2\pi$, you must find all such values of $3x$ where

$$3(0) \le 3x < 3(2\pi)$$
or
$$0 \le 3x < 6\pi$$

There are six such values.

$$3x = \frac{\pi}{6}, \frac{5\pi}{6}, \frac{13\pi}{6}, \frac{17\pi}{6}, \frac{25\pi}{6}, \frac{29\pi}{6}$$

The values of x are found by dividing each of the possible values of $3x$ by 3. Thus, you have

$$x = \frac{\pi}{18}, \frac{5\pi}{18}, \frac{13\pi}{18}, \frac{17\pi}{18}, \frac{25\pi}{18}, \frac{29\pi}{18}$$

Note that the largest of these, $\frac{29\pi}{18}$, is still less than 2π. All six roots do satisfy the given equation. ■

Example 4   Solve $\sin \theta = \cos \theta$ for all θ between $0°$ and $360°$, including $0°$.

Solution   If $\cos \theta \neq 0$, you may divide both sides of the equation by $\cos \theta$.

$$\sin \theta = \cos \theta$$

$$\frac{\sin \theta}{\cos \theta} = 1$$

$$\tan \theta = 1$$

$$\theta = 45°, 225°$$

Both values satisfy the given equation. If $\cos \theta = 0$, then $\theta = 90°$ or $270°$. These values do not satisfy the given equation. Hence, the solutions are $45°$ and $225°$.   ■

Example 5   Solve $\sec^2 x \csc^2 x = \sec^2 x + \csc^2 x$ for x.

Solution   Transform the equation so it involves only sines and cosines. Then, find a common denominator for the fractions on the right and add them. Finally, replace $\sin^2 x + \cos^2 x$ with 1.

$$\sec^2 x \csc^2 x = \sec^2 x + \csc^2 x$$

$$\frac{1}{\cos^2 x} \cdot \frac{1}{\sin^2 x} = \frac{1}{\cos^2 x} + \frac{1}{\sin^2 x}$$

$$\frac{1}{\cos^2 x \sin^2 x} = \frac{\sin^2 x + \cos^2 x}{\sin^2 x \cos^2 x}$$

$$\frac{1}{\cos^2 x \sin^2 x} = \frac{1}{\cos^2 x \sin^2 x}$$

This equation is true for all admissible values of x. Hence, this equation is an identity.   ■

Example 6   Solve the equation $2 \cos^3 \theta = \cos \theta$ for θ, where $0° \leq \theta < 360°$.

Solution   Subtract $\cos \theta$ from both sides of the equation and factor out $\cos \theta$. Then, proceed as follows.

$$2 \cos^3 \theta = \cos \theta$$

$$2 \cos^3 \theta - \cos \theta = 0$$

$$\cos \theta (2 \cos^2 \theta - 1) = 0$$

$$\cos \theta = 0 \quad \text{or} \quad 2 \cos^2 \theta - 1 = 0$$

$$\cos^2 \theta = \frac{1}{2}$$

$$\cos \theta = \pm \frac{1}{\sqrt{2}} = \pm \frac{\sqrt{2}}{2}$$

$$\theta = 90°, 270° \qquad\qquad \theta = 45°, 135°, 225°, 315°$$

All six values satisfy the given equation. Note that, if both sides of the equation had been divided by $\cos \theta$, the solutions 90° and 270° would have been lost.

■

Example 7   Solve $4 \sin^2 \dfrac{x}{2} = 1$ for x, where $0 \leqslant x < 2\pi$.

Solution   Rearrange the terms, factor, and set each factor equal to 0.

$$4 \sin^2 \frac{x}{2} = 1$$

$$4 \sin^2 \frac{x}{2} - 1 = 0$$

$$\left(2 \sin \frac{x}{2} - 1 \right)\left(2 \sin \frac{x}{2} + 1 \right) = 0$$

$$2 \sin \frac{x}{2} - 1 = 0 \qquad \text{or} \qquad 2 \sin \frac{x}{2} + 1 = 0$$

$$\sin \frac{x}{2} = \frac{1}{2} \qquad\qquad\qquad \sin \frac{x}{2} = -\frac{1}{2}$$

$$\frac{x}{2} = \frac{\pi}{6}, \frac{5\pi}{6} \qquad\qquad\qquad \frac{x}{2} = \frac{7\pi}{6}, \frac{11\pi}{6}$$

$$x = \frac{\pi}{3}, \frac{5\pi}{3} \qquad\qquad\qquad x = \frac{7\pi}{3}, \frac{11\pi}{3}$$

Since $7\frac{\pi}{3}$ and $11\frac{\pi}{3}$ are greater than 2π, they must be excluded. The only solutions are $x = \frac{\pi}{3}$ and $x = \frac{5\pi}{3}$. Verify that each of these solutions satisfies the given equation.

■

Example 8   Solve $4 \sin^2 \dfrac{x}{2} = 1$ for x $(0 \leqslant x < 2\pi)$ by using a half-angle identity.

Solution   From the half-angle identity for $\sin \dfrac{x}{2}$, it follows that

$$\sin^2 \frac{x}{2} = \frac{1 - \cos x}{2}$$

Substitute $\dfrac{1 - \cos x}{2}$ for $\sin^2 \dfrac{x}{2}$ in the original equation to get

$$4 \sin^2 \frac{x}{2} = 1$$

$$4\left(\frac{1 - \cos x}{2} \right) = 1$$

$$1 - \cos x = \frac{1}{2}$$

$$\cos x = \frac{1}{2}$$

$$x = \frac{\pi}{3}, \frac{5\pi}{3}$$

Verify that each value satisfies the given equation.

Example 9   Solve $2 \sin x \cos x + \cos x - 2 \sin x - 1 = 0$, where $0 \le x < 2\pi$.

Solution   Use the technique of *factoring by grouping*. The four terms on the left-hand side share no common factors. However, the first two terms share a common factor of cos x, and the last two terms share a common factor of -1. Proceed as follows.

$$2 \sin x \cos x + \cos x - 2 \sin x - 1 = 0$$
$$\cos x(2 \sin x + 1) - 1(2 \sin x + 1) = 0$$
$$(2 \sin x + 1)(\cos x - 1) = 0 \qquad \text{Factor out the common factor of } 2 \sin x + 1.$$

$2 \sin x + 1 = 0$ or	$\cos x - 1 = 0$ Set each factor equal to 0.
$2 \sin x = -1$	$\cos x = 1$
$\sin x = \dfrac{-1}{2}$	
$x = \dfrac{7\pi}{6}, \dfrac{11\pi}{6}$	$x = 0$

Verify that each value satisfies the given equation.

Example 10   Solve $\sin \theta + \cos \theta = 1$ for θ, where $0° \le \theta < 360°$.

Solution   Use the identity

$$A \sin \theta + B \cos \theta = \sqrt{A^2 + B^2} \sin(\theta + \alpha)$$

to write $\sin \theta + \cos \theta$ in the form $k \sin(\theta + \alpha)$.

$$\sin \theta + \cos \theta = 1$$
$$\sqrt{1^2 + 1^2} \sin(\theta + \alpha) = 1$$
$$\sqrt{2} \sin(\theta + \alpha) = 1$$

Remember that α is an angle for which $\sin \alpha = \cos \alpha = \dfrac{1}{\sqrt{2}} = \dfrac{\sqrt{2}}{2}$. Thus, $\alpha = 45°$ and you have

$$\sqrt{2} \sin(\theta + 45°) = 1$$
$$\sin(\theta + 45°) = \frac{1}{\sqrt{2}} = \frac{\sqrt{2}}{2}$$

$\theta + 45° = 45°$ or	$\theta + 45° = 135°$
$\theta = 0°$	$\theta = 90°$

Verify that each value satisfies the given equation.

Exercise 5.1

In Exercises 1–4, solve for all values of the variable. Assume that x is a real number, and that θ is an angle in degrees. **Do not use a calculator or tables.**

1. $\sin \theta = \dfrac{\sqrt{3}}{2}$ **2.** $\cos x = \dfrac{-\sqrt{2}}{2}$

3. $\sin 2x = 0$ **4.** $\cos 2\theta = 1$

In Exercises 5–46, solve for all values of the variable between 0° and 360°, including 0°. **Do not use a calculator or tables.**

5. $\cos \dfrac{\theta}{2} = \dfrac{1}{2}$ **6.** $\sin \dfrac{\theta}{2} = \dfrac{\sqrt{3}}{2}$

7. $\sin \theta \cos \theta = 0$ **8.** $\sin \theta \cos \theta - \sin \theta = 0$

9. $\cos \theta \sin \theta - \cos \theta = 0$ **10.** $\cos \theta \sin \theta + \cos \theta = 0$

11. $\sin \theta \cos \theta + \sin \theta = 0$ **12.** $\cos^2 \theta = 1$

13. $\sin^2 \theta = \dfrac{1}{2}$ **14.** $\sin^2 \theta - \dfrac{1}{2} \sin \theta - \dfrac{1}{2} = 0$

15. $\cos^2 \theta - \dfrac{1}{2} \cos \theta - \dfrac{1}{2} = 0$ **16.** $\sin^2 \theta - 3 \sin \theta + 2 = 0$

17. $4 \sin^2 \theta + 4 \sin \theta + 1 = 0$ **18.** $\sin 2\theta - \cos \theta = 0$

19. $\sin 2A + \cos A = 0$ **20.** $\cos^2 A + 4 \cos A + 3 = 0$

21. $\cos B = \sin B$ **22.** $\cos B = \sqrt{3} \sin B$

23. $\cos \theta = \cos \dfrac{\theta}{2}$ **24.** $\cos \theta = \sin \dfrac{\theta}{2}$

25. $\cos 2\theta = \cos \theta$ **26.** $\cos 2\theta = \sin \theta$

27. $\tan A = -\sin A$ **28.** $\cot A = -\cos A$

29. $\cos^2 C - \sin^2 C = \dfrac{1}{2}$ **30.** $\cos^2 C + \sin^2 C = \dfrac{1}{2}$

31. $\sin \theta + \cos \theta = \sqrt{2}$
(Hint: Square both sides.)   **32.** $\sin \theta - \cos \theta = \sqrt{2}$

33. $\cos 2A = 1 - \sin A$ **34.** $\cos 2A = \cos A - 1$

35. $\cos 2B + \cos B + 1 = 0$ **36.** $\cos^2 B + \sin B - 1 = 0$

37. $\sin 4\theta = \sin 2\theta$ **38.** $\cos 4\theta = \cos 2\theta$

39. $\cos B = 1 + \sqrt{3} \sin B$ **40.** $\tan 2B + \sec 2B = 1$

41. $9 \cos^4 A = \sin^4 A$ **42.** $6 \cos^2 A = -9 \sin A$

43. $\sec C = \tan C + \cos C$ **44.** $1 - \tan C = \sqrt{2} \sec C$

45. $2 \cos x \sin x - \sqrt{2} \cos x = \sqrt{2} \sin x - 1$

46. $4 \sin x \cos x - 2 \sin x = 2 \cos x - 1$

In Exercises 47–52, solve for θ*, where* $0° \leq \theta < 360°$*. Recall that the expression* $a \sin x + b \cos x$ *can be written as* $k \sin(x + \phi)$*.*

47. $\dfrac{\sqrt{3}}{2} \sin \theta + \dfrac{1}{2} \cos \theta = \dfrac{1}{2}$

48. $\sqrt{2} \sin \theta + \sqrt{2} \cos \theta = \sqrt{3}$

49. $\dfrac{1}{2} \sin \theta + \dfrac{\sqrt{3}}{2} \cos \theta = 1$

50. $\cos \theta - \sin \theta = \sqrt{2}$

51. $\cos \theta - \sqrt{3} \sin \theta = 1$

52. $\sin \theta + \cos \theta = -\sqrt{2}$

In Exercises 53–62, solve for x*, where* $0 \leq x < 2\pi$*. Consider using a sum-to-product or a product-to-sum identity.*

53. $\sin x \cos x = \dfrac{1}{2}$

54. $\cos x \sin x = \dfrac{\sqrt{3}}{4}$

55. $2 \sin \dfrac{3x}{2} \cos \dfrac{x}{2} = \sin x$

56. $2 \cos \dfrac{3x}{2} \cos \dfrac{x}{2} = \cos 2x$

57. $\sin 3x \cos x = \dfrac{1}{2} \sin 2x$

58. $\cos 3x \cos x = \cos^2 2x$

59. $\sin 4x = -\sin 2x$

60. $\cos 4x = -\cos 2x$

61. $\cos 9x = -\cos 3x$

62. $\sin 12x = -\sin 4x$

In Exercises 63–76, solve for x*, where* $0 \leq x < 2\pi$*.*

63. $2 \sin^4 x - 9 \sin^2 x + 4 = 0$

64. $2 \sin^3 x + \sin^2 x = \sin x$

65. $4 \cos x \sin^2 x - \cos x = 0$

66. $\tan^2 x = 1 + \sec x$

67. $\csc^4 x = 2 \csc^2 x - 1$

68. $\tan^2 x - 5 \tan x + 6 = 0$

69. $4 \sin^2 x + 4 \sin x + 1 = 0$

70. $2 \sin 5x = 1$

71. $2 \sin^2 5x = 1$

72. $2 \sin^2 2x + \sin 2x - 5 = 0$

73. $\tan^2 x - \tan x = 0$

74. $2 \sin x - \sqrt{2} \tan x = \sqrt{2} \sec x - 2$

75. $\cot^2 x \cos x + \cot^2 x - 3 \cos x - 3 = 0$

76. $\tan x + \cot x = -2$

77. Suppose that $\sin \sqrt{2} \, \pi t = 0$. Show that the values of $\cos \sqrt{2} \, \pi t$ are 1 or -1.

78. Suppose that $\cos \frac{t}{2} = 0$. Show that the values of $\sin \frac{t}{2}$ are 1 or -1.

79. Suppose that $\cos t - \sin t = 0$. Show that the values of $\cos t + \sin t$ are $\sqrt{2}$ and $-\sqrt{2}$.

80. Suppose that $\cos 2t = 1$. Show that the value of $2 \sin t \cos t$ is 0.

5.2 INTRODUCTION TO THE INVERSE TRIGONOMETRIC RELATIONS

The process of solving trigonometric equations often leads to simple equations such as

$$\sin \theta = \frac{1}{2} \quad \text{or} \quad \sin \theta = \frac{\sqrt{3}}{2}$$

These equations are easy to solve for values of θ because of your knowledge of the sine function. If the sine of angle θ is $\frac{1}{2}$, then θ is any angle coterminal with 30° or 150°. Solving a more general equation such as $\sin \theta = y$ for the variable θ requires a new concept called the **inverse sine** of y. The inverse sine of y is denoted either as $\sin^{-1} y$ or arcsin y. In this book, the $\sin^{-1} y$ notation will be used more often, although your teacher may prefer the arcsin y notation. Keep in mind that $\sin^{-1} y$ and arcsin y mean the same thing and that either notation can be used.

Any time you see notation indicating the inverse relation of a trigonometric function, you may translate it into more familiar notation by using the following definition.

(5.1)

Definition.		
$\sin^{-1} y = \theta$	is equivalent to	$y = \sin \theta.$
$\cos^{-1} y = \theta$	is equivalent to	$y = \cos \theta.$
$\tan^{-1} y = \theta$	is equivalent to	$y = \tan \theta.$
$\cot^{-1} y = \theta$	is equivalent to	$y = \cot \theta.$
$\csc^{-1} y = \theta$	is equivalent to	$y = \csc \theta.$
$\sec^{-1} y = \theta$	is equivalent to	$y = \sec \theta.$

Be sure to note that the "-1" in the notation "$\sin^{-1} y$" is *not* an exponent and should never be thought of as one. The expression $\sin^{-1} y$ is not the same as $(\sin y)^{-1}$.

$$\sin^{-1} y = \frac{1}{\sin y}$$

It is often convenient to think of the inverse trigonometric relations as angles. Then, the notation "$\sin^{-1} y$" means "all angles whose sine is y." Thus, "$\sin^{-1} \frac{1}{2}$" represents "all angles whose sine is $\frac{1}{2}$," so $\sin^{-1} \frac{1}{2} = 30°$, 150°, or any angle coterminal with 30° or 150°.

Example 1 Find $\sin^{-1} \dfrac{\sqrt{3}}{2}$.

Solution Translate the notation $\sin^{-1} (\sqrt{3}/2) = \theta$ to $\sin \theta = \sqrt{3}/2$ notation. Then, solve the equation $\sin \theta = \sqrt{3}/2$ as in the previous section.

$$\sin \theta = \frac{\sqrt{3}}{2}$$

$\theta = 60°, 120°,$ and all angles coterminal with 60° or 120°

Note that "to find $\sin^{-1} (\sqrt{3}/2)$" means "to find all angles with a sine of $\sqrt{3}/2$." ∎

Example 2 Find $\tan^{-1} \sqrt{3}$.

Solution Translate the notation $\tan^{-1} \sqrt{3} = \theta$ to $\tan \theta = \sqrt{3}$ notation.

$$\tan \theta = \sqrt{3}$$

$\theta = 60°, 240°,$ and all angles coterminal with 60° or 240° ∎

Example 3 Find θ if $\theta = \cos^{-1}(-1)$ and θ is in radians.

Solution The angles (in radians) whose cosines are -1 are those coterminal with π, such as $\pi, 3\pi, 5\pi, \ldots$ and $-\pi, -3\pi, -5\pi, \ldots$. ∎

Example 4 Find θ if $\theta = \sin^{-1}\dfrac{\pi}{2}$.

Solution Be careful. Don't confuse $\sin^{-1}(\pi/2)$ with $\sin(\pi/2)$; the answer is not 1. You are looking for those angles whose sines are $\pi/2$. Because $\pi/2$ is greater than 1 and the sine function can never be greater than 1, there are no values of θ possible. ∎

Example 5 Find x in radians if $x = \arcsin 0.8330$.

Solution Use your calculator, making sure that it is in the radian mode. Enter .833 and press the $\boxed{\text{INV}}$ and $\boxed{\text{SIN}}$ keys in succession. The answer .9845 is displayed. This is a first-quadrant angle. A second-quadrant angle whose sine is .833 is found by subtracting .9845 from π (.9845 is the reference angle in Figure 5-1). The result, 2.1571, is another possible answer. There are infinitely many more, found by adding multiples of 2π to either 0.9845 or 2.1571. ∎

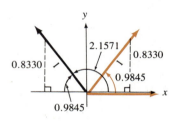

Figure 5-1

Example 6 If $x = \cos^{-1}\dfrac{1}{2}$, find all real numbers x that lie in the interval $0 \leqslant x < 2\pi$.

Solution Think of the equation "$x = \cos^{-1}\frac{1}{2}$" as "all real numbers (or angles in radians) whose cosine is $\frac{1}{2}$." Hence, $\cos x = \frac{1}{2}$. The real numbers (or angles in radians) in the interval $0 \leqslant x < 2\pi$ whose cosines are $\frac{1}{2}$ are $\frac{\pi}{3}$ and $\frac{5\pi}{3}$. ∎

Example 7 If $x = \sec^{-1}(-2)$, find all real numbers x in the interval $0 \leqslant x < 2\pi$.

Solution Think of the equation "$x = \sec^{-1}(-2)$" as "all real numbers (or angles in radians) whose secant is -2." A real number whose secant is -2 is also a number whose cosine is $-\frac{1}{2}$. The numbers in the required interval that satisfy this condition are $\frac{2\pi}{3}$ and $\frac{4\pi}{3}$. ∎

Exercise 5.2

In Exercises 1–12, find all values of θ that lie in the interval $0° \leq \theta < 360°$ without using a calculator or tables.

1. $\sin^{-1} \dfrac{1}{2} = \theta$　　　　　　　　　　**2.** $\cos^{-1} \dfrac{1}{2} = \theta$

3. $\tan^{-1}(-1) = \theta$　　　　　　　　　　　**4.** $\cot^{-1} 1 = \theta$

5. $\cos^{-1} \left(\dfrac{-\sqrt{2}}{2} \right) = \theta$　　　　　　　**6.** $\sin^{-1}(-2) = \theta$

7. $\cot^{-1} \sqrt{3} = \theta$　　　　　　　　　　**8.** $\tan^{-1} \left(\dfrac{-\sqrt{3}}{3} \right) = \theta$

9. $\sin^{-1} 0 = \theta$　　　　　　　　　　　　**10.** $\tan^{-1} 0 = \theta$

11. $\arccos \sqrt{3} = \theta$　　　　　　　　　　**12.** $\text{arccot}(-\sqrt{3}) = \theta$

In Exercises 13–24, find all real numbers x, if any, that lie in the interval $0 \leq x < 2\pi$ without using a calculator or tables.

13. $\cos^{-1} \left(-\dfrac{1}{2} \right) = x$　　　　　　　**14.** $\sin^{-1} \dfrac{\sqrt{2}}{2} = x$

15. $\sin^{-1} \dfrac{1}{2} = x$　　　　　　　　　　**16.** $\cos^{-1} 0 = x$

17. $\tan^{-1} \sqrt{3} = x$　　　　　　　　　　**18.** $\cot^{-1} \dfrac{\sqrt{3}}{3} = x$

19. $\cot^{-1} \left(\dfrac{-\sqrt{3}}{3} \right) = x$　　　　　　**20.** $\tan^{-1} \left(\dfrac{\sqrt{3}}{3} \right) = x$

21. $\arccos \sqrt{3} = x$　　　　　　　　　　**22.** $\arccos \left(\dfrac{\sqrt{2}}{2} \right) = x$

23. $\arcsin 1 = x$　　　　　　　　　　　　**24.** $\arcsin \pi = x$

In Exercises 25–36, use a calculator to find all values of x in radians, if any, that lie in the interval $0 \leq x < 2\pi$. Give all answers to the nearest thousandth.

25. $\sin^{-1} 0.8415 = x$　　　　　　　　　**26.** $\cos^{-1} 0.5403 = x$

27. $\tan^{-1} 1.557 = x$　　　　　　　　　　**28.** $\cot^{-1} 0.6421 = x$

29. $\cos^{-1}(-0.4161) = x$　　　　　　　　**30.** $\sin^{-1} 0.9193 = x$

31. $\cot^{-1}(-0.4577) = x$　　　　　　　　**32.** $\tan^{-1}(-2.1850) = x$

33. $\arcsin 0.1411 = x$　　　　　　　　　**34.** $\arccos(-0.9900) = x$

35. $\text{arccot}(-7.0153) = x$　　　　　　　**36.** $\arctan(-0.1425) = x$

In Exercises 37–48, use a calculator to find all values of θ in degrees, if any, that lie in the interval $0° \leq \theta < 360°$. Give all answers to the nearest degree.

37. $\sin^{-1} 0.8192 = \theta$　　　　　　　　　**38.** $\cos^{-1} 0.2588 = \theta$

39. $\tan^{-1}(-1.428) = \theta$　　　　　　　　**40.** $\cot^{-1}(-1.3270) = \theta$

41. $\cos^{-1}(0.5736) = \theta$

42. $\sin^{-1}(-0.4226) = \theta$

43. $\cot^{-1}(-0.8391) = \theta$

44. $\tan^{-1} 0.0875 = \theta$

45. $\sec^{-1} 1.5557 = \theta$

46. $\csc^{-1} 1.2208 = \theta$

47. $\csc^{-1} 3.8637 = \theta$

48. $\sec^{-1}(-1.0353) = \theta$

5.3   THE INVERSE SINE, COSINE, AND TANGENT FUNCTIONS

A relation in the set R of real numbers is any nonempty set of ordered pairs of real numbers. Certain special relations are called **functions**. Recall that a function is a relation in which to each value of a first component there corresponds exactly one value of a second component.

The set of ordered pairs of real numbers (x, y) defined by the equation $y = \sin x$ is a function because any number x gives a single value of y. Its graph appears in Figure 5-2.

$y = \sin x$

x	y
$-\pi$	0
$-\dfrac{\pi}{2}$	-1
0	0
$\dfrac{\pi}{2}$	1
π	0
$\dfrac{3\pi}{2}$	-1
2π	0

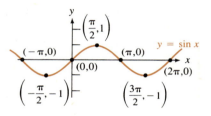

Figure 5-2

Interchanging x and y in the function $y = \sin x$ yields $x = \sin y$, or by the notation used in the previous section, $y = \sin^{-1} x$. The graph in Figure 5-3 is the graph of the relation $y = \sin^{-1} x$.

$x = \sin y$
or
$y = \sin^{-1} x$

x	y
0	$-\pi$
-1	$-\dfrac{\pi}{2}$
0	0
1	$\dfrac{\pi}{2}$
0	π
-1	$\dfrac{3\pi}{2}$
0	2π

$y = \sin^{-1} x$ or $x = \sin y$

Figure 5-3

It is evident from the graph of Figure 5-3 that the relation $y = \sin^{-1} x$ is not a function: it does not pass the vertical line test. However, suppose that we chose the portion of the graph lying between $y = -\frac{\pi}{2}$ and $y = \frac{\pi}{2}$ as shown in Figure 5-4. Then that portion does pass the vertical line test and does determine a function. The colored portion of Figure 5-4 is the graph of the *inverse sine function,* or the *arcsine function.* It is denoted either by $y = \text{Sin}^{-1}x$ or by $y = $ Arcsin x. It is important to note and remember that uppercase letters "S" and "A" are used to denote the inverse sine *function.* As stated earlier, $y = \text{Sin}^{-1}x$ will be used most often in this book.

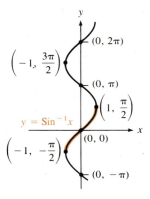

Figure 5-4

The graph of the inverse sine function appears in Figure 5-5.

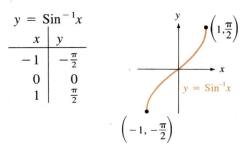

$y = \text{Sin}^{-1}x$

x	y
-1	$-\frac{\pi}{2}$
0	0
1	$\frac{\pi}{2}$

Figure 5-5

The formal definition of the inverse sine function is as follows.

(5.2)

Definition. The **inverse sine** (or arcsine) **function,** denoted by $y = \text{Sin}^{-1} x$ (or $y = $ Arcsin x), has a domain of $\{x \mid -1 \le x \le 1\}$ and a range of $\{y \mid -\frac{\pi}{2} \le y \le \frac{\pi}{2}\}$.

$y = \text{Sin}^{-1} x$ if and only if $x = \sin y$ and $-\frac{\pi}{2} \le y \le \frac{\pi}{2}$

(5.3)

> **Definition.**   The value of $\text{Sin}^{-1} x$ is called the **principal value** of the relation $y = \sin^{-1}x$.

The heavy black curve in Figure 5-6 is that portion of $y = \sin x$ for which $-\frac{\pi}{2} \leqslant x \leqslant \frac{\pi}{2}$. That portion represents an increasing function, which has an inverse function and, therefore, is one-to-one. Its inverse is the red curve of Figure 5-6, the function $y = \text{Sin}^{-1}x$. Note that the black and the red curves are reflections of each other, with the line $y = x$ as the axis of symmetry.

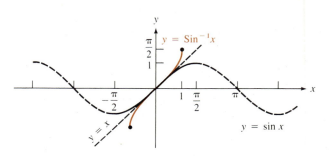

Figure 5-6

Although the domain and range of the inverse sine function are defined to be sets of real numbers, it is often convenient to think of $\text{Sin}^{-1} x$ as the "angle whose sine is x." In this situation, the range of the inverse sine function may be thought of as a set of angles measured in radians.

Example 1   Find   **a.** $\text{Sin}^{-1} \dfrac{1}{2}$,   **b.** $\text{Sin}^{-1}\left(-\dfrac{\sqrt{2}}{2}\right)$,   **c.** $\text{Sin}^{-1} \pi$,   **d.** $\text{Sin}^{-1} \dfrac{\pi}{4}$, and

e. Arcsin 0.8330.

Solution   **a.** The expression $\text{Sin}^{-1} \frac{1}{2}$ represents the number whose sine is $\frac{1}{2}$. Although infinitely many numbers have a sine of $\frac{1}{2}$ ($\frac{5\pi}{6}, \frac{\pi}{6}, -\frac{7\pi}{6}$, and so on), only $\frac{\pi}{6}$ lies in the range of the inverse sine function. The value $\frac{\pi}{6}$ is the principal value of $\sin^{-1} \frac{1}{2}$. Hence, you have

$$\text{Sin}^{-1}\frac{1}{2} = \frac{\pi}{6}$$

b. The expression $\text{Sin}^{-1}\left(-\frac{\sqrt{2}}{2}\right)$ represents the number between $-\frac{\pi}{2}$ and $\frac{\pi}{2}$, inclusive, that has a sine of $-\frac{\sqrt{2}}{2}$. Because $\sin\left(-\frac{\pi}{4}\right) = -\frac{\sqrt{2}}{2}$, you have

$$\text{Sin}^{-1}\left(-\frac{\sqrt{2}}{2}\right) = -\frac{\pi}{4}$$

c. The expression $\text{Sin}^{-1} \pi$ represents the number between $-\frac{\pi}{2}$ and $\frac{\pi}{2}$, inclusive, with a sine of π. Because no number has a sine of π, $\text{Sin}^{-1} \pi$ is undefined.

d. Be careful with $\text{Sin}^{-1} \frac{\pi}{4}$. It is tempting to misread it as $\text{Sin} \frac{\pi}{4}$ and incorrectly answer "$\frac{\sqrt{2}}{2}$." Use a calculator to get the right answer. Divide π by 4. Make sure that your calculator is set for radian measure, and then press the $\boxed{\text{INV}}$ and $\boxed{\text{SIN}}$ keys in that order. The angle whose sine is $\frac{\pi}{4}$ is approximately 0.9033 radians as displayed on the calculator.

e. To find Arcsin 0.8330, use your calculator. Make sure that it is set for radian measure. Enter .833 and press the $\boxed{\text{INV}}$ and $\boxed{\text{SIN}}$ keys in succession. The angle whose sine is 0.8330 is approximately 0.9845 radians as displayed on the calculator.   ◼

Because $y = \sin x$ (restricted to $-\frac{\pi}{2} \le x \le \frac{\pi}{2}$) and $y = \text{Sin}^{-1} x$ are inverse functions of each other, the effect of performing the functions in succession is worth considering. Because $\text{Sin}^{-1} x$ means "the principal angle whose sine is x," $\sin(\text{Sin}^{-1} x)$ means "the sine of the angle whose sine is x." The obvious answer to this "Who's buried in Grant's tomb?" question is "x."

(5.4)   **If $-1 \le x \le 1$, then $\sin(\text{Sin}^{-1} x) = x$**

Similarly:

(5.5)   **If $-\dfrac{\pi}{2} \le x \le \dfrac{\pi}{2}$, then $\text{Sin}^{-1}(\sin x) = x$**

However, if x is not restricted to the proper interval, then $\text{Sin}^{-1}(\sin x)$ might *not* be x. For example, we have

$$\text{Sin}^{-1}\left(\sin \frac{5\pi}{6}\right) = \text{Sin}^{-1}\left(\frac{1}{2}\right) = \frac{\pi}{6}$$

Similar considerations produce inverse functions of the remaining five trigonometric functions.

(5.6)

> **Definition.**   The **inverse cosine** (or **arccosine**) **function,** denoted by $y = \text{Cos}^{-1} x$ (or $y = \text{Arccos } x$), has a domain of $\{x \mid -1 \le x \le 1\}$ and a range of $\{y \mid 0 \le y \le \pi\}$.
>
> $y = \text{Cos}^{-1} x$ if and only if $x = \cos y$ and $0 \le y \le \pi$.

The graph of $y = \text{Cos}^{-1} x$, shown in Figure 5-7, is the reflection of a portion of the cosine curve with the line $y = x$ as an axis of symmetry.

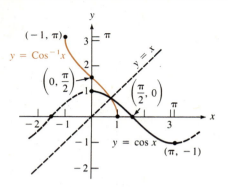

Figure 5-7

(5.7)

> **Definition.** The **inverse tangent** (or arctangent) **function,** denoted by $y = \text{Tan}^{-1} x$ (or $y = \text{Arctan } x$), has a domain of the real numbers and a range of $\{y \mid -\frac{\pi}{2} < y < \frac{\pi}{2}\}$.
>
> $y = \text{Tan}^{-1} x$ if and only if $x = \tan y$ and $-\frac{\pi}{2} < y < \frac{\pi}{2}$.

The graph of $y = \text{Tan}^{-1} x$ shown in Figure 5-8 is the reflection of a portion of the tangent curve with the line $y = x$ as an axis of symmetry.

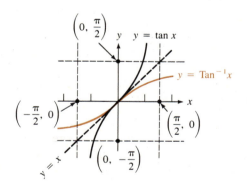

Figure 5-8

Example 2   Find **a.** $\text{Tan}^{-1}(-1)$, **b.** $\text{Cos}^{-1}\left(\dfrac{\sqrt{3}}{2}\right)$, and **c.** $\text{Tan}^{-1}\left(\dfrac{\pi}{4}\right)$.

Solution   **a.** Because $\tan\left(-\frac{\pi}{4}\right) = -1$ and because $-\frac{\pi}{4}$ is in the interval $-\frac{\pi}{2} \leq x \leq \frac{\pi}{2}$, you have $\text{Tan}^{-1}(-1) = -\frac{\pi}{4}$.

b. Because $\cos\left(\frac{\pi}{6}\right) = \frac{\sqrt{3}}{2}$ and because $0 \leq \frac{\pi}{6} \leq \pi$, you have $\text{Cos}^{-1}\frac{\sqrt{3}}{2} = \frac{\pi}{6}$.

c. To evaluate $\text{Tan}^{-1}\frac{\pi}{4}$, use a calculator set in radian mode. Divide π by 4. Then, press the $\boxed{\text{INV}}$ and $\boxed{\text{TAN}}$ keys in that order to get

$$\text{Tan}^{-1}\left(\frac{\pi}{4}\right) \approx 0.6658$$

∎

Example 3   Find $\cos(\text{Sin}^{-1} 1)$.

Solution   Because $\frac{\pi}{2}$ is the angle whose sine is 1, it follows that

$$\cos(\text{Sin}^{-1} 1) = \cos\left(\frac{\pi}{2}\right) = 0$$

∎

Example 4   Find $\cos(\text{Sin}^{-1} x)$ given that $0 < x \leqslant 1$.

Solution   Because the value of x is unknown, you don't know what angle has a sine of x. However, you do know that $\text{Sin}^{-1} x$ is a first- or fourth-quadrant angle, because $-\frac{\pi}{2} \leqslant \text{Sin}^{-1} x \leqslant \frac{\pi}{2}$. In Figure 5-9, the angle denoted by $\text{Sin}^{-1} x$ has a sine of x ($x > 0$), because the side opposite $\text{Sin}^{-1} x$ is labeled x and the hypotenuse is 1. Use the Pythagorean Theorem to find that the remaining side of the triangle has length of $\sqrt{1 - x^2}$. You can then read the value of the cosine of $\text{Sin}^{-1} x$ from the figure.

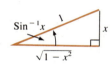

Figure 5-9

$$\cos(\text{Sin}^{-1} x) = \frac{\text{adjacent side}}{\text{hypotenuse}}$$

$$= \frac{\sqrt{1 - x^2}}{1}$$

$$= \sqrt{1 - x^2}$$

Because $-\frac{\pi}{2} \leqslant \text{Sin}^{-1} x \leqslant \frac{\pi}{2}$, the expression $\cos(\text{Sin}^{-1} x)$ is never negative and the positive value of the radical is correct. ∎

Example 5   Find $\tan(2\,\text{Cos}^{-1} x)$, where $-1 \leqslant x \leqslant 1$.

Solution   In this exercise, you must consider two cases: when $0 \leqslant x \leqslant 1$ and when $-1 \leqslant x < 0$.

If x is positive or zero, then $\text{Cos}^{-1} x$ is an acute angle that may be drawn in standard position as in Figure 5-10. Because $\text{Cos}^{-1} x$ represents the angle whose cosine is x, the side adjacent to the angle may be labeled x and the hypotenuse labeled 1. The remaining side of the right triangle is determined by the Pythagorean Theorem.

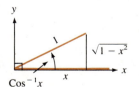

Figure 5-10

$$\tan(\text{Cos}^{-1} x) = \frac{\sqrt{1 - x^2}}{x}$$

The tangent of an acute angle is positive. Because x is also positive, the radical was chosen to be positive.

Now consider the case when x is negative. If x is negative, then $\text{Cos}^{-1} x$ is a second-quadrant angle that may be drawn in standard position as in Figure 5-11. Because $\text{Cos}^{-1} x$ is the angle whose cosine is x, the adjacent side may be labeled x and the hypotenuse labeled 1. Use the Pythagorean Theorem to determine that the remaining side has length of $\sqrt{1 - x^2}$.

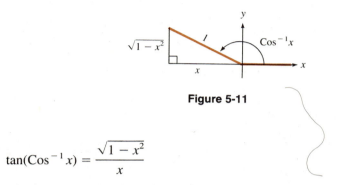

Figure 5-11

$$\tan(\text{Cos}^{-1} x) = \frac{\sqrt{1 - x^2}}{x}$$

The tangent of a second-quadrant angle is negative. Because x is negative, the radical must again be chosen positive. Taken together, the results of these two cases determine that, for all x in the interval $-1 \leqslant x \leqslant 1$, we have

$$\tan(\text{Cos}^{-1} x) = \frac{\sqrt{1 - x^2}}{x}$$

To find $\tan(2\,\text{Cos}^{-1} x)$, substitute the value of $\tan(\text{Cos}^{-1} x)$ into the right side of the double-angle identity for $\tan 2\theta$.

$$\tan(2\,\text{Cos}^{-1} x) = \frac{2\tan(\text{Cos}^{-1} x)}{1 - \tan^2(\text{Cos}^{-1} x)}$$

$$= \frac{2\dfrac{\sqrt{1 - x^2}}{x}}{1 - \dfrac{1 - x^2}{x^2}}$$

$$= \frac{\dfrac{2\sqrt{1 - x^2}}{x}}{\dfrac{x^2 - 1 + x^2}{x^2}}$$

$$= \frac{2x\sqrt{1 - x^2}}{2x^2 - 1}$$

Example 6   Write $\sin(\text{Sin}^{-1} x + \text{Sin}^{-1} y)$ as an algebraic expression in the variables x and y.

Solution   Note that both $\text{Sin}^{-1} x$ and $\text{Sin}^{-1} y$ are angles, and use the identity $\sin(A + B) = \sin A \cos B + \cos A \sin B$ to remove parentheses.

$$\sin(\text{Sin}^{-1} x + \text{Sin}^{-1} y) = \sin(\text{Sin}^{-1} x) \cos(\text{Sin}^{-1} y)$$
$$+ \cos(\text{Sin}^{-1} x) \sin(\text{Sin}^{-1} y)$$

Because $\sin(\text{Sin}^{-1} z) = z$ and $\cos(\text{Sin}^{-1} z) = \sqrt{1 - z^2}$ (see Example 4), it follows that

$$\sin(\text{Sin}^{-1} x + \text{Sin}^{-1} y) = x \sqrt{1 - y^2} + \sqrt{1 - x^2} \cdot y$$
$$= x \sqrt{1 - y^2} + y \sqrt{1 - x^2} \qquad ■$$

Example 7   Verify the identity $\text{Sin}^{-1} x + \text{Cos}^{-1} x = \dfrac{\pi}{2}$.

Solution   You are asked to verify that the sum of two angles is equal to a third angle. Normally, you would verify an identity by showing that one side may be transformed into the other. Here, however, verify the identity by showing that the sine of the left side equals the sine of the right side.

$$\text{Sin}^{-1} x + \text{Cos}^{-1} x = \frac{\pi}{2}$$

$$\sin(\text{Sin}^{-1} x + \text{Cos}^{-1} x) = \sin \frac{\pi}{2}$$

$$\sin(\text{Sin}^{-1} x) \cos(\text{Cos}^{-1} x) + \cos(\text{Sin}^{-1} x) \sin(\text{Cos}^{-1} x) = 1$$
$$x \cdot x + \sqrt{1 - x^2} \cdot \sqrt{1 - x^2} = 1$$
$$x^2 + 1 - x^2 = 1$$
$$1 = 1$$

Because $-\frac{\pi}{2} \le \text{Sin}^{-1} x \le \frac{\pi}{2}$ and $0 \le \text{Cos}^{-1} x \le \pi$, you know by adding these inequalities that

$$-\frac{\pi}{2} \le \text{Sin}^{-1} x + \text{Cos}^{-1} x \le \frac{3\pi}{2}$$

Because the left side of the identity is an angle between $-\frac{\pi}{2}$ and $\frac{3\pi}{2}$, and because the sine of only one angle $\left(\frac{\pi}{2}\right)$ in that interval attains the value of 1, the identity is verified.   ■

Exercise 5.3

In Exercises 1–12, find the value of x, if any. **Do not use a calculator or tables.** *Note that each answer should be a real number.*

1. $\text{Sin}^{-1} \dfrac{1}{2} = x$

2. $\text{Cos}^{-1} \dfrac{\sqrt{3}}{2} = x$

3. $\text{Cos}^{-1} 0 = x$

4. $\text{Sin}^{-1} \sqrt{3} = x$

5. $\text{Tan}^{-1} 1 = x$

6. $\text{Tan}^{-1} 0 = x$

7. $\text{Sin}^{-1} 3 = x$

8. $\text{Tan}^{-1} (-\sqrt{3}) = x$

9. $\text{Cos}^{-1} \left(-\dfrac{\sqrt{2}}{2} \right) = x$

10. $\text{Sin}^{-1} \dfrac{\sqrt{2}}{2} = x$

11. $\text{Arcsin} \dfrac{\sqrt{3}}{2} = x$

12. $\text{Arccos} \dfrac{1}{2} = x$

In Exercises 13–24, find the sine, cosine, and tangent of θ. **Do not use a calculator or tables.**

13. $\text{Sin}^{-1} \dfrac{1}{2} = \theta$

14. $\text{Cos}^{-1} \dfrac{\sqrt{3}}{2} = \theta$

15. $\text{Tan}^{-1} 0 = \theta$

16. $\text{Tan}^{-1} 1 = \theta$

17. $\text{Cos}^{-1} \left(-\dfrac{\sqrt{3}}{2} \right) = \theta$

18. $\text{Sin}^{-1} \left(-\dfrac{\sqrt{3}}{2} \right) = \theta$

19. $\text{Arcsin} 1 = \theta$

20. $\text{Arctan} 1 = \theta$

21. $\text{Cos}^{-1} (-1) = \theta$

22. $\text{Sin}^{-1} \dfrac{\sqrt{2}}{2} = \theta$

23. $\text{Arccos} \dfrac{\sqrt{2}}{2} = \theta$

24. $\text{Arcsin} 0 = \theta$

In Exercises 25–36, find the value of x. **Do not use a calculator or tables.**

25. $\sin \left(\text{Sin}^{-1} \dfrac{1}{2} \right) = x$

26. $\cos \left(\text{Cos}^{-1} \dfrac{1}{2} \right) = x$

27. $\tan(\text{Tan}^{-1} 1) = x$

28. $\sin(\text{Sin}^{-1} 0) = x$

29. $\cos(\text{Cos}^{-1} 1) = x$

30. $\tan(\text{Tan}^{-1} 0) = x$

31. $\sin \left(\text{Cos}^{-1} \dfrac{\sqrt{3}}{2} \right) = x$

32. $\cos \left(\text{Sin}^{-1} \dfrac{1}{2} \right) = x$

33. $\tan \left[\text{Sin}^{-1} \left(-\dfrac{\sqrt{3}}{2} \right) \right] = x$

34. $\cot \left(\text{Cos}^{-1} \dfrac{\sqrt{2}}{2} \right) = x$

35. $\cos(\text{Arctan} 1) = x$

36. $\sin(\text{Arctan} 0) = x$

In Exercises 37–48, evaluate each expression without using a calculator or tables.

37. $\cos \left(\text{Sin}^{-1} \dfrac{4}{5} \right)$

38. $\sin \left(\text{Cos}^{-1} \dfrac{3}{5} \right)$

39. $\sin \left(\text{Cos}^{-1} \dfrac{5}{13} \right)$

40. $\cos \left[\text{Sin}^{-1} \left(-\dfrac{5}{13} \right) \right]$

41. $\tan \left[\text{Sin}^{-1} \left(-\dfrac{4}{5} \right) \right]$

42. $\tan \left(\text{Cos}^{-1} \dfrac{3}{5} \right)$

43. $\tan \left(\text{Cos}^{-1} \dfrac{5}{13} \right)$

44. $\tan \left[\text{Sin}^{-1} \left(-\dfrac{12}{13} \right) \right]$

45. $\cos\left[\text{Tan}^{-1}\left(-\dfrac{5}{12}\right)\right]$

46. $\sin\left(\text{Tan}^{-1}\dfrac{12}{5}\right)$

47. $\sin\left(\text{Arccos}\dfrac{9}{41}\right)$

48. $\cos\left(\text{Arcsin}\dfrac{40}{41}\right)$

In Exercises 49–60, find the required value without using a calculator or tables.

49. $\sin\left(\text{Sin}^{-1}\dfrac{1}{2} + \text{Cos}^{-1}\dfrac{1}{2}\right)$

50. $\sin\left(\text{Sin}^{-1}\dfrac{1}{2} - \text{Cos}^{-1}\dfrac{1}{2}\right)$

51. $\cos\left(\text{Sin}^{-1}\dfrac{1}{2} - \text{Cos}^{-1}\dfrac{1}{2}\right)$

52. $\cos\left(\text{Sin}^{-1}\dfrac{1}{2} + \text{Cos}^{-1}\dfrac{1}{2}\right)$

53. $\sin 2\left(\text{Sin}^{-1}\dfrac{\sqrt{2}}{2}\right)$

54. $\cos 2\left(\text{Sin}^{-1}\dfrac{\sqrt{2}}{2}\right)$

55. $\tan 2\left(\text{Sin}^{-1}\dfrac{\sqrt{2}}{2}\right)$

56. $\cot 2\left(\text{Sin}^{-1}\dfrac{\sqrt{2}}{2}\right)$

57. $\sin\dfrac{1}{2}\left(\text{Cos}^{-1}\dfrac{1}{2}\right)$

58. $\cos\dfrac{1}{2}\left(\text{Cos}^{-1}\dfrac{1}{2}\right)$

59. $\tan\dfrac{1}{2}\left(\text{Arccos}\dfrac{1}{2}\right)$

60. $\cot\dfrac{1}{2}\left(\text{Arcsin}\dfrac{1}{2}\right)$

In Exercises 61–76, rewrite each value as an algebraic expression in the variable x.

61. $\sin(\text{Tan}^{-1}x)$

62. $\cos(\text{Tan}^{-1}x)$

63. $\tan(\text{Sin}^{-1}x)$

64. $\tan(\text{Cos}^{-1}x)$

65. $\sin(\text{Cos}^{-1}x)$

66. $\cos(\text{Sin}^{-1}x)$

67. $\sin(2\,\text{Arcsin}\,x)$

68. $\cos(2\,\text{Arccos}\,x)$

69. $\tan(2\,\text{Arctan}\,x)$

70. $\sin(2\,\text{Arccos}\,x)$

71. $\cos(2\,\text{Sin}^{-1}x)$

72. $\sin\left(\dfrac{1}{2}\text{Sin}^{-1}x\right)$

73. $\cos\left(\dfrac{1}{2}\text{Cos}^{-1}x\right)$

74. $\tan\left(\dfrac{1}{2}\text{Tan}^{-1}\dfrac{\sqrt{1-x^2}}{x}\right)$

75. $\sin\left(\dfrac{1}{2}\text{Arccos}\,x\right)$

76. $\cos\left(\dfrac{1}{2}\text{Arcsin}\,x\right)$

In Exercises 77–80, evaluate each of the following expressions.

77. $\text{Sin}^{-1}\left(\sin\dfrac{11\pi}{6}\right)$

78. $\text{Cos}^{-1}\left(\cos\dfrac{7\pi}{6}\right)$

79. $\text{Cos}^{-1}\left(\sin\dfrac{3\pi}{4}\right)$

80. $\text{Sin}^{-1}\left(\cos\dfrac{5\pi}{4}\right)$

In Exercises 81–86, verify each of the identities.

81. $\text{Tan}^{-1}y + \text{Tan}^{-1}\dfrac{1}{y} = \dfrac{\pi}{2}$ $(y > 0)$

82. $\text{Sin}^{-1}x = \dfrac{\pi}{2} - \text{Cos}^{-1}x$

83. $\text{Sin}^{-1}(-x) = -\text{Sin}^{-1}x$

84. $\text{Cos}^{-1}(-x) + \text{Cos}^{-1}x = \pi$

85. $\cos(\text{Arctan }x) = \dfrac{\sqrt{1+x^2}}{1+x^2}$ $(x > 0)$

86. $\sin\dfrac{1}{2}(\text{Arccos }2x) = \sqrt{\dfrac{1-2x}{2}}$

5.4   THE INVERSE COTANGENT, SECANT, AND COSECANT FUNCTIONS (OPTIONAL)

If we interchange the x- and y-coordinates of the ordered pairs of the relation defined by the equation $y = \cot x$, we obtain the inverse relation $y = \cot^{-1}x$. A graph of the equation $y = \cot^{-1}x$ is shown in Figure 5-12.

$$x = \cot y \text{ or } y = \cot^{-1}x$$

x	y
0	$-\dfrac{\pi}{2}, \dfrac{\pi}{2}, \dfrac{3\pi}{2}$
1	$-\dfrac{3\pi}{4}, \dfrac{\pi}{4}, \dfrac{5\pi}{4}$
-1	$-\dfrac{\pi}{4}, \dfrac{3\pi}{4}, \dfrac{7\pi}{4}$

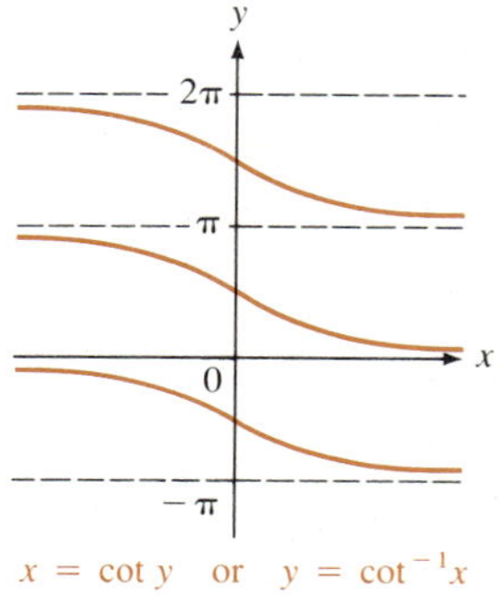

$$x = \cot y \quad \text{or} \quad y = \cot^{-1}x$$

Figure 5-12

If we restrict the range of the relation defined by $y = \cot^{-1}x$ to the interval $0 < y < \pi$, the equation $y = \cot^{-1}x$ defines a function. This function, graphed in Figure 5-13, is called the **inverse cotangent function,** or the **arccotangent function,** and is denoted by either $y = \text{Cot}^{-1}x$ or $y = \text{Arccot }x$.

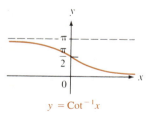

$$y = \text{Cot}^{-1}x$$

Figure 5-13

(5.8)

> **Definition.**   The **inverse cotangent** (or **arccotangent**) **function,** denoted by $y = \text{Cot}^{-1}x$ (or $y = \text{Arccot }x$), is a function with the real numbers for its domain and $\{y \mid 0 < y < \pi\}$ for its range.
>
> $y = \text{Cot}^{-1}x$ is equivalent to $x = \cot y$ and $0 < y < \pi$.

Example 1   Find   **a.** $\text{Cot}^{-1}1$,   **b.** $\text{Cot}^{-1}0$,   **c.** $\text{Cot}^{-1}(-\sqrt{3})$,   and **d.** $\text{Cot}^{-1}1.576$.

Solution   **a.** $\text{Cot}^{-1}1$ is that number between 0 and π whose cotangent is 1. Hence, you have

$$\text{Cot}^{-1}1 = \frac{\pi}{4}$$

Note that $\cot \dfrac{\pi}{4} = 1$ and that $0 < \dfrac{\pi}{4} < \pi$.

b. $\text{Cot}^{-1}0$ is that number between 0 and π whose cotangent is 0. Hence, you have

$$\text{Cot}^{-1}0 = \frac{\pi}{2}$$

Note that $\cot \dfrac{\pi}{2} = 0$ and that $0 < \dfrac{\pi}{2} < \pi$.

c. $\text{Cot}^{-1}(-\sqrt{3})$ is that number between 0 and π whose cotangent is $-\sqrt{3}$. Hence, you have

$$\text{Cot}^{-1}(-\sqrt{3}) = \frac{5\pi}{6}$$

Note that $\cot \dfrac{5\pi}{6} = -\sqrt{3}$ and that $0 < \dfrac{5\pi}{6} < \pi$.

d. To find $\text{Cot}^{-1}1.576$, use your calculator set in radian mode. Enter 1.576 and find its reciprocal by pressing the $\boxed{1/x}$ key. Then press the $\boxed{\text{INV}}$ and $\boxed{\text{TAN}}$ keys in that order to obtain the number 0.56541429. Thus, you have

$$\text{Cot}^{-1}1.576 \approx 0.5654$$

Note that $\cot 0.5654 \approx 1.576$ and that $0 < 0.5654 < \pi$.   ■

If we restrict the ranges of the relations defined by $y = \sec^{-1} x$ and $y = \csc^{-1} x$ to appropriate numbers, the equations $y = \sec^{-1} x$ and $y = \csc^{-1} x$ define functions also.

(5.9)

> **Definition.**   The **inverse secant** (or **arcsecant**) **function,** denoted by $y = \text{Sec}^{-1} x$ (or $y = \text{Arcsec } x$), has a domain of $\{x \mid x \leqslant -1 \text{ or } x \geqslant 1\}$ and a range of $\{y \mid 0 \leqslant y \leqslant \pi \text{ and } y \neq \frac{\pi}{2}\}$.*
>
> $y = \text{Sec}^{-1} x$ is equivalent to $x = \sec y$ and $0 \leqslant y \leqslant \pi$ and $y \neq \dfrac{\pi}{2}$.

---

*Some books restrict y to the interval $-\pi \leqslant y < -\dfrac{\pi}{2}$ or $0 \leqslant y < \dfrac{\pi}{2}$.

The graphs of $y = \sec^{-1} x$ and $y = \text{Sec}^{-1} x$ appear in Figure 5-14.

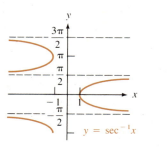
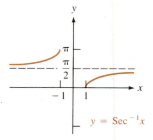

Figure 5-14

(5.10)

> **Definition.** The **inverse cosecant** (or **arccosecant**) **function,** denoted by $y = \text{Csc}^{-1} x$ (or $y = \text{Arccsc } x$), has a domain of $\{x \mid x \leq -1 \ \text{ or } \ x \geq 1\}$ and a range of $\{y \mid -\frac{\pi}{2} \leq y \leq \frac{\pi}{2} \text{ and } y \neq 0\}.$[†]
>
> $y = \text{Csc}^{-1} x$ is equivalent to $x = \csc y$ and $-\dfrac{\pi}{2} \leq y \leq \dfrac{\pi}{2}$ and $y \neq 0.$

The graphs of $y = \csc^{-1} x$ and $y = \text{Csc}^{-1} x$ appear in Figure 5-15.

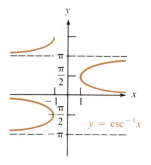
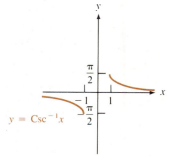

Figure 5-15

Example 2   Find   **a.** $\text{Sec}^{-1}2$,   **b.** $\text{Sec}^{-1}1.75$,   **c.** $\text{Csc}^{-1}\left(-\dfrac{2\sqrt{3}}{3}\right)$, and   **d.** $\text{Csc}^{-1}\dfrac{1}{2}$.

Solution   **a.** $\text{Sec}^{-1}2$ is that number between 0 and π whose secant is 2 (or whose cosine is $\frac{1}{2}$). Hence, you have

---

[†]Some books restrict y to the interval $-\pi < y \leq -\dfrac{\pi}{2}$ or $0 < y \leq \dfrac{\pi}{2}$.

$$\text{Sec}^{-1}2 = \frac{\pi}{3}$$

Note that $\sec \dfrac{\pi}{3} = 2$ and that $0 \le \dfrac{\pi}{3} \le \pi$.

b. To find $\text{Sec}^{-1}1.75$, set your calculator to radian mode, enter 1.75, and find its reciprocal by pressing the $\boxed{1/x}$ key. Then press the $\boxed{\text{INV}}$ and $\boxed{\text{COS}}$ keys to obtain the number 0.96255075. Hence, you have

$$\text{Sec}^{-1}1.75 \approx 0.9625$$

Note that $\sec 0.9625 \approx 1.75$ and that $0 \le 0.9625 \le \pi$.

c. $\text{Csc}^{-1}\left(-\frac{2\sqrt{3}}{3}\right)$ is that number between $-\frac{\pi}{2}$ and $\frac{\pi}{2}$ whose cosecant is $-\frac{2\sqrt{3}}{3}$ (or whose sine is $-\frac{3}{2\sqrt{3}}$, or $-\frac{\sqrt{3}}{2}$). Hence, you have

$$\text{Csc}^{-1}\left(-\frac{2\sqrt{3}}{3}\right) = -\frac{\pi}{3}$$

Note that $\csc\left(-\dfrac{\pi}{3}\right) = -\dfrac{2\sqrt{3}}{3}$ and that $-\dfrac{\pi}{2} \le -\dfrac{\pi}{3} < \dfrac{\pi}{2}$.

d. $\text{Csc}^{-1}\frac{1}{2}$ is that number whose cosecant is $\frac{1}{2}$ (or whose sine is 2). Because no such number exists, the expression $\text{Csc}^{-1}\frac{1}{2}$ is not defined.   ■

Example 3   Find $\sin\left(\text{Csc}^{-1}\dfrac{5}{4} + \text{Sec}^{-1}\dfrac{13}{5}\right)$.

Solution   Draw two right triangles with acute angles of $\text{Csc}^{-1}\frac{5}{4}$ and $\text{Sec}^{-1}\frac{13}{5}$. See Figure 5-16. These triangles will help you compute the sine and cosine of each angle. To evaluate $\sin(\text{Csc}^{-1}\frac{5}{4} + \text{Sec}^{-1}\frac{13}{5})$, use the identity for the sine of the sum of two angles and refer to Figure 5-16.

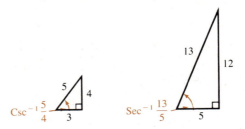

Figure 5-16

$$\sin\left(\text{Csc}^{-1}\frac{5}{4} + \text{Sec}^{-1}\frac{13}{5}\right)$$

$$= \sin\left(\text{Csc}^{-1}\frac{5}{4}\right)\cos\left(\text{Sec}^{-1}\frac{13}{5}\right) + \cos\left(\text{Csc}^{-1}\frac{5}{4}\right)\sin\left(\text{Sec}^{-1}\frac{13}{5}\right)$$

$$= \frac{4}{5} \cdot \frac{5}{13} + \frac{3}{5} \cdot \frac{12}{13}$$

$$= \frac{20}{65} + \frac{36}{65}$$

$$= \frac{56}{65}$$

Example 4 Express $\cot(\mathrm{Csc}^{-1} x)$ as an algebraic expression in the variable x.

Solution The expression $\mathrm{Csc}^{-1} x$ can represent an angle between $-\frac{\pi}{2}$ and $\frac{\pi}{2}$ whose cosecant is x. If $x \geqslant 1$, then $\mathrm{Csc}^{-1} x$ is the first-quadrant angle shown in Figure 5-17. The ratio of the hypotenuse to the side opposite $\mathrm{Csc}^{-1} x$ is $\frac{x}{1}$, or x, as required. The remaining side of the triangle is $\sqrt{x^2 - 1}$, as determined by the Pythagorean Theorem.

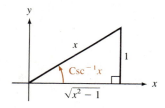

Figure 5-17

The cotangent of $\mathrm{Csc}^{-1} x$ is the ratio of the adjacent side to the opposite side. Thus, you have

$$\cot(\mathrm{Csc}^{-1} x) = \sqrt{x^2 - 1} \qquad \text{for } x \geqslant 1$$

If $x \leqslant -1$, then the expression $\mathrm{Csc}^{-1} x$ is a fourth-quadrant angle, and its cotangent is negative. Thus, you have

$$\cot(\mathrm{Csc}^{-1} x) = -\sqrt{x^2 - 1} \qquad \text{for } x \leqslant -1$$

Exercise 5-4

In Exercises 1–6, evaluate each expression.

1. $\mathrm{Cot}^{-1}\sqrt{3}$

2. $\mathrm{Cot}^{-1}\left(-\dfrac{\sqrt{3}}{3}\right)$

3. $\mathrm{Sec}^{-1}(-1)$

4. $\mathrm{Sec}^{-1}(-2)$

5. $\mathrm{Csc}^{-1}(-1)$

6. $\mathrm{Csc}^{-1}2$

In Exercises 7–14, simplify each expression.

7. $\sec(\mathrm{Sec}^{-1} 3)$

8. $\csc(\mathrm{Csc}^{-1}5)$

9. $\sec\left(\text{Csc}^{-1}\dfrac{5}{4}\right)$

10. $\csc\left(\text{Sec}^{-1}\dfrac{25}{20}\right)$

11. $\sin 2\left(\text{Sec}^{-1}\dfrac{5}{3}\right)$

12. $\cos 2\left(\text{Csc}^{-1}\dfrac{13}{5}\right)$

13. $\cos\left(\text{Cot}^{-1}\dfrac{3}{4} + \text{Sec}^{-1}\dfrac{13}{5}\right)$

14. $\sin(\text{Tan}^{-1}1 + \text{Cot}^{-1}1)$

In Exercises 15–20, rewrite each value as an algebraic expression in the variable x. Make sure that you consider all numbers in the domain of x.

15. $\sin(\text{Csc}^{-1}x)$

16. $\cos(\text{Cot}^{-1}x)$

17. $\sec(\text{Cot}^{-1}x)$

18. $\csc(\text{Sec}^{-1}x)$

19. $\tan(\text{Sec}^{-1}x)$

20. $\tan(\text{Cot}^{-1}x)$

5.5 APPLICATIONS OF TRIGONOMETRIC EQUATIONS AND INVERSE FUNCTIONS

Trigonometric equations and inverse trigonometric functions are used in several areas besides pure mathematics. Here are four applications from carpentry, optics, and mathematics.

1. Cutting rafters

If the peak of a roof is h ft above the ceiling joists (see Figure 5-18) and the distance between the wall and the king post is r ft, then the **slope** of the roof is defined as h/r.

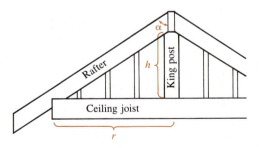

Figure 5-18

Example 1   At what angle α should the rafter in Figure 5-18 be cut?

Solution   The tangent of angle α is $\dfrac{r}{h}$. Hence, you have

$$\tan\alpha = \frac{r}{h}$$

$$\alpha = \text{Tan}^{-1}\frac{r}{h}$$

The rafter should be cut at an angle of $\text{Tan}^{-1}\dfrac{r}{h}$.

Note that $\alpha = \text{Tan}^{-1}\left(\dfrac{1}{\text{slope}}\right)$. ■

2. Optics: Total internal reflection

Snell's law (see Section 2.8) indicates the relationship between the angle of incidence i and the angle of refraction r of a beam of light as it enters a substance such as glass, with an index of refraction n. The relationship is

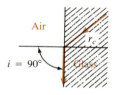

$$n = \frac{\sin i}{\sin r}$$

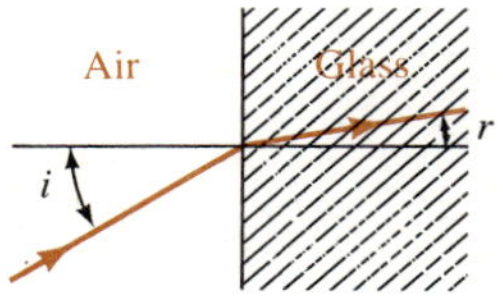

Figure 5-19

See Figure 5-19. The ray of light is reversible: a beam passing right to left—from the glass into the air—will follow the same path. As angle r increases, so does angle i, until angle i becomes $90°$ and no light emerges from the glass. See Figure 5-20.

The angle r_c is called the **critical angle.** It is found by substituting $i = 90°$ into Snell's law and then solving for $r = r_c$.

$$\frac{\sin i}{\sin r} = n$$

$$\frac{\sin 90°}{\sin r_c} = n$$

$$\frac{1}{\sin r_c} = n$$

$$\sin r_c = \frac{1}{n}$$

$$r_c = \text{Sin}^{-1}\frac{1}{n}$$

Figure 5-20

Although values of the inverse sine function are real numbers (or angles measured in radians), it is common in some applications to express answers in degrees.

Example 2   Given that the index of refraction for glass is 1.6, find its critical angle.

Solution   The critical angle is

$$r_c = \text{Sin}^{-1}\frac{1}{1.6}$$

$$= \text{Sin}^{-1}0.625$$

$$\approx 38.68°$$

Any light striking the surface of the glass (from the inside) at an angle greater than $38.68°$ will not exit from the glass but will be internally reflected. ■

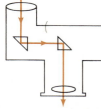

Glass prisms in binoculars use total internal reflection to increase the optical distance between the objective lens and the eyepiece. See Figure 5-21.

Rhinestones glitter because some of their surfaces are coated with a mirrorlike substance. With their higher index of refraction, real diamonds glitter because of internal reflections—they need no mirror coatings.

Figure 5-21

3. Polarized light

Light energy is transmitted as an electromagnetic wave, just like a radio wave. Most light waves behave like the "wave" of a jump rope that is spun very rapidly. See Figure 5-22. A point on the rope moves in a circular pattern and has both an up/down and a side-to-side motion. If the rope is threaded through a picket fence, however, the side-to-side motion is filtered out and only the up/down motion continues. If two boards are nailed horizontally to the pickets, one above the rope and one below, the up/down motion would also stop, eliminating the jump rope wave completely.

Figure 5-22

The optical equivalent of a picket fence is a substance called **polaroid,** developed by Edwin H. Land in 1935. If two sheets of polaroid are oriented with their optical "pickets" at an angle θ with respect to each other, the intensity of the light transmitted by the second polaroid filter is $\cos^2 \theta$ times the intensity of the light entering that filter.

Example 3   At what angle θ will one-half of the light be transmitted through the second filter in Figure 5-23?

Solution   Solve the equation $I \cos^2 \theta = \frac{1}{2} I$ for θ.

$$I \cos^2 \theta = \frac{1}{2} I$$

$$\cos^2 \theta = \frac{1}{2}$$

$$\cos \theta = \frac{1}{\sqrt{2}}$$

$$\cos \theta = \frac{\sqrt{2}}{2}$$

$$\theta = \cos^{-1} \frac{\sqrt{2}}{2}$$

$$\theta = 45°$$

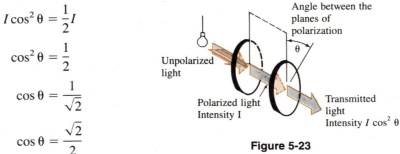

Angle between the planes of polarization

Unpolarized light

Polarized light Intensity I

Transmitted light Intensity $I \cos^2 \theta$

Figure 5-23

Figure 5-24

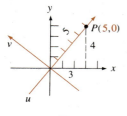

Figure 5-25

Light reflected from nonmetallic surfaces is partially polarized. Polaroid sunglasses reduce glare by partially blocking reflected light.

4. Rotation of axes

Point P of Figure 5-24 has x and y coordinates of (3, 4). The coordinates of P with respect to a different set of axes would no longer be (3, 4). In Figure 5-25, for example, the unit distances measured on the u- and v-axes are the same as they are on the x- and y-axes, but the u- and v-axes are tilted at an angle to the x- and y-axes so that the positive u-axis passes through point P. With respect to the u, v coordinate system, P has u and v coordinates of (5, 0).

More generally, suppose we have two coordinate systems, as in Figure 5-26, one determined by the x- and y-axes, and one determined by the u- and v-axes. Suppose further that these two systems have a common origin, the same unit of measure on each axis, and that the u, v system is tilted at an angle θ with respect to the x, y system. A point P has coordinates of x and y when referred to the x, y system, and coordinates of u and v when referred to the u, v system. We will now determine how u and v are related to x and y.

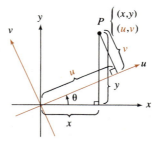

Figure 5-26

If perpendicular lines are drawn from P to both the x-axis and u-axis, and lines CQ and CB are drawn as in Figure 5-27, then the relationship between the x and y coordinates of P and the u and v coordinates of P can be found. Note that

$$\cos \theta = \frac{OB}{OC} = \frac{OB}{u} \quad \text{or} \quad OB = u \cos \theta$$

$$\sin \theta = \frac{QC}{PC} = \frac{AB}{v} \quad \text{or} \quad AB = v \sin \theta$$

$$\sin \theta = \frac{BC}{OC} = \frac{BC}{u} \quad \text{or} \quad BC = u \sin \theta$$

and

$$\cos \theta = \frac{QP}{PC} = \frac{QP}{v} \quad \text{or} \quad QP = v \cos \theta$$

Now we can express x and y in terms of u, v, and θ.

$$x = OA \qquad\qquad y = AP$$
$$= OB - AB \qquad = BC + QP$$
$$= u \cos \theta - v \sin \theta \qquad = u \sin \theta + v \cos \theta$$

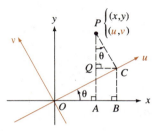

Figure 5-27

The equations relating the coordinates of P in the x, y system to the coordinates of P in the u, v system are as follows.

(5.11)

> **Equations of Rotation.**
>
> $$x = u \cos \theta - v \sin \theta$$
> $$y = u \sin \theta + v \cos \theta$$

Example 4   In Figure 5-28, the u, v coordinate system is rotated by $30°$ from the x, y system. The u and v coordinates of P are $(\sqrt{3} + 1, \sqrt{3} - 1)$. What are the x and y coordinates of point P?

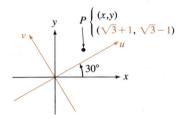

Figure 5-28

Solution   Use the equations

$$\begin{cases} x = u \cos \theta - v \sin \theta \\ y = u \sin \theta + v \cos \theta \end{cases}$$

with $u = \sqrt{3} + 1$, $v = \sqrt{3} - 1$, and $\theta = 30°$.

$$x = u \cos 30° - v \sin 30° \qquad\qquad y = u \sin 30° + v \cos 30°$$

$$x = (\sqrt{3} + 1)\frac{\sqrt{3}}{2} - (\sqrt{3} - 1)\frac{1}{2} \qquad\qquad y = (\sqrt{3} + 1)\frac{1}{2} + (\sqrt{3} - 1)\frac{\sqrt{3}}{2}$$

$$= \frac{3}{2} + \frac{\sqrt{3}}{2} - \frac{\sqrt{3}}{2} + \frac{1}{2} \qquad\qquad = \frac{\sqrt{3}}{2} + \frac{1}{2} + \frac{3}{2} - \frac{\sqrt{3}}{2}$$

$$= 2 \qquad\qquad\qquad\qquad = 2$$

With respect to the x, y system, the coordinates of point P are $(2, 2)$. ■

Example 5 At what angle θ should the u-axis be tilted from the x-axis to transform the equation $xy = 1$ into an equation with no term involving the product uv?

Solution Use the rotation equations to transform the equation $xy = 1$ into a new equation with variables u and v.

$$xy = 1$$
$$(u \cos \theta - v \sin \theta)(u \sin \theta + v \cos \theta) = 1$$
$$u^2 \cos \theta \sin \theta + uv \cos^2 \theta - uv \sin^2 \theta - v^2 \sin \theta \cos \theta = 1$$

1. $\qquad u^2 \cos \theta \sin \theta + uv(\cos^2 \theta - \sin^2 \theta) - v^2 \sin \theta \cos \theta = 1$

This equation is to be free of any term involving the product uv. Hence, set the coefficient of uv in Equation 1 equal to 0, and solve for θ.

$$\cos^2 \theta - \sin^2 \theta = 0$$
$$\cos 2\theta = 0$$
$$2\theta = \text{Cos}^{-1} 0 = 90°$$
$$\theta = 45°$$

If the u, v coordinate system is rotated by 45° from the x, y system, then Equation 1 becomes

$$u^2 \cos 45° \sin 45° - v^2 \sin 45° \cos 45° = 1$$

$$u^2 \left(\frac{\sqrt{2}}{2} \cdot \frac{\sqrt{2}}{2}\right) - v^2 \left(\frac{\sqrt{2}}{2} \cdot \frac{\sqrt{2}}{2}\right) = 1$$

$$\frac{u^2}{2} - \frac{v^2}{2} = 1$$

Note that if $xy = 1$ is graphed on the x, y coordinate system, and

$$\frac{u^2}{2} - \frac{v^2}{2} = 1$$

is graphed on the u, v coordinate system, the exact same curve results. See Figure 5-29.

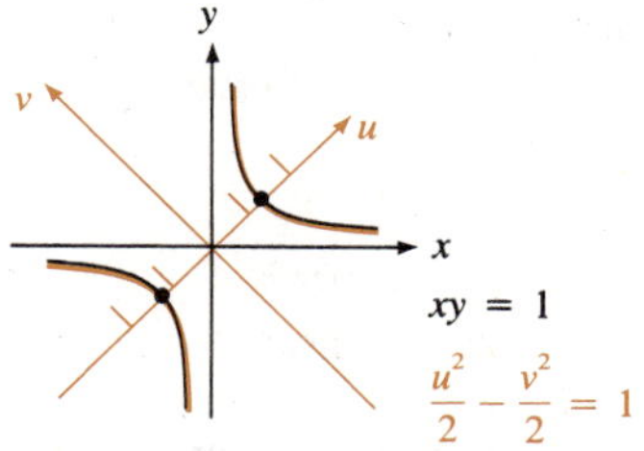

Figure 5-29

Exercise 5.5

1. What is the critical angle for water with an index of refraction of 1.33?

2. What is the critical angle for a diamond with an index of refraction of 2.4?

3. At what angle α should the rafters for a house be cut if the slope of the roof is $\frac{5}{4}$?

4. If the rafters for a house were cut at an angle of 80°, what is the slope of the roof?

5. At what angle θ should two polarizing filters be oriented to transmit one-third of the intensity of the light striking the second filter?

6. One polarizing filter reduces the intensity of light by one-half. At what angle should the two filters be placed to reduce by 90% the intensity of the light passing through both filters?

In Exercises 7–12, the u and v coordinates of a point P and an angle of rotation are given. Find the x and y coordinates of point P relative to the x, y system.

7. $(2\sqrt{2}, 0)$, $\theta = 45°$

8. $\left(-\dfrac{\sqrt{2}}{2}, \dfrac{\sqrt{2}}{2}\right)$, $\theta = 45°$

9. $(1, \sqrt{3})$, $\theta = 60°$

10. $\left(\dfrac{\sqrt{3}}{2} - 2, -\dfrac{1}{2} - 2\sqrt{3}\right)$, $\theta = 30°$

11. $(1 + \sqrt{3}, 1 - \sqrt{3})$, $\theta = 60°$

12. $(\sqrt{3}, 1)$, $\theta = 60°$

Transform each of the following equations in x and y into equations containing the variables u and v, but containing no terms involving the product uv. What is the angle of rotation between the x- and u-axes?

13. $xy = 2$

14. $xy = \sqrt{2}$

15. $x^2 + 3xy + y^2 = 2$

16. $7x^2 + 2\sqrt{3}\, xy + 5y^2 = 8$

CHAPTER SUMMARY

Key Words

critical angle   (5.5)

inverse cosecant function   (5.4)

inverse cosine function   (5.3)

inverse tangent function   (5.3)

polarization of light   (5.5)

principal value   (5.3)

inverse cotangent function   (5.4)   *rotation of coordinate axes*   (5.5)
inverse secant function   (5.4)   *Snell's law*   (5.5)
inverse sine function   (5.3)

Key Ideas

(5.1)   Solving trigonometric equations.

(5.2)   The inverse trigonometric relations.

$\sin^{-1} y = \theta$ is equivalent to $y = \sin\theta$.
$\cos^{-1} y = \theta$ is equivalent to $y = \cos\theta$.
$\tan^{-1} y = \theta$ is equivalent to $y = \tan\theta$.
$\cot^{-1} y = \theta$ is equivalent to $y = \cot\theta$.
$\csc^{-1} y = \theta$ is equivalent to $y = \csc\theta$.
$\sec^{-1} y = \theta$ is equivalent to $y = \sec\theta$.

(5.3)   If $y = \text{Sin}^{-1} x$, then $-\dfrac{\pi}{2} \leq y \leq \dfrac{\pi}{2}$, and $-1 \leq x \leq 1$.

If $y = \text{Cos}^{-1} x$, then $0 \leq y \leq \pi$, and $-1 \leq x \leq 1$.

If $y = \text{Tan}^{-1} x$, then $-\dfrac{\pi}{2} < y < \dfrac{\pi}{2}$, and x is a real number.

Figure 5-30 shows the graphs of the above inverse trigonometric functions.

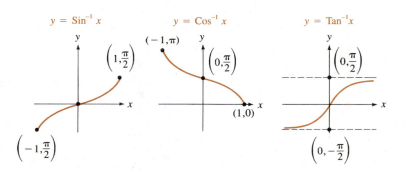

Figure 5-30

(5.4)   If $y = \text{Cot}^{-1} x$, then $0 < y < \pi$, and x is any real number.
(optional)

If $y = \text{Sec}^{-1} x$, then $0 \leq y \leq \pi$, $y \neq \dfrac{\pi}{2}$, and $x \leq -1$ or $x \geq 1$.

If $y = \text{Csc}^{-1} x$, then $-\dfrac{\pi}{2} \leq y \leq \dfrac{\pi}{2}$, $y \neq 0$, and $x \leq -1$ or $x \geq 1$.

Figure 5-31 shows the graphs of the inverse cotangent, inverse secant, and inverse cosecant functions.

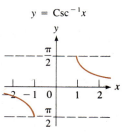

$y = \text{Cot}^{-1}x$
$y = \text{Sec}^{-1}x$
$y = \text{Csc}^{-1}x$

Figure 5-31

REVIEW EXERCISES

In Review Exercises 1–24, solve for all values of the variable between $0°$ and $360°$, including $0°$. **Do not use a calculator or tables.**

1. $\cos \theta = \dfrac{\sqrt{3}}{2}$

2. $\tan \theta = -\dfrac{\sqrt{3}}{3}$

3. $\sin 2\theta = 1$

4. $\cos 2\theta = -1$

5. $\sin \dfrac{\theta}{2} = -\dfrac{1}{2}$

6. $\cos \dfrac{\theta}{2} = -\dfrac{\sqrt{3}}{2}$

7. $\sin \theta \tan \theta + \tan \theta = 0$

8. $\cos \theta \tan \theta - \tan \theta = 0$

9. $\sin A \cos A - \cos A - \sin A + 1 = 0$

10. $2 \sin B \cos B + 2 \sin B + \cos B + 1 = 0$

11. $\sin \theta + 1 = \tan \theta + \cos \theta$

12. $\sin 2\theta = \sin \theta$

13. $2 \cos^2 A = -\sin A + 2$

14. $\cos 2B + \sin B = 0$

15. $\sin^2 \dfrac{C}{2} = \cos^2 C$

16. $\cos^2 \dfrac{A}{2} = 1 - \sin^2 A$

17. $4 \cos^2 \dfrac{\theta}{2} = 1$

18. $\tan \alpha - \cot \alpha = 0$

19. $\cos^2 \theta - \sin^2 \theta = 0$

20. $\csc \theta \sec \theta = \cot \theta + \tan \theta$

21. $\csc \beta \sec \beta = \sec \beta + \cot \beta$

22. $\sin^2 \theta - 2 = \cos \theta + \cos^2 \theta - 2$

23. $\sqrt{2} \sin \theta + \sqrt{2} \cos \theta = 2$

24. $\sqrt{3} \cos \theta - \sin \theta = 2$

In Review Exercises 25–36, find all values of x, if any, that lie in the interval $0 \leqslant x < 2\pi$. Express all answers in radians. **Do not use a calculator or tables.**

25. $\sin^{-1} \dfrac{\sqrt{3}}{2} = x$

26. $\tan^{-1} \dfrac{\sqrt{3}}{3} = x$

27. $\cos^{-1} \left(-\dfrac{1}{2} \right) = x$

28. $\sin^{-1} \left(-\dfrac{\sqrt{3}}{2} \right) = x$

29. $\tan^{-1}(-\sqrt{3}) = x$

30. $\sin^{-1} \sqrt{2} = x$

31. $\cos^{-1} \sqrt{3} = x$

32. $\cot^{-1} \sqrt{3} = x$

33. $\cot^{-1} 0 = x$

34. $\csc^{-1} 2 = x$

35. $\csc^{-1}(0.5) = x$

36. $\sec^{-1}(-2) = x$

In Review Exercises 37–48, use a calculator to find all values of x, if any, that lie in the interval $0 \le x < 2\pi$. Give all answers in radians and to the nearest thousandth.

37. $\sin^{-1} 0.588 = x$

38. $\cos^{-1} 1.732 = x$

39. $\tan^{-1} 75 = x$

40. $\cot^{-1} 75 = x$

41. $\cos^{-1} 3.2 = x$

42. $\sin^{-1} 4.5 = x$

43. $\cot^{-1} 75.3 = x$

44. $\tan^{-1} 0.003 = x$

45. $\csc^{-1} 0.5 = x$

46. $\sec^{-1} 5 = x$

47. $\sec^{-1} 2 = x$

48. $\csc^{-1} 0.1 = x$

In Review Exercises 49–60, find the values of x, if any. **Do not use a calculator or tables.** *Express all answers in radians.*

49. $\mathrm{Sin}^{-1}\left(-\dfrac{1}{2}\right) = x$

50. $\mathrm{Cos}^{-1}\left(-\dfrac{1}{2}\right) = x$

51. $\mathrm{Tan}^{-1}\dfrac{\sqrt{3}}{3} = x$

52. $\mathrm{Tan}^{-1}\left(-\dfrac{\sqrt{3}}{3}\right) = x$

53. $\mathrm{Cos}^{-1} 0 = x$

54. $\mathrm{Sin}^{-1}(-1) = x$

55. $\mathrm{Arctan}(-1) = x$

56. $\mathrm{Arctan}\ \sqrt{3} = x$

57. $\mathrm{Arcsin}\ 0 = x$

58. $\mathrm{Arccos}\ 5 = x$

59. $\mathrm{Arccos}\ 1 = x$

60. $\mathrm{Arcsin}\ 1 = x$

In Review Exercises 61–72, find all real numbers x, if any, without using a calculator or tables.

61. $\sin\left(\mathrm{Sin}^{-1}\dfrac{4}{5}\right) = x$

62. $\cos\left(\mathrm{Cos}^{-1}\dfrac{1}{3}\right) = x$

63. $\sin\left(\mathrm{Cos}^{-1}\dfrac{1}{2}\right) = x$

64. $\cos\left(\mathrm{Sin}^{-1}\dfrac{1}{2}\right) = x$

65. $\tan(\mathrm{Tan}^{-1} 17) = x$

66. $\tan\left[\mathrm{Sin}^{-1}\left(-\dfrac{\sqrt{3}}{2}\right)\right] = x$

67. $\tan\left(\mathrm{Cos}^{-1}\dfrac{\sqrt{2}}{2}\right) = x$

68. $\sin(\mathrm{Tan}^{-1} 1) = x$

69. $\cos[\mathrm{Tan}^{-1}(-1)] = x$

70. $\sec\left[\mathrm{Sin}^{-1}\left(-\dfrac{1}{2}\right)\right] = x$

71. $\csc\left(\mathrm{Arcsin}\dfrac{1}{2}\right) = x$

72. $\csc\left(\mathrm{Arcsin}\dfrac{\sqrt{2}}{2}\right) = x$

In Review Exercises 73–84, find the required value without using a calculator or tables.

73. $\sin\left[\mathrm{Sin}^{-1}\left(-\dfrac{1}{2}\right) + \mathrm{Cos}^{-1}\left(-\dfrac{1}{2}\right)\right]$

74. $\sin\left(\mathrm{Sin}^{-1}\dfrac{\sqrt{3}}{2} - \mathrm{Cos}^{-1}\dfrac{\sqrt{3}}{2}\right)$

75. $\cos\left(\operatorname{Sin}^{-1}\dfrac{\sqrt{3}}{2} - \operatorname{Cos}^{-1}\dfrac{\sqrt{3}}{2}\right)$

76. $\cos\left[\operatorname{Sin}^{-1}\left(-\dfrac{1}{2}\right) + \operatorname{Cos}^{-1}\left(-\dfrac{1}{2}\right)\right]$

77. $\sin 2\left(\operatorname{Sin}^{-1}\dfrac{1}{2}\right)$

78. $\cos 2\left(\operatorname{Sin}^{-1}\dfrac{1}{2}\right)$

79. $\tan 2\left(\operatorname{Sin}^{-1}\dfrac{1}{2}\right)$

80. $\cot 2\left(\operatorname{Cos}^{-1}\dfrac{1}{2}\right)$

81. $\sin\dfrac{1}{2}\left[\operatorname{Arccos}\left(-\dfrac{1}{2}\right)\right]$

82. $\cos\dfrac{1}{2}(\operatorname{Arccos} 0)$

83. $\tan\dfrac{1}{2}(\operatorname{Arccos} 1)$

84. $\cot\dfrac{1}{2}\left[\operatorname{Arccos}\left(-\dfrac{1}{2}\right)\right]$

In Review Exercises 85–96, rewrite each value as an algebraic expression in the variable u.

85. $\sin(\operatorname{Cos}^{-1} u)$

86. $\cos(\operatorname{Sin}^{-1} u)$

87. $\tan(\operatorname{Sin}^{-1} u)$

88. $\sin(\operatorname{Tan}^{-1} u)$

89. $\cos(\operatorname{Sin}^{-1} u)$

90. $\tan(\operatorname{Cos}^{-1} u)$

91. $\sin(2\operatorname{Sin}^{-1} u)$

92. $\cos(2\operatorname{Cos}^{-1} u)$

93. $\tan(2\operatorname{Tan}^{-1} u)$

94. $\sin(2\operatorname{Cos}^{-1} u)$

95. $\cos(2\operatorname{Arcsin} u)$

96. $\cos\left(\dfrac{1}{2}\operatorname{Arccos} u\right)$

In Review Exercises 97–98, evaluate each expression. These exercises are from an optional section.

97. $\operatorname{Cot}^{-1}\left(\dfrac{\sqrt{3}}{3}\right)$

98. $\operatorname{Sec}^{-1}\sqrt{2}$

In Review Exercises 99–100, simplify each expression. These exercises are from an optional section.

99. $\cos\left(\operatorname{Csc}^{-1}\dfrac{5}{3}\right)$

100. $\sin\left(\operatorname{Cot}^{-1}\dfrac{5}{12}\right)$

CHAPTER SIX

OBLIQUE TRIANGLES

Given two sides—or one acute angle and one side—of a *right* triangle, it is always possible to compute the remaining parts. If the given triangle is *not* a right triangle, it is called an **oblique triangle,** and the triangle-solving techniques that were previously discussed no longer apply. The solution of oblique triangles is the topic of this chapter.

In the discussion of oblique triangles, we shall use the following convention: capital letters will be used to name the vertices of a triangle, and the corresponding lowercase letters will name the sides opposite those vertices. Thus, side a is opposite angle A, side b is opposite angle B, and side c is opposite angle C.

In elementary geometry, you studied congruency of triangles and learned conditions that "fix" the size and shape of a triangle. For example, if two sides and the included angle of one triangle are equal to two sides and the included angle of a second triangle, then the two triangles are congruent. This fact (often abbreviated as SAS) indicates that if two sides and the angle between them are given, the triangle is uniquely determined. Because the triangle is "fixed," it is possible to compute the other side and the remaining two angles.

Similarly, if the three sides of a triangle are given, the triangle is determined (SSS) and the three angles of the triangle can be computed. If two angles and any side of a triangle are known (ASA or AAS), the triangle is "fixed" and all remaining parts can be calculated.

The law of cosines is useful in solving oblique triangles when the given information is in SSS or SAS form.

6.1   THE LAW OF COSINES

Place triangle ABC in a coordinate system with vertex A at the origin, as indicated in Figure 6-1. Because angle A is in standard position and point B is on its terminal side, the cosine of A is the ratio of the x-coordinate of B to the distance that B is from the origin. Thus, we have

$$\cos A = \frac{x\text{-coordinate of } B}{c}$$

or

$$x\text{-coordinate of } B = c \cos A$$

Similarly, we have

$$\sin A = \frac{y\text{-coordinate of } B}{c}$$

or

$$y\text{-coordinate of } B = c \sin A$$

Thus, the coordinates of point B are $(c \cos A, c \sin A)$, the coordinates of A are $(0, 0)$, and the coordinates of C are $(b, 0)$.

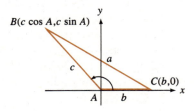

Figure 6-1

We can now use the distance formula to compute a^2.

$$
\begin{aligned}
a^2 &= (c \cos A - b)^2 + (c \sin A - 0)^2 \\
&= c^2 \cos^2 A - 2bc \cos A + b^2 + c^2 \sin^2 A \\
&= c^2(\cos^2 A + \sin^2 A) + b^2 - 2bc \cos A \\
&= c^2 + b^2 - 2bc \cos A
\end{aligned}
$$

1. $a^2 = b^2 + c^2 - 2bc \cos A$

Although the triangle in Figure 6-1 is obtuse, this derivation is valid for any triangle. Also, it need not be vertex A that is placed at the origin. If point B had been placed at the origin instead of point A, a different but similar formula would have been derived.

2. $b^2 = c^2 + a^2 - 2ca \cos B$

If point C had been placed at the origin, a third and equally valid formula would have been found.

3. $c^2 = a^2 + b^2 - 2ab \cos C$

These three formulas are called the **law of cosines.**

Note that a cyclical change of the letters in any one of these three formulas produces another: let a become b, let b become c, let c become a (and similarly for the capital letters).

(6.1)

The Law of Cosines.   The square of any side of any triangle is equal to the sum of the squares of the remaining two sides, minus twice the product of these two sides and the cosine of the angle between them.

$$
\begin{aligned}
a^2 &= b^2 + c^2 - 2bc \cos A \\
b^2 &= c^2 + a^2 - 2ca \cos B \\
c^2 &= a^2 + b^2 - 2ab \cos C
\end{aligned}
$$

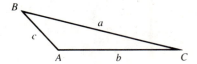

You will be asked to prove that the Pythagorean Theorem is a special case of the law of cosines in the exercises.

Note that the law of cosines is useful if you are given two sides and the included angle (SAS) or three sides (SSS) of a triangle. The first two examples illustrate these cases.

Example 1   Oblique triangle ABC has $b = 27$, $c = 14$, and $A = 43°$. Find side a.

Solution   Sides b and c and the included angle A are given (SAS). See Figure 6-2. Use the law of cosines.

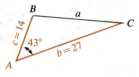

$$a^2 = b^2 + c^2 - 2bc \cos A$$
$$= 27^2 + 14^2 - 2(27)(14) \cos 43°$$
$$\approx 729 + 196 - 552.90$$
$$\approx 372.10$$
$$a \approx 19.29$$
$$\approx 19$$

Figure 6-2

The value of a is approximately 19 units.

Example 2   In the oblique triangle ABC in Figure 6-3, $a = 5.2$, $b = 3.7$, and $c = 7.1$ units. Find angle B.

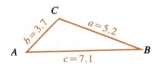

Figure 6-3

Solution   Use the form of the law of cosines that involves angle B.

$$b^2 = a^2 + c^2 - 2ac \cos B$$

Solve this formula for $\cos B$, substitute the values for a, b, and c, and calculate angle B.

$$\cos B = \frac{a^2 + c^2 - b^2}{2ac}$$

$$\cos B = \frac{5.2^2 + 7.1^2 - 3.7^2}{2(5.2)(7.1)}$$

$$\cos B \approx 0.8635$$

$$B \approx 30.29°$$

To the nearest degree, angle $B = 30°$.

Example 3   A farmer uses two horses and two ropes to pull a tractor out of the mud. When pulled tight, the ropes form an angle of 27°, and each horse exerts a pull of 950 lb. What force is applied to the tractor?

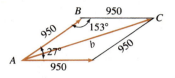

Figure 6-4

Solution   By the parallelogram law for adding vectors, the combined force is length b in the diagram of Figure 6-4. Because opposite sides of a parallelogram are equal, all four sides are 950 units in length. Because consecutive angles of a parallelogram are supplementary, the obtuse angle at B is $180° - 27° = 153°$. Apply the law of cosines to triangle ABC.

$$b^2 = a^2 + c^2 - 2ac \cos B$$
$$b^2 = 950^2 + 950^2 - 2(950)(950) \cos 153°$$
$$\approx 3,413,266.8$$
$$b \approx 1847.5$$
$$\approx 1800$$

The combined pull of the horses is approximately 1800 lb.   ∎

Example 4   An airplane flies N 86.2° W for a distance of 143 km. The pilot, experiencing some engine problems, alters course and flies 79.5 km in the direction S 32.7° E in an attempt to find a place to land. He crash lands safely in a cornfield. How far, and in what direction, will the rescue team need to travel?

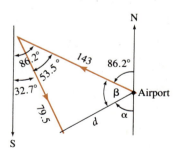

Figure 6-5

Solution   The course diagram appears in Figure 6-5. To find distance d, use the law of cosines.

$$d^2 = 79.5^2 + 143^2 - 2(79.5)(143) \cos 53.5°$$
$$\approx 13,244.8$$

$$d \approx 115.1$$
$$\approx 115$$

The pilot is about 115 km from the airport.

As a first step toward finding the direction in which the rescue team must travel, use the law of cosines again to find angle β.

$$79.5^2 = 143^2 + 115^2 - 2(143)(115) \cos \beta$$

$$\cos \beta = \frac{143^2 + 115^2 - 79.5^2}{2(143)(115)}$$

$$\cos \beta \approx 0.8317$$

$$\beta \approx 33.7°$$

Angle α is $180° - 86.2° - 33.7°$, or $60.1°$. The rescue team must bear S $60.1°$ W.

Example 5   A 100-lb weight is suspended by two cables as in Figure 6-6. The tension on the left cable is 55 lb and on the right cable, 75 lb. What angle does each cable make with the horizontal?

Solution   The forces in the two ropes are such that their resultant force is 100 lb, directed upward to exactly counter the 100-lb downward pull. The force diagram appears in Figure 6-7. The angles that the cables make with the horizontal are the complements of angles α and β. Angles α and β can be found by using the law of cosines.

Figure 6-6

$$55^2 = 75^2 + 100^2 - 2(75)(100) \cos \alpha$$

$$\cos \alpha = \frac{75^2 + 100^2 - 55^2}{2(75)(100)}$$

$$\cos \alpha = 0.8400$$

$$\alpha \approx 32.9°$$

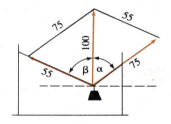

Figure 6-7

Similarly, you have

$$75^2 = 55^2 + 100^2 - 2(55)(100) \cos \beta$$

$$\cos \beta = \frac{55^2 + 100^2 - 75^2}{2(55)(100)}$$

$$\cos \beta \approx 0.6727$$

$$\beta \approx 47.7°$$

The left cable makes an angle of $90° - 47.7° \approx 42°$ with the horizontal. The right cable is $90° - 32.9° \approx 57°$ from the horizontal.

Exercise 6.1

In Exercises 1–12, refer to Illustration 1 and find the value requested. Use a calculator.

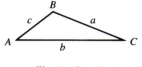

Illustration 1

1. a = 42.9 cm, c = 37.2 cm, B = 99.0°; find b.

2. b = 192 m, c = 86.9 m, A = 21.2°; find a.

3. a = 2730 km, b = 3520 km, C = 21.7°; find c.

4. b = 2.1 km, c = 1.3 km, A = 14°; find a.

5. a = 91.1 cm, c = 87.6 cm, B = 43.2°; find b.

6. a = 107 cm, b = 205 cm, C = 86.5°; find c.

7. a = 19 km, b = 23 km, c = 18 km; find A.

8. a = 14.3 km, b = 29.7 km, c = 21.3 km; find B.

9. a = 30 ft, b = 40 ft, c = 50 ft; find C.

10. a = 130 mi, b = 50 mi, c = 120 mi; find A.

11. a = 1580, b = 2137, c = 3152; find B.

12. a = 0.0031, b = 0.0047, c = 0.0093; find C.

13. Two men are pulling on ropes attached to the bumper of a car that is stuck in a snowdrift. If one man pulls with a force of 114 lb and the other with a force of 97 lb and the angle between the ropes is 13°, what is the force exerted on the car?

14. Two forces, one of 75 lb and the other of 90 lb, are exerted at an angle of 102° from each other. What is the magnitude of the resultant force?

15. In Exercise 13, what is the angle between the resultant force and the rope pulled by the stronger man?

16. In Exercise 14, what is the angle between the resultant force and the direction of the 90-lb force?

17. A donkey and a horse are tied to a large stone. The horse pulls with a force of 950 lb; the donkey lazily tugs with a force of 150 lb. The angle between their tethers is 19.5°. With what force do they pull on the stone?

18. A ship sails 21.2 nautical miles in a direction of N 42.0° W and then turns onto a course of S 15.0° E and sails 19.0 nautical miles. How far is the ship from its starting point?

19. A ship sails 14.3 nautical miles in a direction of S 28.0° W and then turns onto a course of S 52.0° W and sails 23.2 nautical miles. How far is the ship from its starting point?

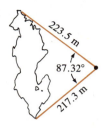

Illustration 2

20. To measure the length of a lake, a surveyor determines the measurements shown in Illustration 2. How long is the lake?

21. To estimate the cost of building a tunnel, a surveyor must find the distance through a hill. The surveyor determines the measurements shown in Illustration 3. How long must the tunnel be to pass through the hill?

22. The three circles in Illustration 4 have radii of 4.0 cm, 7.0 cm, and 9.0 cm. If the circles are externally tangent to each other, what are the angles of the triangle that joins their centers?

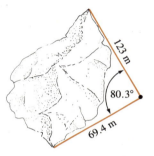

Illustration 3

Illustration 4

23. The three circles in Illustration 4 have radii of 21.2 cm, 19.3 cm, and 31.2 cm. If the circles are externally tangent to each other, what are the angles of the triangle that joins their centers?

24. Show that the Pythagorean Theorem is a special case of the law of cosines.

25. Consider triangle ABC in Figure 6-1. Rotate triangle ABC so that point B is at the origin. Prove that $b^2 = c^2 + a^2 - 2ac \cos B$.

26. Consider triangle ABC in Figure 6-1. Rotate triangle ABC so that point C is at the origin. Prove that $c^2 = a^2 + b^2 - 2ab \cos C$.

27. To determine whether two interior walls meet at a right angle, carpenters often mark a point 3 ft from the corner on one wall and a point (at the same height) on the other wall 4 ft from the corner. If the straight-line distance between those points is 5 ft, the walls are square. At what angle do the walls meet if the distance measures 4 ft 10 in.?

28. To build a counter top for a kitchen, a cabinetmaker must determine the angle with which two walls meet. The method of Exercise 27 is used. What is the angle between the walls if the measured distance is 5 ft 3 in.?

29. Triangle ABC is formed by points $A(3, 4)$, $B(1, 5)$, and $C(5, 9)$. Find angle A to the nearest tenth of a degree.

30. In Exercise 29, find angle B to the nearest tenth of a degree.

31. In Illustration 5, D is the midpoint of BC. Find angle 2 to the nearest tenth of a degree.

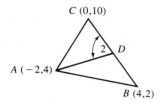

Illustration 5

32. In Exercise 31, find angle *DAB* to the nearest tenth of a degree.

33. A lighthouse is 15.0 nautical miles N 23.0° W of a dock. A ship leaves the dock heading due east at 26.3 knots. How long will it take for the ship to reach a distance of 35.0 nautical miles from the lighthouse?

6.2 THE LAW OF SINES

The law of cosines can be used to solve oblique triangles when the given information fits the SAS or SSS forms. However, the law of cosines is not useful in the ASA or AAS case. Another set of formulas called the **law of sines** is required.

In the triangle of Figure 6-8 and the triangle of Figure 6-9, we draw a line segment *CD* perpendicular to side *AB* (or to an extension of *AB*). Let *h* be the length of segment *CD*. In the two right triangles of Figure 6-8, the following formulas are true.

$$h = b \sin A$$

and

$$h = a \sin B$$

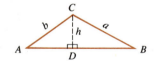

Figure 6-8

Because both $b \sin A$ and $a \sin B$ are equal to h, they are equal to each other.

$$b \sin A = a \sin B$$

or

$$\frac{a}{\sin A} = \frac{b}{\sin B}$$

In the two right triangles of Figure 6-9, these relations hold:

$$h = a \sin B$$

and

$$h = b \sin(180° - A) = b \sin A$$

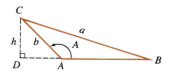

Figure 6-9

Because $a \sin B$ and $b \sin A$ both equal h, they are equal to each other.

$$b \sin A = a \sin B$$

or

$$\frac{a}{\sin A} = \frac{b}{\sin B}$$

In either of these triangles, the perpendicular need not be drawn from C to AB. If drawn from another vertex, similar reasoning yields

$$\frac{a}{\sin A} = \frac{c}{\sin C}$$

or

$$\frac{b}{\sin B} = \frac{c}{\sin C}$$

Because of the transitive law of equality, it follows that

$$\frac{a}{\sin A} = \frac{c}{\sin C} = \frac{b}{\sin B}$$

These results are summed up in a formula called the **law of sines.**

(6.2)

> **The Law of Sines.** The sides in any triangle are proportional to the sines of the angles opposite those sides.
>
> $$\frac{a}{\sin A} = \frac{b}{\sin B} = \frac{c}{\sin C}$$
>
>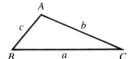

There are three equations implicit in the law of sines, each providing a relation between two angles of a triangle and the sides opposite those angles. If any three of these are known, the fourth can be calculated.

Example 1   In Figure 6-10, $a = 14$ km, $A = 21°$, and $B = 35°$. Find side b.

Solution   Note that the given information, in the pattern AAS, fills three of the four spots in the law of sines.

$$\frac{a}{\sin A} = \frac{b}{\sin B}$$

$$\frac{14}{\sin 21°} = \frac{b}{\sin 35°}$$

$$b \approx 22.4$$

$$\approx 22$$

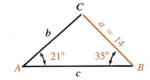

Figure 6-10

Thus, b is approximately 22 km.

Example 2   In Figure 6-11, $a = 29$ meters, $B = 42°$, and $C = 31°$. Find side c.

Solution   Note that the given information is of the pattern ASA. The law of sines is not convenient to use unless you know angle A. Because A is the "rest of" 180°, you have

$$A = 180° - B - C$$
$$= 180° - 42° - 31°$$
$$= 107°$$

Now use the law of sines.

$$\frac{a}{\sin A} = \frac{c}{\sin C}$$

$$\frac{29}{\sin 107°} = \frac{c}{\sin 31°}$$

$$c \approx 15.62$$

$$\approx 16$$

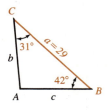

Figure 6-11

Thus, c is approximately 16 meters.

Example 3   A ship is sailing due east. The skipper observes a lighthouse with a bearing of N 37.5° E. After the ship has sailed 4.70 nautical miles, the bearing to the lighthouse is N 9.0° W. How close to the lighthouse did the ship pass?

Solution   The information of the problem is contained in Figure 6-12. If you can find distance b first, then the required distance d can be obtained by solving right triangle ACD. The law of sines provides a way to compute b.

$$\frac{b}{\sin B} = \frac{c}{\sin C}$$

$$b = \sin 81° \left(\frac{4.70}{\sin 46.5°} \right)$$

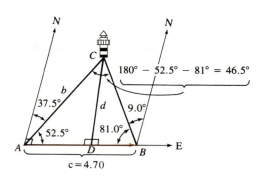

Figure 6-12

$$b \approx 6.3996$$

$$\approx 6.40$$

In right triangle ACD, it follows that

$$d = b \sin A$$

$$\approx 6.40 \sin 52.5°$$

$$\approx 5.0775$$

$$\approx 5.08$$

The ship's closest approach to the lighthouse is 5.08 nautical miles. ■

Example 4 On a bright moonlit night, the moon is directly overhead in city A. In city B, 2500 mi to the north, the moonlight strikes a pole at an angle of 36.57°. If the radius of the earth is 4000 mi (to two significant digits), find the distance from the earth to the moon.

Solution Refer to Figure 6-13. Because 2500 mi is $\frac{1}{10}$ of the earth's circumference, angle BOA is $\frac{1}{10}$ of a complete revolution, or 36°. Angle OBM is supplementary to 36.57°. Hence, angle OBM is 143.43°. Because the sum of the angles in any triangle must be 180°, angle M is 0.57°. Use the law of sines to set up the following proportion and solve for x.

$$\frac{x}{\sin 143.43°} = \frac{4000}{\sin 0.57°}$$

$$x \approx \frac{0.5958(4000)}{0.00995}$$

$$x \approx 239{,}518$$

$$x \approx 240{,}000$$

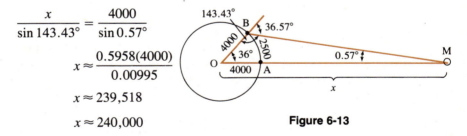

Figure 6-13

The estimate of the distance to the moon is $240{,}000 - 4000 = 236{,}000$ mi. To two significant digits, the distance is 240,000 mi. ■

Example 5 A pilot wishes to fly in the direction of 20.0° east of north, against a 45.0° mph wind blowing from the east. The airspeed of the plane is to be 185 mph. What should be the pilot's heading, and what will be the plane's groundspeed?

Solution Refer to Figure 6-14. The pilot's intended direction of travel is represented by the vector OA. The direction of vector OA is 20.0° east of north. The length of vector OH represents the plane's airspeed of 185 mph. To find the pilot's heading, you must find angle θ, which is the direction of vector OH.

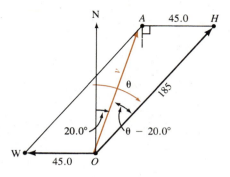

Figure 6-14

In triangle OAH, side AH has a length of 45.0 mi. Angle OAH is $90.0°$ + $20.0°$ or $110.0°$. Use the law of sines on triangle OAH to find $\theta - 20.0°$. You can then find θ.

$$\frac{185}{\sin 110.0°} = \frac{45.0}{\sin(\theta - 20.0°)}$$

$$\sin(\theta - 20.0°) = \frac{45.0 \sin 110.0°}{185}$$

$$\sin(\theta - 20.0°) \approx 0.2286$$

$$\theta - 20.0° \approx 13.2°$$

$$\theta \approx 33.2°$$

To attain his intended direction of travel, the pilot must set a heading of $33.2°$.

In the figure, the groundspeed of the plane is represented by the length, v, of vector OA. To determine this length, calculate angle H and use the law of cosines. Note that

$$\text{angle } H \approx 180.0° - 13.2° - 110.0°$$

$$\approx 56.8°$$

By the law of cosines, you get

$$v^2 \approx 45^2 + 185^2 - 2(45)(185)\cos 56.8°$$

$$\approx 27133$$

$$v \approx 164.7$$

$$\approx 165$$

The plane's groundspeed is approximately 165 mph.

Exercise 6.2

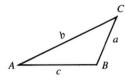

Illustration 1

In Exercises 1–16, refer to Illustration 1. Use the law of sines and your calculator to find the required value.

1. $A = 12°$, $B = 97°$, $a = 14$ km; find b.

2. $A = 19°$, $C = 102°$, $c = 37$ ft; find a.

3. $A = 21.3°$, $B = 19.2°$, $a = 143$ m; find c.

4. $A = 28.8°$, $C = 9.3°$, $c = 135$ m; find b.

5. $B = 8.6°$, $C = 9.2°$, $c = 2.73$ m; find b.

6. $A = 86.3°$, $C = 7.6°$, $a = 43.0$ km; find c.

7. $A = 99.8°$, $C = 43.2°$, $b = 186$ m; find a.

8. $A = 14.61°$, $B = 87.10°$, $c = 1437$ ft; find b.

9. $A = \dfrac{\pi}{7}$, $C = \dfrac{\pi}{2}$, $b = 44.3$ cm; find c.

10. $A = \dfrac{2\pi}{9}$, $B = \dfrac{\pi}{11}$, $c = 56.7$ cm; find a.

11. $B = 107°$, $C = 11°$, $a = 0.96$ mi; find b.

12. $A = 141°$, $B = 5°$, $c = 0.037$; find a.

13. $B = 32.0°$, $b = 120$, $a = c$; find a.

14. $C = 12°$, $c = 5.4$, $b = a$; find b.

15. $A = x°$, $C = 2x°$, $B = 3x°$, $b = 7.93$; find a.

16. $A = x°$, $C = 3x°$, $B = 5x°$, $c = 12.5$; find a.

Illustration 2

17. To measure the distance up a steep hill, Mary determines the measurements shown in Illustration 2. What is the distance d?

18. A ship sails 3.2 nautical miles on a bearing of N 33° E. After reaching a lighthouse, the ship turns and sails 6.7 nautical miles to a position that is due east of the starting point. What is the bearing of the lighthouse from the ship's current position?

19. Points A and B are on opposite sides of a river. See Illustration 3. A tree at point C is 310 ft from point A. Angle A measures 125°, and angle C measures 32°. How wide is the river?

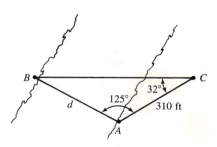

Illustration 3

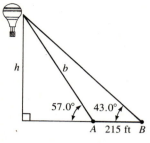

Illustration 4

20. Observers at points A and B are directly in line with a hot-air balloon and are themselves 215 ft apart. See Illustration 4. The angles of elevation of the balloon from A and B are as shown in the illustration. How high is the balloon? (Hint: First use the law of sines to find b.)

21. A radio tower 175 ft high is located on top of a hill. At a point 800 ft down the hill, the angle of elevation to the top of the tower is 19.0°. What angle does the hill make with the horizontal? See Illustration 5.

22. Two children are on one river bank, 120 ft apart. They each sight the same tree on the opposite bank. See Illustration 6. Angle A measures 79°, and angle B measures 63°. How wide is the river?

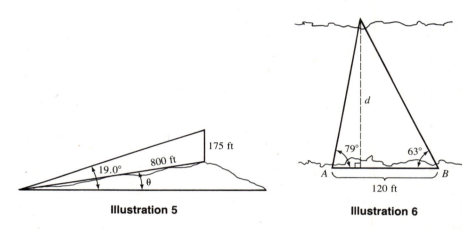

Illustration 5   **Illustration 6**

23. A ship sails due north at 3.2 knots (nautical miles per hour). At 2:00 PM, the skipper sights a lighthouse in the direction of N 43° W. One hour later, the lighthouse bears S 78° W. How close to the lighthouse did the ship sail?

24. A ship sails on a course bearing N 21° E at a speed of 14 knots (nautical miles per hour). At 12:00 noon, the first mate sights an island in the direction N 35° E and one hour later sights the same island due east. If the ship continues on its course, how close will it approach the island?

25. In Exercise 23, at what time was the ship closest to the lighthouse?

26. In Exercise 24, at what time will the ship be nearest the island?

6.3 THE AMBIGUOUS CASE: SSA

A triangle is uniquely determined if two sides and the included angle are given. This situation is handled by the law of cosines. However, a triangle may or may not be determined if two sides and a nonincluded angle are given. There may even be no triangle at all! SSA is called the **ambiguous case.**

Consider the problem of constructing a triangle with sides of 1 and 2 in. and a nonincluded angle of 20°, as indicated in Figure 6-15. For the 20° angle to not be the included angle, the 1-inch side must hang on to point A. There are two possible positions for it, and consequently, two possible triangles. The two possible positions are determined by the two intersections of a 1-in. radius circle with the third side.

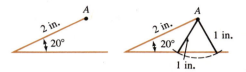

Figure 6-15

Figure 6-16

If the angle is given as 30°, there is only one possible triangle formed because the 1-in. radius circle is tangent to the third side. The triangle is a 30°–60°-right triangle. (In a 30°–60°-right triangle, the side opposite the 30° angle is one-half the hypotenuse.) See Figure 6-16.

If the angle is greater than 30°, the 1-in. side is not long enough to "close up" the triangle, and no triangle is formed. See Figure 6-17.

Here are some generalizations about the ambiguous case of SSA. If side *b* is given and angle *A* is acute, the length of side *a* determines the possible number of triangles. Side *a* could be too short as in Figure 6-18*i*, and no triangle would be formed. Side *a* could be "just right," and one right triangle would be formed (Figure 6-18*ii*). As side *a* continues to get longer, the two possibilities of Figure 6-18*iii* develop, but if *a* is longer than *b*, only one triangle is possible (Figure 6-18*iv*).

Figure 6-17

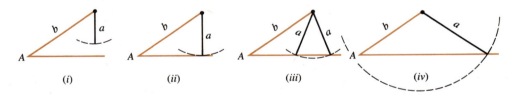

Figure 6-18

If side *b* is given and angle *A* is obtuse, there are two more cases to consider. If *a* is less than or equal to *b*, no triangle is formed, as in Figure 6-19*i*. If *a* is greater than *b*, as in Figure 6-19*ii*, one triangle is possible.

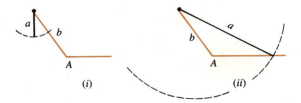

Figure 6-19

Memorizing all of these possibilities is confusing and often unnecessary. If you are given information in the SSA form, sketch a careful scale drawing of the triangle. Then, use your common sense to determine whether two, one, or no triangles are expected. If a scale drawing indicates a situation that is too close to call from the diagram alone, let the law of sines decide the outcome. This will be illustrated in Example 2.

Example 1   Three parts of a triangle are exactly $a = 1$, $b = 2$, and $A = 20°$. Find angles B and C to the nearest tenth of a degree and side c to the nearest hundredth.

Solution   A sketch indicates that there are two possible triangles. See Figure 6-20. By the law of sines, it follows that

$$\frac{a}{\sin A} = \frac{b}{\sin B}$$

$$\sin B = \frac{b \sin A}{a}$$

$$\sin B = \frac{2 \sin 20°}{1}$$

$$\sin B \approx 0.6840$$

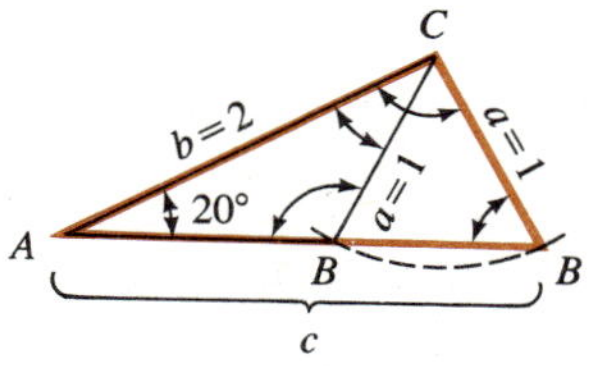

Figure 6-20

There are two possible values of B, one acute (a first-quadrant angle) and the other obtuse (a second-quadrant angle).

$$B \approx 43.2°$$

or

$$B \approx 180° - 43.2° = 136.8°$$

The third angle C has two possibilities also:

$$C = 180° - A - \mathbf{B}$$
$$C \approx 180° - 20° - \mathbf{43.2}°$$
$$C \approx 116.8°$$

or

$$C \approx 180° - 20° - \mathbf{136.8}°$$
$$C \approx 23.2°$$

The third side c can also be found by using the law of sines. Side c has two possibilities:

$$\frac{c}{\sin C} = \frac{a}{\sin A}$$

$$c \approx \frac{1 \cdot \sin 116.8°}{\sin 20°} \approx 2.61$$

or

$$c \approx \frac{1 \cdot \sin 23.2°}{\sin 20°} \approx 1.15$$

The two triangles are shown in Figure 6-21.

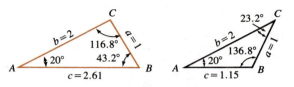

Figure 6-21

Example 2   In the triangle of Figure 6-22, $a = 1$, $b = 2$, and $A = 30°$. Solve the triangle.

Solution   A careful scale drawing indicates a situation too close to call from the diagram alone, so let the law of sines determine the number of possible triangles. By the law of sines, it follows that

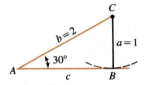

Figure 6-22

$$\frac{a}{\sin A} = \frac{b}{\sin B}$$

$$\frac{1}{\sin 30°} = \frac{2}{\sin B}$$

$$\sin B = 2\left(\frac{1}{2}\right)$$

$$\sin B = 1$$

$$B = 90°$$

There is only one triangle possible—a right triangle. Because the right triangle has a 30° angle, the remaining angle must be 60° and the remaining side must be $\sqrt{3}$.

Example 3   In triangle ABC, $a = 1$, $b = 2$, and $A = 45°$. If possible, solve the triangle.

Solution   By the law of sines, it follows that

$$\frac{a}{\sin A} = \frac{b}{\sin B}$$

$$\frac{1}{\sin 45°} = \frac{2}{\sin B} \quad \left(\frac{1}{2}\right)$$

$$\sin B = 2\sin 45°$$

$$= 2\left(\frac{\sqrt{2}}{2}\right)$$

$$= \sqrt{2}$$

Because $\sqrt{2}$ is greater than 1, and $\sin B$ cannot be greater than 1, there is no triangle that satisfies the given conditions. ■

Example 4 A vertical tower 255 ft tall stands on a hill. From a point 700 ft down the hill, the angle between the hill and an observer's line of sight to the top of the tower is 12.0°. What is the angle of inclination of the hill (the angle the ground makes with the horizontal)?

Solution The given information is used to draw Figure 6-23. By the law of sines, it follows that

$$\frac{255}{\sin 12.0°} = \frac{700}{\sin B}$$

$$\sin B = \frac{700 \cdot \sin 12.0°}{255}$$

$$\approx 0.5707$$

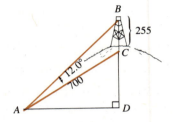

Figure 6-23

There are two possibilities for angle B—one acute and one obtuse. However, from the diagram, you want only the acute angle. Because $\sin B \approx 0.5707$, you have

$$B \approx 34.8°$$

Once you know the measure of angle B, it is easy to compute angle C (angle ACB).

$$C = 180° - A - B$$
$$\approx 180° - 12.0° - 34.8°$$
$$\approx 133.2°$$

The angle the hill makes with the horizontal is $C - 90°$, which is approximately 43.2°. ■

Exercise 6.3

In Exercises 1–16, calculate all possibilities for the indicated value. If no triangle is possible, so indicate.

1. $A = 42.0°$, $a = 123$ ft, $b = 96.0$ ft; find B.

2. $B = 56.2°$, $b = 13.5$ yd, $c = 15.3$ yd; find C.

3. $C = 98.6°$, $a = 42.1$ cm, $c = 47.3$ cm; find A.

4. $B = 17.5°$, $a = 0.063$ m, $b = 0.152$ m; find A.

5. $A = 98.6°$, $a = 42.1$ in., $c = 47.3$ in.; find C.

6. $C = 86°$, $b = 20$ ft, $c = 19$ ft; find B.

7. $C = 23°$, $b = 48$, $c = 52$; find A.

8. $B = 7.35°$, $b = 2.683$, $c = 4.752$; find A.

9. $A = 9.86°$, $a = 3761$, $b = 5293$; find C.

10. $B = 78.3°$, $a = 0.057$, $b = 0.093$; find C.

11. $A = 102.0°$, $a = 13.9$, $c = 15.0$; find B.

12. $C = 47.6°$, $a = 10.5$, $c = 7.35$; find B.

13. $A = 57°$, $b = 13$ m, $a = 12$ m; find c.

14. $C = 48°$, $b = 29$ km, $c = 26$ km; find a.

15. $B = 87°$, $a = 35$ cm, $b = 32$ cm; find c.

16. $B = 38°$, $a = 12$ cm, $b = 40$ cm; find c.

17. The triangular piece of land owned by farmer Brown is bounded by three straight highways. The angle between two of them—U.S. 45 and county M—is 43°. Brown's property runs for 2500 ft along county M, and for 2000 ft along the third highway—scenic Silo Drive. How much land might Brown own fronting on U.S. 45?

18. From the roof of farmer Brown's barn, the angle of elevation to the top of a ranger lookout tower is 17°. From the barn's ground level 43 ft below, the angle of elevation of the tower is 21°. How far above ground level is the top of the tower?

19. A 210-ft television tower stands on the top of an office building. From a point on level ground, the angles of elevation to the top and base of the tower are 25.2° and 21.1°. How tall is the office building?

20. A pilot leaves point A and flies 800 km with a heading of 320° to point B. From B, she flies due south to a point C, which is 700 km from point A. How long is the BC leg of the trip?

6.4   MOLLWEIDE'S EQUATIONS AND THE LAW OF TANGENTS (OPTIONAL)

There are formulas, named after Karl Brandan Mollweide (1774–1825),* that involve all six parts of a triangle. For this reason, they are useful to verify that a triangle has been solved correctly.

Consider any triangle ABC. By the law of sines, we have

$$\frac{a}{c} = \frac{\sin A}{\sin C} \quad \text{and} \quad \frac{b}{c} = \frac{\sin B}{\sin C}$$

Adding these equations together and using a sum-to-product identity yields

$$\frac{a + b}{c} = \frac{\sin A + \sin B}{\sin C}$$

---

*These formulas were published and used as early as 1746, 28 years before Mollweide was born.

1. $\dfrac{a+b}{c} = \dfrac{2\sin\frac{1}{2}(A+B)\cos\frac{1}{2}(A-B)}{2\sin\frac{1}{2}C\cos\frac{1}{2}C}$

Note that in any triangle ABC

$$\sin\frac{1}{2}(A+B) = \sin\frac{1}{2}(180° - C)$$

$$= \sin\left(90° - \frac{1}{2}C\right)$$

$$= \cos\frac{1}{2}C$$

Substituting $\cos\frac{1}{2}C$ for $\sin\frac{1}{2}(A+B)$ in Equation 1 gives

$$\frac{a+b}{c} = \frac{2\cos\frac{1}{2}C\cos\frac{1}{2}(A-B)}{2\sin\frac{1}{2}C\cos\frac{1}{2}C}$$

After removing the common factor of $2\cos\frac{1}{2}C$ in the numerator and the denominator of the fraction on the right side, we have

(6.3) $\dfrac{a+b}{c} = \dfrac{\cos\frac{1}{2}(A-B)}{\sin\frac{1}{2}C}$

By a similar argument, we have

(6.4) $\dfrac{a-b}{c} = \dfrac{\sin\frac{1}{2}(A-B)}{\cos\frac{1}{2}C}$

These two results, together with all formulas obtained by a cyclic change of the letters a, b, c and A, B, C, are called **Mollweide's equations.**

Example 1   A lazy student is asked to measure the six parts of a triangle. He claims that $A = 43°$, $B = 40°$, $C = 97°$ and $a = 13.5$, $b = 11.1$, $c = 14.7$. Is he reasonably close?

Solution   As indicated, Mollweide's equations, which involve all six parts of a triangle, can be used to check these measurements. Begin by substituting the indicated values into Mollweide's second equation.

$$\frac{a - b}{c} \overset{?}{=} \frac{\sin \frac{1}{2}(A - B)}{\cos \frac{1}{2}C}$$

$$\frac{13.5 - 11.1}{14.7} \overset{?}{=} \frac{\sin \frac{1}{2}(43° - 40°)}{\cos \frac{1}{2}(97°)}$$

$$0.1633 \overset{?}{=} 0.0395$$

Because 0.1633 is not "close" to 0.0395, no triangle exists with these measurements. If you had obtained equality, the existence of a triangle would still be uncertain. The data must then be substituted into Mollweide's first equation. Only when the data satisfy both equations is the existence of the triangle guaranteed. ■

Mollweide's equations also provide a way to develop additional formulas that can be used to solve oblique triangles. These new formulas are called the **law of tangents.**

If one of Mollweide's equations,

$$\frac{a - b}{c} = \frac{\sin \frac{1}{2}(A - B)}{\cos \frac{1}{2}C}$$

is divided by another,

$$\frac{a + b}{c} = \frac{\cos \frac{1}{2}(A - B)}{\sin \frac{1}{2}C}$$

the result is

$$\frac{a - b}{a + b} = \frac{\sin \frac{1}{2}(A - B)}{\cos \frac{1}{2}(A - B)} \cdot \frac{\sin \frac{1}{2}C}{\cos \frac{1}{2}C}$$

2. $$\frac{a - b}{a + b} = \tan \frac{1}{2}(A - B) \cdot \tan \frac{1}{2}C$$

This can be simplified by noting that

$$\tan \frac{1}{2}C = \tan \frac{1}{2}[180° - (A + B)]$$

$$= \tan\left[90° - \frac{1}{2}(A + B)\right]$$

$$= \cot\frac{1}{2}(A + B)$$

$$= \frac{1}{\tan\dfrac{1}{2}(A + B)}$$

After substituting in Equation 2, we get

$$\frac{a - b}{a + b} = \frac{\tan\dfrac{1}{2}(A - B)}{\tan\dfrac{1}{2}(A + B)}$$

(6.5) | **The Law of Tangents.** The formula

$$\frac{a - b}{a + b} = \frac{\tan\dfrac{1}{2}(A - B)}{\tan\dfrac{1}{2}(A + B)}$$

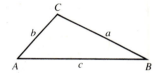

and all formulas produced by cyclic changes of the letters are collectively called the **law of tangents.**

The law of tangents can be used to solve triangles that fit the AAS and ASA cases.

Example 2 Suppose that $A = 42°$, $B = 17°$, and $a = 13$ in a triangle ABC. Use the law of tangents to find b.

Solution By the law of tangents, it follows that

$$\frac{a - b}{a + b} = \frac{\tan\dfrac{1}{2}(A - B)}{\tan\dfrac{1}{2}(A + B)}$$

$$\frac{13 - b}{13 + b} = \frac{\tan\dfrac{1}{2}(42° - 17°)}{\tan\dfrac{1}{2}(42° + 17°)}$$

$$\frac{13 - b}{13 + b} \approx 0.3918$$

Clear the fractions and solve for b.

$$13 - b \approx 0.3918(13 + b)$$
$$b \approx 5.68$$
$$b \approx 5.7 \, \text{units}$$

■

Exercise 6.4

In Exercises 1–10, use Mollweide's equations to decide whether triangles exist with the six given parts.

1. $a = 20.0, b = 41.0, c = 27.2, A = 25.0°, B = 120.0°, C = 35.0°$
2. $a = 97.8, b = 93.0, c = 112.2, A = 56.0°, B = 52.0°, C = 72.0°$
3. $a = 13.2, b = 15.7, c = 10.6, A = 56.2°, B = 82.0°, C = 41.8°$
4. $a = 48.3, b = 112, c = 96.1, A = 25.4°, B = 96.0°, C = 58.6°$
5. $a = 30.4, b = 122.5, c = 133.6, A = 12.7°, B = 62.3°, C = 105.0°$
6. $a = 40.9, b = 23.0, c = 62.2, A = 18.0°, B = 10.0°, C = 152.0°$
7. $a = 1.73, b = 1.00, c = 2.00, A = 30.0°, B = 60.0°, C = 90.0°$
8. $a = 1.41, b = 1.41, c = 2.00, A = 50.0°, B = 50.0°, C = 80.0°$
9. $a = 16, b = 17, c = 29, A = 20°, B = 40°, C = 120°$
10. $a = 35.4, b = 28.1, c = 68.6, A = 60.0°, B = 50.0°, C = 70.0°$

In Exercises 11–22, use the law of tangents to solve for the value indicated.

11. $A = 14.3°, B = 46.8°, a = 25.1$; find b.
12. $A = 19.2°, B = 39.7°, b = 123$; find a.
13. $A = 27°, C = 29°, a = 47$; find b.
14. $A = 94.0°, C = 10.8°, b = 863$; find a.
15. $B = 7.83°, C = 56.37°, c = 271.3$; find b.
16. $B = 9.27°, C = 63.81°, b = 4.276$; find c.
17. $A = 47.9°, B = 32.2°, c = 13.7$; find a.
18. $A = 29.8°, B = 49.3°, c = 24.2$; find b.
19. $A = 56°, C = 72°, b = 18$; find a.
20. $A = 73.1°, C = 98.6°, b = 158$; find c.
21. $B = 99°, C = 12°, a = 800$; find b.
22. $B = 102°, C = 62°, a = 56$; find c.
23. Derive the following alternative form of the law of tangents,

$$\tan \frac{1}{2}(A - B) \tan \frac{1}{2}C = \frac{a - b}{a + b}$$

[*Hint*:   Because $A + B + C = 180°$, $\frac{1}{2}(A + B) = \frac{1}{2}(180° - C)$.]

6.5 AREAS OF TRIANGLES

The area of any triangle is given by the formula

(6.6) $A = \dfrac{1}{2} bh$

where b is the base and h is the height of the triangle. In any of the triangles of Figure 6-24, the base is 6 and the height is 3. The area of each triangle is

$$A = \frac{1}{2} bh$$

$$= \frac{1}{2} (6)(3)$$

$$= 9 \text{ sq units}$$

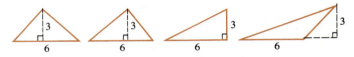

Figure 6-24

The height of an oblique triangle is never one of the six parts of the triangle because it is not a side. Is it possible to compute the area of an oblique triangle when only parts of the triangle are given? If we are given three parts that uniquely determine the triangle, the answer is "yes."

If two sides and the included angle are given (SAS), the height can be calculated. In triangle ABC of Figure 6-25, assume that b, a, and angle C are given. The height h is $a \sin C$, and the area K is

$$K = \frac{1}{2} bh$$

Hence, we have

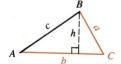

Figure 6-25

(6.7) $K = \dfrac{1}{2} ba \sin C$

If c, b, and angle A are given, a similar argument gives

(6.8) $K = \dfrac{1}{2} cb \sin A$

Finally, if a, c, and angle B are given, we have

(6.9) $K = \dfrac{1}{2} ac \sin B$

Example 1   Find the area of the triangle in Figure 6-26.

   Solution   Use Formula (6.9).

$$K = \frac{1}{2} ac \sin B$$

$$= \frac{1}{2} (15)(17) \sin 20°$$

$$\approx 43.6076$$

$$\approx 44 \text{ sq units}$$

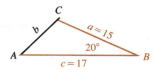

Figure 6-26

Example 2   What is the area of an equilateral triangle of side s?

   Solution   Each angle of an equilateral triangle is 60°. Hence, we have

$$K = \frac{1}{2} ab \sin C$$

$$= \frac{1}{2} ss \sin 60°$$

$$= \frac{\sqrt{3}}{4} s^2$$

   If two angles and a side are given (AAS or ASA), the formulas (6.7) through (6.9) can be adjusted to provide the area. In the triangle of Figure 6-26, we have

$$K = \frac{1}{2} cb \sin A$$

By the law of sines, we have

$$b = \frac{c \sin B}{\sin C}$$

Substituting for b, we have

$$K = \frac{1}{2} c \frac{c \sin B}{\sin C} \sin A$$

or

(6.10) $$K = \frac{c^2 \sin A \sin B}{2 \sin C}$$

A similar argument produces two more formulas.

(6.11) $$K = \frac{a^2 \sin B \sin C}{2 \sin A}$$

and

(6.12) $K = \dfrac{b^2 \sin C \sin A}{2 \sin B}$

Note the cyclic change of the letters in the above three formulas.

Example 3   What is the area of the triangle in Figure 6-27?

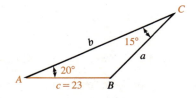

Figure 6-27

Solution   Angle $B = 180° - 20° - 15° = 145°$, and

$K = \dfrac{c^2 \sin A \sin B}{2 \sin C}$

$= \dfrac{23^2 \sin 20° \sin 145°}{2 \sin 15°}$

≈ 200.4806

$\approx 200 \text{ sq units}$

Example 4   Find the area of the isosceles triangle in Figure 6-28.

Solution   The vertex angle B is $180° - 2\theta$, and

$K = \dfrac{b^2 \sin A \sin C}{2 \sin B}$

$= \dfrac{b^2 \sin \theta \sin \theta}{2 \sin(180° - 2\theta)}$

$= \dfrac{b^2 \sin^2 \theta}{2 \sin 2\theta}$

$= \dfrac{b^2 \sin^2 \theta}{2(2 \sin \theta \cos \theta)}$

$= \dfrac{b^2 \sin \theta}{4 \cos \theta}$

$= \dfrac{b^2}{4} \tan \theta$

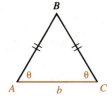

Figure 6-28

Finally, if three sides of a triangle are given (SSS), the triangle is determined, and the area can be calculated by a formula attributed to Heron (Hero) of Alexandria (circa 250 B.C.).

The plan for deriving Heron's formula is not difficult, but the details are messy. Here's the plan.

By the law of cosines, it follows that

$$\cos A = \frac{b^2 + c^2 - a^2}{2bc}$$

We will compute $1 - \cos A$ and $1 + \cos A$; their product is $\sin^2 A$. This will be substituted into $K^2 = \frac{1}{4} b^2 c^2 \sin^2 A$, the square of the area formula $K = \frac{1}{2} cb \sin A$. The area of the given triangle will be K, the square root of $\frac{1}{4} b^2 c^2 \sin^2 A$.

Here are the details.

$$\cos A = \frac{b^2 + c^2 - a^2}{2bc}$$

$$1 - \cos A = 1 - \frac{b^2 + c^2 - a^2}{2bc}$$

$$= \frac{2bc - b^2 - c^2 + a^2}{2bc} \qquad \text{Find a common denominator and add.}$$

$$= \frac{a^2 - (b - c)^2}{2bc} \qquad \text{Factor } 2bc - b^2 - c^2 \text{ as } -(b - c)^2.$$

$$= \frac{(a + b - c)(a - b + c)}{2bc} \qquad \text{Factor the numerator, which is the difference of two squares.}$$

In a similar fashion, we have

$$1 + \cos A = 1 + \frac{b^2 + c^2 - a^2}{2bc}$$

$$= \frac{2bc + b^2 + c^2 - a^2}{2bc} \qquad \text{Find a common denominator and add.}$$

$$= \frac{(b + c)^2 - a^2}{2bc} \qquad \text{Factor } 2bc + b^2 + c^2 \text{ as } (b + c)^2.$$

$$= \frac{(b + c - a)(b + c + a)}{2bc} \qquad \text{Factor the numerator, which is a difference of two squares.}$$

The product of these two results is $\sin^2 A$:

$$\sin^2 A = 1 - \cos^2 A$$

$$= (1 - \cos A)(1 + \cos A)$$

$$= \frac{(a + b - c)(a - b + c)(b + c - a)(b + c + a)}{4b^2 c^2}$$

This result can be substituted into the formula $K^2 = \frac{1}{4}b^2c^2 \sin^2 A$.

$$K^2 = \frac{1}{4}b^2c^2 \sin^2 A$$

$$= \frac{1}{4}b^2c^2 \frac{(a + b - c)(a - b + c)(b + c - a)(b + c + a)}{4b^2c^2}$$

$$= \frac{(a + b - c)(a - b + c)(b + c - a)(b + c + a)}{16}$$

Instead of dividing the product in the numerator by 16, we give each of the four factors its own divisor of 2. Also, each of the four factors is written in a slightly different form; you should verify that they are equivalent.

$$K^2 = \left(\frac{a + b + c - 2c}{2}\right)\left(\frac{a + b + c - 2b}{2}\right)\left(\frac{a + b + c - 2a}{2}\right)\left(\frac{a + b + c}{2}\right)$$

$$= \left(\frac{a + b + c}{2} - c\right)\left(\frac{a + b + c}{2} - b\right)\left(\frac{a + b + c}{2} - a\right)\left(\frac{a + b + c}{2}\right)$$

The expression

$$\frac{a + b + c}{2}$$

is one-half of the perimeter of the triangle. This is often called the **semiperimeter** and is denoted by the letter s. Substituting s for

$$\frac{a + b + c}{2}$$

in the preceding equation gives

$$K^2 = (s - c)(s - b)(s - a)s$$

We take the square root of both sides of this equation and rearrange the terms to obtain **Heron's formula.**

(6.13)

> **Heron's Formula.**   If a, b, and c are the three sides of a triangle and
>
> $$s = \frac{a + b + c}{2}$$
>
> then the area of the triangle is given by
>
> $$K = \sqrt{s(s - a)(s - b)(s - c)}$$

Example 5   A triangle has sides that are exactly 5, 7, and 10 cm. What is its area?

Solution   Let $a = 5$, $b = 7$, and $c = 10$. Then

$$s = \frac{5 + 7 + 10}{2} = 11$$

Use Heron's formula to find the area.

$$K = \sqrt{s(s-a)(s-b)(s-c)}$$
$$= \sqrt{11(11-5)(11-7)(11-10)}$$
$$= \sqrt{11 \cdot 6 \cdot 4 \cdot 1}$$
$$= \sqrt{264}$$
$$\approx 16.248$$

The area is approximately 16.248 sq cm.   ■

Exercise 6.5

In Exercises 1–16, find the area of the triangle whose parts are given, if possible.

1. $b = 23$ ft, $a = 17$ ft, $C = 80°$

2. $c = 1.7$ yd, $b = 3.5$ yd, $A = 60°$

3. $a = 32.3$ cm, $c = 21.5$ cm, $B = 120.0°$

4. $B = 33.2°$, $a = 101$ km, $c = 97.3$ km

5. $a = 3.0$, $b = 5.0$, $c = 7.0$

6. $a = 2.1$, $b = 3.2$, $c = 5.7$

7. $a = 3$, $b = 4$, $c = 5$

8. $a = 1.2$, $b = 2.3$, $c = 3.4$

9. $A = 55°$, $B = 45°$, $c = 12$

10. $A = 102°$, $C = 47°$, $b = 82$

11. $B = 15°$, $A = 70°$, $b = 23$

12. $C = 41°$, $B = 62°$, $c = 17$

13. $a = 0.06$ mm, $b = 0.05$ mm, $c = 0.07$ mm

14. $a = 0.017$ cm, $b = 0.032$ cm, $c = 0.055$ cm

15. $a = 976$ km, $b = 728$ km, $c = 543$ km

16. $a = 1860$ m, $b = 2150$ m, $c = 1590$ m

Illustration 1

17. To find the area of a triangular lot, the owner starts at one corner and walks due west 205 ft to a second corner. After turning through an angle of 87.3°, he walks 307 ft to the third corner. What is the area of the lot in square feet?

18. A painter wishes to estimate the area of the gable end of a house. What is its area in square feet if the triangle has dimensions as shown in Illustration 1?

19. A printer wishes to make a sign in the form of an isosceles triangle with base angles of 70° and a side of 15 m. Find the area of the triangle.

20. Point C has a bearing of N 20° E from point A and a bearing of N 10° E from point B. What is the area of triangle ABC if B is due east of A and 17 km from C?

21. Three circles with radii of exactly 3, 5, and 9 cm are externally tangent. What is the area of the triangle joining their centers?

22. Three circles have diameters of 7.8, 5, and 11.4 cm. If they are tangent externally, what is the area of the triangle joining their centers?

23. A Boy Scout walks 520 ft, turns and walks 490 ft, turns again and walks 670 ft, returning to his starting point. What area did his walk encompass?

24. A Girl Scout hikes 523 m, turns and jogs 412 m, turns again, and runs 375 m, returning to her starting point. What area did her trip encompass?

25. Prove that in any triangle,

$$\cos^2 \frac{A}{2} = \frac{s(s-a)}{bc}$$

where s is one-half of the perimeter.

26. Prove that in any triangle,

$$\sin^2 \frac{A}{2} = \frac{(s-b)(s-c)}{bc}$$

where s is one-half of the perimeter.

27. Prove that the area of a parallelogram is one-half the product of the diagonals and the sine of the angle between the diagonals.

Illustration 2

28. Three externally tangent circles have radii of 5, 7, and 8, as shown in Illustration 2. What is the area of the curve-sided "triangle" they enclose?

29. Find the area of an isosceles triangle with vertex angle α and base b.

30. Find the area of an isosceles triangle with vertex angle α and one of the equal sides of length a.

31. Find the area of an isosceles triangle if the base b is one-half the length of one of the equal sides.

32. Derive a formula for the area of the segment of the circle (the shaded area) in Illustration 3.

Illustration 3

6.6 MORE ON VECTORS (OPTIONAL)

Some physical quantities, called **scalar quantities,** can be described completely by their numerical values, or magnitudes. Some examples of scalar quantities are temperature, distance, area, volume, and elapsed time. Other quantities, called **vector quantities,** have both magnitude and direction. Examples of vector quantities are force, velocity, acceleration, and displacement. Although we have discussed vectors previously, we now examine some of their properties in more detail.

Recall that a **vector** is a directed line segment. A vector can be denoted by a boldface letter such as **V.** However, in handwritten work, we often express a vector as a letter with an arrow above it, such as $\vec{V}$. If a vector starts at point A and ends at point B, that vector can be denoted either as **AB** or as $\overrightarrow{AB}$.

(6.14)

> **Definition.** Two **vectors** are equal if and only if they have the same length and the same direction.

Figure 6-29

In Figure 6-29, note that vector **AB** could also be denoted as $\overrightarrow{AB}$ or as $\overrightarrow{V}$. Furthermore, because vectors **AB** and **CD** have the same length and direction, it is true that **AB** = **CD**. The length of a vector **V** is called its **magnitude** or its **norm**. Thus, because the norm of a vector is a numerical value, it is a scalar quantity. This scalar is denoted by $|\mathbf{V}|$. By the distance formula, the vector **OA** in Figure 6-30 has a norm of 5, and $|\mathbf{OA}| = 5$. Similarly, $|\mathbf{OB}| = 5$ and $|\mathbf{OC}| = \sqrt{2}$. Note that $|\mathbf{OA}| = |\mathbf{OB}|$, but **OA** ≠ **OB**.

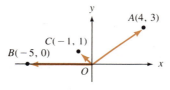

Figure 6-30

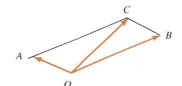

Figure 6-31

Recall that vectors can be added by using the **parallelogram law.** The sum of the two vectors **OA** and **OB** shown in Figure 6-31 is found by constructing the parallelogram *OACB* with vectors **OA** and **OB** forming two of its adjacent sides. The **sum,** or **resultant,** of vectors **OA** and **OB** is the parallelogram's diagonal **OC**.

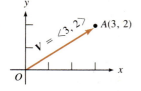

Figure 6-32

Vectors are easier to handle mathematically if they are placed on a coordinate system. The vector **V,** for example, in Figure 6-32 is placed on a coordinate system so that it starts at the origin and ends at the point (3, 2). If we assume that *all* vectors start at the origin, then each is completely determined by its endpoint. Thus, we can denote the vector shown in Figure 6-32 by the ordered pair of numbers ⟨3, 2⟩. We use corner brackets ⟨ ⟩ to distinguish the vector from the point (3, 2). Note that the distance formula can be used to find the norm of **V.**

$$|\mathbf{V}| = \sqrt{3^2 + 2^2}$$
$$= \sqrt{13}$$

In general, we have

(6.15)

> **Theorem.** If **V** is a vector placed on a coordinate system and **V** is represented by ⟨a, b⟩, then the norm of **V** is given by
>
> $$|\mathbf{V}| = \sqrt{a^2 + b^2}$$

Example 1 If vector $\mathbf{V} = \langle 3, -5 \rangle$, find $|\mathbf{V}|$.

Solution
$$\begin{aligned} |\mathbf{V}| &= \sqrt{3^2 + (-5)^2} \\ &= \sqrt{9 + 25} \\ &= \sqrt{34} \end{aligned}$$

The addition of vectors is easy using this ordered-pair notation. The coordinates of point C in the parallelogram of Figure 6-33 are found by adding the corresponding coordinates of points A and B. Thus, if $\mathbf{OA} = \langle -3, 1 \rangle$ and $\mathbf{OB} = \langle 6, 1 \rangle$, then

$$\begin{aligned} \mathbf{OA} + \mathbf{OB} &= \langle -3 + 6, \ 1 + 1 \rangle \\ &= \langle 3, 2 \rangle \\ &= \mathbf{OC} \end{aligned}$$

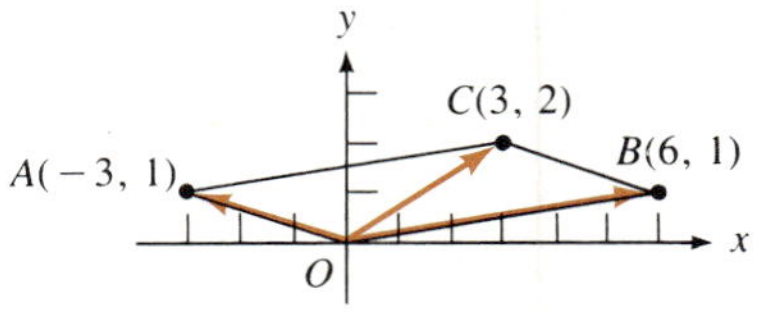

Figure 6-33

The previous example suggests the following definition.

(6.16)

> **Definition.** If vector $\mathbf{V} = \langle a, b \rangle$ and vector $\mathbf{W} = \langle c, d \rangle$, then
> $$\mathbf{V} + \mathbf{W} = \langle a + c, b + d \rangle$$

Example 2 If $\mathbf{V} = \langle 3, -5 \rangle$ and $\mathbf{W} = \langle 1, 2 \rangle$, find $\mathbf{V} + \mathbf{W}$.

Solution
$$\begin{aligned} \mathbf{V} + \mathbf{W} &= \langle 3, -5 \rangle + \langle 1, 2 \rangle \\ &= \langle 3 + 1, -5 + 2 \rangle \\ &= \langle 4, -3 \rangle \end{aligned}$$

If k is a real number and $\mathbf{V}$ is a vector, we can define a type of multiplication called **scalar multiplication.**

(6.17)

> **Definition.** If k is a scalar and $\mathbf{V}$ is the vector $\langle a, b \rangle$, then
> $$k\mathbf{V} = k\langle a, b \rangle = \langle ka, kb \rangle$$

Example 3 If $\mathbf{V} = \langle 3, -5 \rangle$, find **a.** $2\mathbf{V}$ and **b.** $-8\mathbf{V}$.

Solution **a.**
$$\begin{aligned} 2\mathbf{V} &= 2\langle 3, -5 \rangle \\ &= \langle 6, -10 \rangle \end{aligned}$$

b.
$$\begin{aligned} -8\mathbf{V} &= -8\langle 3, -5 \rangle \\ &= \langle -24, 40 \rangle \end{aligned}$$

Note that, if k is a real number and **V** is a vector, then the product k**V** is also a vector. Its norm is $|k|$ times the norm of **V** itself. If k is a positive real number, then k**V** has the same direction as **V**. If k is a negative real number, then k**V** has the opposite direction of **V**. See Figure 6-34. The vector k**V** is called a **scalar multiple** of **V**.

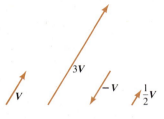

Figure 6-34

Let **i** be the vector $\langle 1, 0 \rangle$. Note that **i** is a vector of length 1 unit, pointing in the positive x direction. Similarly, let **j** be the vector $\langle 0, 1 \rangle$. The vector **j** is of length 1 unit, and it points in the positive y direction. Any vector can be written as the sum of scalar multiples of these vectors **i** and **j**. For example, the vector $\langle 5, 2 \rangle$ can be written in this form by proceeding as follows.

$$\langle 5, 2 \rangle = \langle 5, 0 \rangle + \langle 0, 2 \rangle$$
$$= 5\langle 1, 0 \rangle + 2\langle 0, 1 \rangle$$
$$= 5\mathbf{i} + 2\mathbf{j}$$

The two vectors $5\mathbf{i}$ and $2\mathbf{j}$ are called the **x-** and the **y-components** of the vector $\langle 5, 2 \rangle$. See Figure 6-35.

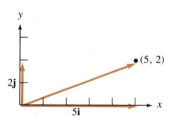

Figure 6-35

(6.18)   **Definition.**   The vectors $\mathbf{i} = \langle 1, 0 \rangle$ and $\mathbf{j} = \langle 0, 1 \rangle$ are called **unit coordinate vectors.**

Any vector $\langle x, y \rangle$ is **resolved** into its **x-** or **horizontal component** $x\mathbf{i}$, and its **y-** or **vertical component** $y\mathbf{j}$ when it is written in the form $x\mathbf{i} + y\mathbf{j}$.

Example 4   Let $V = 2i + 3j$ and $W = 4i - j$. Calculate **a.** $5V + 3W$ and **b.** $V - W$.

Solution   **a.** $5V + 3W = 5(2i + 3j) + 3(4i - 1j)$

$$= (10i + 15j) + (12i - 3j)$$

$$= 22i + 12j$$

b. $V - W = V + (-1)W$

$$= (2i + 3j) + (-1)(4i - 1j)$$

$$= (2i + 3j) + (-4i + 1j)$$

$$= -2i + 4j$$

The definition of scalar multiplication provides the way to multiply a vector by a real number. We now define a way, called the **dot product,** to multiply one vector by another.

(6.19)

> **Definition.**   The **dot product** of vectors V and W is the *scalar*
>
> $V \cdot W = |V| \, |W| \cos \theta$
>
> where θ is the angle between V and W.

It is not convenient to calculate the dot product of two vectors by using Definition (6.19) directly. It is easy, however, to calculate a dot product by using a theorem which we state here without proof.

(6.20)

> **Theorem.**   Let $V = ai + bj$ and $W = ci + dj$. Then
>
> $V \cdot W = ac + bd$

Example 5   If $A = 2i + 3j$ and $B = 5i - 4j$, calculate $A \cdot B$.

Solution $A \cdot B = (2i + 3j) \cdot (5i - 4j)$

$$= 2 \cdot 5 + 3 \cdot (-4)$$

$$= -2$$

Example 6   Find the angle between $A = 3i + 4j$ and $B = 5i - 12j$.

Solution   See Figure 6-36. By definition, $A \cdot B = |A| \, |B| \cos \theta$, where θ is the angle between the vectors. Solve for $\cos \theta$, and proceed as follows:

$$\mathbf{A} \cdot \mathbf{B} = |\mathbf{A}| \, |\mathbf{B}| \cos \theta$$

$$\cos \theta = \frac{\mathbf{A} \cdot \mathbf{B}}{|\mathbf{A}| \, |\mathbf{B}|}$$

$$= \frac{(3\mathbf{i} + 4\mathbf{j}) \cdot (5\mathbf{i} - 12\mathbf{j})}{\sqrt{3^2 + 4^2} \, \sqrt{5^2 + (-12)^2}}$$

$$= \frac{15 - 48}{5 \cdot 13}$$

$$\cos \theta = \frac{-33}{65}$$

$$\theta = \cos^{-1}\left(\frac{-33}{65}\right)$$

$$\theta = 120.5°$$

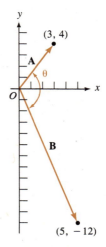

Figure 6-36

If the dot product, $|\mathbf{A}| \, |\mathbf{B}| \cos \theta$, of two nonzero vectors **A** and **B** is 0, then $\cos \theta$ must be 0. If $\cos \theta = 0$, then $\theta = 90°$, and the two vectors must be perpendicular. Thus, the dot product provides a test for the perpendicularity of two vectors.

(6.21)   **Theorem.**   Two nonzero vectors are perpendicular if and only if their dot product is zero.

Example 7   Are the vectors $\mathbf{A} = 6\mathbf{i} - 2\mathbf{j}$ and $\mathbf{B} = \mathbf{i} + 3\mathbf{j}$ perpendicular?

Solution   Calculate $\mathbf{A} \cdot \mathbf{B}$. If $\mathbf{A} \cdot \mathbf{B} = 0$, the vectors are perpendicular. If $\mathbf{A} \cdot \mathbf{B} \neq 0$, the vectors are not perpendicular.

$$\mathbf{A} \cdot \mathbf{B} = (6\mathbf{i} - 2\mathbf{j}) \cdot (\mathbf{i} + 3\mathbf{j})$$
$$= 6 \cdot 1 + (-2)(+3)$$
$$= 6 - 6$$
$$= 0$$

Because the dot product is 0, the vectors **A** and **B** are perpendicular.

Example 8   What are the horizontal and vertical components of a 2.0-lb force that makes an angle of 30° with the x-axis? Express the 2.0-lb force in $a\mathbf{i} + b\mathbf{j}$ form.

Solution   The horizontal component of the given force is vector **OA**, which is one leg of the right triangle *OAC*. See Figure 6-37.

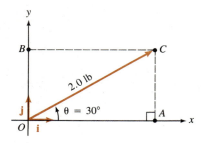

Figure 6-37

The norm of **OA** is found as follows.

$$|\mathbf{OA}| = |\mathbf{OC}| \cos \theta$$
$$= 2.0(\cos 30°)$$
$$= 2.0 \frac{\sqrt{3}}{2}$$
$$= \sqrt{3}$$

The horizontal component is $\sqrt{3}$ lb.

Similarly, the vertical component **OB** is found as follows.

$$|\mathbf{OB}| = |\mathbf{OC}| \sin \theta$$
$$= 2.0(\sin 30°)$$
$$= 1.0$$

The vertical component is 1.0 lb.

Thus, you have $\mathbf{OC} = \sqrt{3}\, \mathbf{i} + \mathbf{j}$. ■

Example 9 A force of 2.0 lb makes an angle of 30° with the horizontal. What is the component of this force in the direction $\mathbf{OB} = 12\mathbf{i} + 5\mathbf{j}$?

Solution You must find the component of the given force in a direction other than that of an axis. See Figure 6-38.

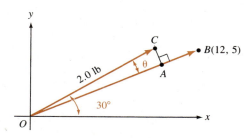

Figure 6-38

You must find the norm of vector **OA**. In right triangle OAC, the following relation holds.

$$|\mathbf{OA}| = |\mathbf{OC}| \cos \theta$$

Use the dot product to make this calculation. Proceed as follows.

$$
\begin{aligned}
|\mathbf{OA}| &= |\mathbf{OC}| \cos \theta \\
&= \frac{|\mathbf{OC}||\mathbf{OB}| \cos \theta}{|\mathbf{OB}|} \qquad \text{Multiply and divide by } |\mathbf{OB}|. \\
&= \frac{\mathbf{OC} \cdot \mathbf{OB}}{|\mathbf{OB}|} \qquad \text{Use the definition of dot product.} \\
&= \frac{(\sqrt{3}\mathbf{i} + \mathbf{j}) \cdot (12\mathbf{i} + 5\mathbf{j})}{\sqrt{12^2 + 5^2}} \qquad \text{Use the result from Example 8, and} \\
&\qquad\qquad\qquad\qquad\qquad\qquad \mathbf{OB} = 12\mathbf{i} + 5\mathbf{j}. \\
|\mathbf{OA}| &= \frac{12\sqrt{3} + 5}{13}
\end{aligned}
$$

The component of the 2.0-lb force in the direction of **OB** is $\dfrac{12\sqrt{3} + 5}{13}$. ■

Exercise 6.6

In Exercises 1–12, let $\mathbf{U} = \langle 2, -3 \rangle$, $\mathbf{V} = \langle 5, -2 \rangle$, *and* $\mathbf{W} = \langle -1, 1 \rangle$. *Calculate each quantity.*

1. $\mathbf{U} + \mathbf{V}$
2. $\mathbf{V} + \mathbf{W}$
3. $3\mathbf{U}$
4. $5\mathbf{V}$
5. $2\mathbf{U} + \mathbf{V}$
6. $3\mathbf{U} - \mathbf{W}$
7. $|\mathbf{U}|$
8. $|3\mathbf{U}|$
9. $|\mathbf{U} + \mathbf{W}|$
10. $|\mathbf{V} - \mathbf{W}|$
11. $|\mathbf{U}| + |\mathbf{W}|$
12. $|3\mathbf{V} - 5\mathbf{W}|$

In Exercises 13–18, resolve each vector into its horizontal and its vertical components by writing each vector in $a\mathbf{i} + b\mathbf{j}$ *form.*

13. $\langle 3, 5 \rangle + \langle 5, 3 \rangle$

14. $\langle -2, 7 \rangle + \langle 2, 3 \rangle$

15. A vector of length 10, making an angle of 30° with the x-axis.

16. A vector of length 10, making an angle of 45° with the x-axis.

17. A vector of length 23.3, making an angle of 37.2° with the x-axis.

18. A vector of length 19.1, making an angle of 183.7° with the x-axis.

In Exercises 19–24, find the dot product of the two given vectors.

19. $\langle 2, -3 \rangle$ and $\langle 3, -1 \rangle$
20. $\langle 1, -5 \rangle$ and $\langle 5, 1 \rangle$
21. $2\mathbf{i} + 5\mathbf{j}$ and $\mathbf{i} + \mathbf{j}$
22. $3\mathbf{i} - 3\mathbf{j}$ and $2\mathbf{i} + \mathbf{j}$
23. $\mathbf{i}$ and $\mathbf{j}$
24. $2\mathbf{i}$ and $3\mathbf{i}$

In Exercises 25–30, find the angle between the given vectors.

25. $\langle 2, 2 \rangle$ and $\langle 5, 0 \rangle$ **26.** $\langle \sqrt{3}, 1 \rangle$ and $\langle 3, 3 \rangle$

27. $\langle \sqrt{3}, -1 \rangle$ and $\langle -1, \sqrt{3} \rangle$ **28.** $2\mathbf{i} + 3\mathbf{j}$ and $-3\mathbf{i} + 2\mathbf{j}$

29. $3\mathbf{i} - \mathbf{j}$ and $3\mathbf{i} + \mathbf{j}$ **30.** $3\mathbf{i} - 4\mathbf{j}$ and $5\mathbf{i} + 12\mathbf{j}$

In Exercises 31–36, indicate whether the given vectors are perpendicular.

31. $\langle 2, 3 \rangle$ and $\langle -3, 2 \rangle$ **32.** $\langle 2, 3 \rangle$ and $\langle 3, 2 \rangle$

33. $5\mathbf{i} + \mathbf{j}$ and $\mathbf{i} + \mathbf{j}$ **34.** $6\mathbf{i} - 2\mathbf{j}$ and $\mathbf{i} + 3\mathbf{j}$

35. $\mathbf{i}$ and $\mathbf{j}$ **36.** $-\mathbf{i}$ and $3\mathbf{i}$

In Exercises 37–40, find the component of the first vector in the direction of the second vector.

37. $\langle 3, 4 \rangle$, $\langle 5, 12 \rangle$ **38.** $\langle 1, 1 \rangle$, $\langle 3, 2 \rangle$

39. $6\mathbf{i} + 8\mathbf{j}$, $4\mathbf{i} - 3\mathbf{j}$ **40.** $\mathbf{i}$, $\mathbf{i} + \mathbf{j}$

41. Find an example to illustrate that $|\mathbf{U} + \mathbf{V}| \neq |\mathbf{U}| + |\mathbf{V}|$.

42. Find an example to support the distributive law, $(a + b)\mathbf{V} = a\mathbf{V} + b\mathbf{V}$.

43. Find an example to support the distributive law, $a(\mathbf{V} + \mathbf{W}) = a\mathbf{V} + a\mathbf{W}$.

44. Find an example to support the associative law, $a(\mathbf{V} \cdot \mathbf{W}) = (a\mathbf{V}) \cdot \mathbf{W}$.

45. Let $\mathbf{V} = a\mathbf{i} + b\mathbf{j}$. Prove that $\mathbf{V} \cdot \mathbf{V} = |\mathbf{V}|^2$.

CHAPTER SUMMARY

Key Words

ambiguous case (6.3)   *norm of a vector* (6.6)
dot product (6.6)   *oblique triangles* (6.1)
Heron's formula (6.5)   *scalars* (6.6)
law of cosines (6.1)   *scalar multiplication* (6.6)
law of sines (6.2)   *semiperimeter* (6.5)
law of tangents (6.4)   *unit vectors* (6.6)
Mollweide's equations (6.4)   *vectors* (6.6)

Key Ideas

(6.1)   The law of cosines.
In triangle *ABC* with sides of *a*, *b*, and *c*,

$$a^2 = b^2 + c^2 - 2bc \cos A$$
$$b^2 = c^2 + a^2 - 2ca \cos B$$

and

$$c^2 = a^2 + b^2 - 2ab \cos C$$

The law of cosines can be used in the SAS and SSS cases.

(6.2) The law of sines.
In triangle ABC with sides of a, b, and c,

$$\frac{a}{\sin A} = \frac{b}{\sin B} = \frac{c}{\sin C}$$

The law of sines can be used in the AAS, ASA, and SSA cases.

(6.3) When given information in the ambiguous SSA case, sketch a careful scale drawing to determine whether two, one, or no triangles exist.

(6.4) Mollweide's equations.
(Optional)

$$\frac{a+b}{c} = \frac{\cos\frac{1}{2}(A-B)}{\sin\frac{1}{2}C}$$

$$\frac{a-b}{c} = \frac{\sin\frac{1}{2}(A-B)}{\cos\frac{1}{2}C}$$

The law of tangents.

$$\frac{a-b}{a+b} = \frac{\tan\frac{1}{2}(A-B)}{\tan\frac{1}{2}(A+B)}$$

(6.5) Formulas for the area K of a triangle with vertices at A, B, and C and sides of a, b, and c.

$$K = \frac{1}{2}bh \qquad \text{Where } b \text{ is the base and } h \text{ is the altitude of the triangle}$$

$$K = \frac{1}{2}ba\sin C$$

$$K = \frac{1}{2}cb\sin A$$

$$K = \frac{1}{2}ac\sin B$$

$$K = \frac{c^2\sin A\sin B}{2\sin C}$$

$$K = \frac{a^2\sin B\sin C}{2\sin A}$$

$$K = \frac{b^2 \sin C \sin A}{2 \sin B}$$

$$K = \sqrt{s(s-a)(s-b)(s-c)}$$ With s equal to half the triangle's perimeter (s is the semiperimeter)

(6.6)
(Optional)   Two vectors are equal if and only if they have the same length and the same direction.

If $\mathbf{V} = \langle a, b \rangle$, $\mathbf{W} = \langle c, d \rangle$, and k is a real number, then

$$|\mathbf{V}| = \sqrt{a^2 + b^2}$$
$$\mathbf{V} + \mathbf{W} = \langle a + c, b + d \rangle$$
$$k\mathbf{V} = \langle ka, kb \rangle$$
$$\mathbf{V} \cdot \mathbf{W} = |\mathbf{V}||\mathbf{W}| \cos \theta = ac + bd$$
$$\mathbf{i} = \langle 1, 0 \rangle$$
$$\mathbf{j} = \langle 0, 1 \rangle$$

Two nonzero vectors are perpendicular if and only if their dot product is zero.

REVIEW EXERCISES

In Review Exercises 1–3, state the indicated law in your own words.

1. The law of cosines.

2. The law of sines.

3. Heron's formula for the area of a triangle.

4. Explain why SSA is called the *ambiguous case*.

In Review Exercises 5–16, consider the given parts of triangle ABC. Use your calculator to solve for the required value, if possible. If more than one value is possible, give both.

5. $a = 12$, $c = 15$, $B = 30°$; find b.

6. $b = 23$, $a = 13$, $C = 125°$; find c.

7. $c = 0.5$, $b = 0.8$, $A = 50°$; find a.

8. $a = 28.7$, $b = 37.8$, $C = 11.2°$; find c.

9. $a = 12$, $c = 18$, $C = 40°$; find A.

10. $b = 17$, $a = 12$, $A = 25°$; find B.

11. $c = 31.5$, $b = 27.5$, $B = 16.2°$; find C.

12. $a = 315.2$, $b = 457.8$, $A = 32.51°$; find B.

13. $A = 24.3°$, $B = 56.8°$, $a = 32.3$; find b.

14. $B = 10.3°$, $C = 59.4°$, $c = 341$; find b.

15. $b = 17$, $c = 21$, $B = 42°$; find A.

16. $c = 189$, $a = 150$, $C = 85.3°$; find B.

In Review Exercises 17–18, use Mollweide's equations to verify that a triangle exists with the given parts. These exercises come from an optional section.

17. $A = 30°, B = 60°, C = 90°, a = 28.9, b = 50.1, c = 57.8$

18. $A = 22°, B = 54°, C = 104°, a = 75, b = 160, c = 194.3$

In Review Exercises 19–26, find the area of the triangle with the given parts, if possible.

19. $a = 32, C = 47°, b = 55$ **20.** $b = 29, A = 96°, c = 85$

21. $B = 33°, C = 25°, a = 17$ **22.** $A = 85°, C = 80°, b = 7.5$

23. $A = 130°, B = 20°, a = 3.5$ **24.** $C = 15°, A = 110°, c = 91$

25. $a = 57, b = 85, c = 110$ **26.** $a = 8.50, b = 17.4, c = 22.3$

27. Two airplanes leave an airport at 2:00 PM, one with a heading of 30.0° and ground-speed of 425 mph and the other with a heading of 85.0° and groundspeed of 375 mph. How far apart are they at 3:30 PM?

28. Find the angles of the triangle with vertices of (0, 0), (5, 0), and (7, 8).

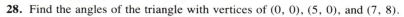

29. From a point 542 ft away from the base of the leaning tower of Pisa, the angle of elevation to its top is 17.9°. If the tower makes an angle with the ground of 94.5° (see Illustration 1), how tall is the tower?

Illustration 1

30. From a point 312 ft from the base of the great pyramid of Khufu (Cheops) at Gizeh, the angle of elevation to its top is 25.5°. If the pyramid makes an angle with the ground of 141.8° (see Illustration 2), find its slant height.

31. Find the area of a triangular lot with sides of 21, 32, and 47 m.

32. Find the area of the triangle joining points with coordinates of (0, 0), (0, 8), and (6, 14).

In Review Exercises 33–36, assume that $\mathbf{V} = \langle 3, 7 \rangle$ and $\mathbf{W} = \langle -2, 5 \rangle$. Calculate each quantity. These exercises come from an optional section.

33. $2\mathbf{V} + 3\mathbf{W}$ **34.** $|3\mathbf{V}| - |\mathbf{V}|$

35. $5(\mathbf{V} - \mathbf{W})$ **36.** $|3\mathbf{V} - \mathbf{W}|$

In Review Exercises 37–40, find the angle between the two given vectors. These exercises come from an optional section.

Illustration 2

37. $\langle 0, 5 \rangle$ and $\langle 2, 0 \rangle$ **38.** $\langle 8, 2 \rangle$ and $\langle 4, 1 \rangle$

39. $\sqrt{3}\,\mathbf{i} + \mathbf{j}$ and $\mathbf{i} - \mathbf{j}$ **40.** $2\mathbf{i} - 2\sqrt{3}\mathbf{j}$ and $\mathbf{i} + \sqrt{3}\mathbf{j}$

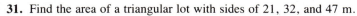

CHAPTER SEVEN

COMPLEX NUMBERS AND POLAR COORDINATES

All of the numbers used thus far in this book have been real numbers. However, some situations in mathematics, such as solving certain quadratic equations, require a new set of numbers. This new set, called the set of **complex numbers,** is the main topic of this chapter.

7.1 COMPLEX NUMBERS

If we solve the equation $x^2 = 9$ for x, we find that the equation has two solutions.

$$x^2 = 9$$
$$x = \sqrt{9} \quad \text{or} \quad x = -\sqrt{9}$$
$$= 3 \quad \quad = -3$$

If we solve the equation $x^2 = -1$ in a similar way, we would expect it to have two solutions also.

$$x^2 = -1$$
$$x = \sqrt{-1} \quad \text{or} \quad x = -\sqrt{-1}$$

Each proposed solution of the equation $x^2 = -1$ involves the symbol $\sqrt{-1}$. The symbol $\sqrt{-1}$ cannot represent a real number, because no real number has a square that is equal to -1. For years, mathematicians believed that square roots of negative numbers, denoted by symbols such as $\sqrt{-3}$, $\sqrt{-4}$, and $\sqrt{-5}$, were nonsense. Even the great English mathematician Sir Isaac Newton (1642–1727) called them "impossible numbers." In the 17th century, these symbols were termed **imaginary numbers** by René Descartes (1596–1650). However, mathematicians no longer think of imaginary numbers as impossible. In fact, imaginary numbers have important uses, such as describing the behavior of alternating current in electronics.

Powers of i
The imaginary number $\sqrt{-1}$ occurs often enough to warrant a special symbol: the letter i is used to denote $\sqrt{-1}$. Because i is used to denote $\sqrt{-1}$, it follows that $i^2 = -1$. The powers of i with natural-number exponents produce an interesting pattern.

$$i = \sqrt{-1} = \boldsymbol{i} \qquad\qquad i^5 = i^4 \cdot i = 1 \cdot i = \boldsymbol{i}$$
$$i^2 = \sqrt{-1}\,\sqrt{-1} = \boldsymbol{-1} \qquad i^6 = i^4 \cdot i^2 = 1(-1) = \boldsymbol{-1}$$
$$i^3 = i^2 \cdot i = -1 \cdot i = \boldsymbol{-i} \qquad i^7 = i^4 \cdot i^3 = 1(-i) = \boldsymbol{-i}$$
$$i^4 = i^2 \cdot i^2 = (-1)(-1) = \boldsymbol{1} \qquad i^8 = i^4 \cdot i^4 = 1(1) = \boldsymbol{1}$$

The pattern continues $\boldsymbol{i}$, $\boldsymbol{-1}$, $\boldsymbol{-i}$, $\boldsymbol{1}$,

We can simplify powers of i involving negative integral exponents by rationalizing denominators.

$$i^{-1} = \frac{1}{i} = \frac{1}{i} \cdot \frac{\boldsymbol{i}}{\boldsymbol{i}} = \frac{i}{-1} = -i$$

$$i^{-2} = \frac{1}{i^2} = \frac{1}{-1} = -1$$

$$i^{-3} = \frac{1}{i^3} = \frac{1}{i^3} \cdot \frac{i}{i} = \frac{i}{1} = i$$

$$i^{-4} = \frac{1}{i^4} = \frac{1}{1} = 1$$

If we define i^0 to be 1, then the familiar pattern for the powers of i carries over to integral exponents.

$$\vdots$$

$$i^{-3} = i$$
$$i^{-2} = -1$$
$$i^{-1} = -i$$
$$i^0 = 1$$
$$i^1 = i$$
$$i^2 = -1$$
$$i^3 = -i$$
$$i^4 = 1$$

$$\vdots$$

The sequence i, -1, $-i$, 1 repeats endlessly. Because the powers of i form a repeating sequence, it is easy to compute large powers of i.

Example 1   Simplify i^{365}.

Solution $i^{365} = (i^4)^{91} \cdot i^1 = 1^{91} \cdot i = i$

Note that the number 365 leaves a remainder of 1 when it is divided by 4.   ■

Example 2   Suppose that i is raised to the nth power, where n is a natural number. Discuss the possible values for i^n.

Solution   If the exponent n of i^n is a multiple of 4, then $i^n = 1$.
If n leaves a remainder of 1 when divided by 4, then $i^n = i$.
If n leaves a remainder of 2 when divided by 4, then $i^n = -1$.
If n leaves a remainder of 3 when divided by 4, then $i^n = -i$.   ■

Recall that, if at least one of two numbers x and y is not negative, then $\sqrt{xy} = \sqrt{x}\,\sqrt{y}$. This allows us to write the square root of any negative number in the form bi, where b is a real number and $i = \sqrt{-1}$. For example, we have

$$\sqrt{-4} = \sqrt{4(-1)} = \sqrt{4}\,\sqrt{-1} = 2i$$
$$\sqrt{-9} = \sqrt{9(-1)} = \sqrt{9}\,\sqrt{-1} = 3i$$

and

$$\sqrt{-2} = \sqrt{2(-1)} = \sqrt{2}\sqrt{-1} = \sqrt{2}\,i$$

Because a square root of a number is one of two equal factors of that number, we know that $\sqrt{-3}\sqrt{-3} = -3$. It is incorrect to apply the rule $\sqrt{xy} = \sqrt{x}\sqrt{y}$ in this case: $\sqrt{-3}\sqrt{-3} \neq \sqrt{(-3)(-3)} = \sqrt{9} = 3$. This example illustrates that, if x and y are both negative, then the rule $\sqrt{xy} = \sqrt{x}\sqrt{y}$ is not valid.

Expressions such as $3 + 4i$, $-5 + 7i$, and $-1 + 9i$ indicate the sum of a real number and an imaginary number. Such expressions form a new set of numbers called the set of **complex numbers.**

(7.1) | **Definition.** A **complex number** is any number that can be expressed in the form $a + bi$, where a and b are real numbers and $i = \sqrt{-1}$. The number a is called the **real part** and the number b is called the **imaginary part** of the complex number $a + bi$.

If $b = 0$, the complex number $a + bi$ is the real number a. Thus, any real number is a complex number with a zero imaginary part. If $a = 0$ and $b \neq 0$, the complex number $a + bi$ is the imaginary number bi. Thus, any imaginary number is a complex number with a zero real part. It follows that the real-number set and the imaginary-number set are both subsets of the complex-number set.

We must accept some definitions before doing any arithmetic with complex numbers.

(7.2) | **Equality of Complex Numbers.** Two complex numbers are equal if and only if their real parts are equal and their imaginary parts are equal. Thus, if $a + bi$ and $c + di$ are two complex numbers, then

$a + bi = c + di$ if and only if $a = c$ and $b = d$

Example 3   For what real-number values of x and y is $3x + 4i = (2y + x) + xi$?

Solution   Because the imaginary parts of these two complex numbers must be equal, $x = 4$. Because the real parts of these two complex numbers must be equal, $3x = 2y + x$. You can solve the system of equations

$$\begin{cases} x = 4 \\ 3x = 2y + x \end{cases}$$

by substituting 4 for x in the equation $3x = 2y + x$ and solving for y. You will find that $y = 4$. Hence, the solution is $x = 4$ and $y = 4$. ∎

Complex numbers can be added and multiplied as if they were simple binomials.

(7.3)

> **Addition of Complex Numbers.**   Two complex numbers such as $a + bi$ and $c + di$ are added as if they were algebraic binomials:
>
> $$(a + bi) + (c + di) = (a + c) + (b + d)i$$

Note that Definition (7.3) implies that the sum of two complex numbers is another complex number.

Example 4   Find the sum of $3 + 4i$ and $2 + 7i$.

Solution
$$(3 + 4i) + (2 + 7i) = 3 + 4i + 2 + 7i$$
$$= 3 + 2 + 4i + 7i$$
$$= 5 + 11i$$

(7.4)

> **Multiplication of Complex Numbers.**   Two complex numbers such as $a + bi$ and $c + di$ are multiplied as if they were algebraic binomials, with $i^2 = -1$:
>
> $$(a + bi)(c + di) = (ac - bd) + (ad + bc)i$$

Note that Definition (7.4) implies that the product of two complex numbers is another complex number.

Example 5   Find the product of $3 + 4i$ and $2 + 7i$.

Solution
$$(3 + 4i)(2 + 7i) = 6 + 21i + 8i + 28i^2$$
$$= 6 + 21i + 8i - 28$$
$$= -22 + 29i$$

When working with complex numbers, always express all numbers in $a + bi$ form before attempting any algebraic manipulations. This procedure will help you avoid making errors in determining the sign of the result.

Example 6   Find the product of $-2 + \sqrt{-16}$ and $4 - \sqrt{-9}$.

Solution   First, change each complex number to $a + bi$ form.
$$-2 + \sqrt{-16} = -2 + \sqrt{16}\,\sqrt{-1}$$
$$= -2 + 4i$$
$$4 - \sqrt{-9} = 4 - \sqrt{9}\,\sqrt{-1}$$
$$= 4 - 3i$$

Then, find the product of $-2 + 4i$ and $4 - 3i$.

$$(-2 + 4i)(4 - 3i) = -8 + 6i + 16i - 12i^2$$
$$= -8 + 6i + 16i + 12$$
$$= 4 + 22i$$

The following definition is useful in finding the quotient of two complex numbers.

(7.5)
> **Definition.**   The complex numbers $a + bi$ and $a - bi$ are called **conjugates** of each other.

The conjugate of $3 + 4i$ is $3 - 4i$, and the conjugate of $-\frac{1}{2} - 4i$ is $-\frac{1}{2} + 4i$.

Example 7   Write the number $\dfrac{3}{2 + i}$ in $a + bi$ form.

Solution   To write the given number in $a + bi$ form, multiply both the numerator and the denominator by the conjugate of the denominator, and simplify.

$$\frac{3}{2 + i} = \frac{3}{2 + i} \cdot \frac{2 - i}{2 - i}$$

$$= \frac{6 - 3i}{4 - 2i + 2i - i^2}$$

$$= \frac{6 - 3i}{4 + 1}$$

$$= \frac{6}{5} - \frac{3}{5}i$$

$$= \frac{6}{5} + \left(-\frac{3}{5}\right)i$$

It is common practice to accept $\frac{6}{5} - \frac{3}{5}i$ as a substitute for $\frac{6}{5} + \left(-\frac{3}{5}\right)i$.

Example 8   Write the number $\dfrac{2 - \sqrt{-64}}{3 + \sqrt{-1}}$ in $a + bi$ form.

Solution
$$\frac{2 - \sqrt{-64}}{3 + \sqrt{-1}} = \frac{2 - 8i}{3 + i}$$

$$= \frac{2 - 8i}{3 + i} \cdot \frac{3 - i}{3 - i}$$

$$= \frac{6 - 2i - 24i + 8i^2}{9 - 3i + 3i - i^2}$$

$$= \frac{-2 - 26i}{9 + 1}$$

$$= \frac{2(-1 - 13i)}{10}$$

$$= -\frac{1}{5} - \frac{13}{5}i$$

A generalization of the process involved in Examples 7 and 8 shows that the quotient of two complex numbers is another complex number.

The solutions of certain quadratic equations are complex numbers. The following example shows how the quadratic formula can be used to find these solutions.

Example 9   Use the quadratic formula to solve the quadratic equation $x^2 - 4x + 5 = 0$.

Solution   Substitute **1** for a, **− 4** for b, and **5** for c in the quadratic formula, and simplify.

$$x = \frac{-b \pm \sqrt{b^2 - 4ac}}{2a}$$

$$= \frac{-(-4) \pm \sqrt{(-4)^2 - 4(1)(5)}}{2(1)}$$

$$= \frac{4 \pm \sqrt{16 - 20}}{2}$$

$$= \frac{4 \pm \sqrt{-4}}{2}$$

$$= \frac{4 \pm 2i}{2}$$

$$= 2 \pm i$$

The two solutions of the equation $x^2 - 4x + 5 = 0$ are $x = 2 + i$ and $x = 2 - i$.

Exercise 7.1

In Exercises 1–12, simplify each expression.

1. i^9 　　　　　　　　　　　　　　　**2.** i^{27}

3. i^{38} 　　　　　　　　　　　　　　**4.** i^{99}

5. $\dfrac{1}{i^5}$

6. $\dfrac{1}{i^{10}}$

7. $\dfrac{1}{i^{99}}$

8. $\dfrac{1}{i^{1776}}$

9. i^{-1984}

10. i^{-1492}

11. i^{-1111}

12. i^{-2001}

In Exercises 13–18, solve for the real numbers x and y. You may have to solve a system of equations after using the definition of equality of complex numbers.

13. $x + (x + y)i = y - i$

14. $x + 3y + 2xi + yi = 5x + 2 + 2yi$

15. $x + iy = y + 2xi$

16. $x + y + (x - y)i = x + iy$

17. $3x - 2yi = 2 + (x + y)i$

18. $\begin{cases} 2 + (x + y)i = 2 - i \\ \quad x + 3i = 2 + 3i \end{cases}$

In Exercises 19–46, perform any indicated operations, and express the final answer in a + bi form.

19. $(2 - 7i) + (3 + i)$

20. $(-7 + 2i) + (2 - 8i)$

21. $(5 - 6i) - (7 + 4i)$

22. $(11 + 2i) - (13 - 5i)$

23. $(14i + 2) + (2 - 4i)$

24. $(5 + 8i) - (23i - 32)$

25. $(2 + \sqrt{-9})(-3 - \sqrt{-16})$

26. $(-2 - \sqrt{-16})(4 + \sqrt{-25})$

27. $(-11 + \sqrt{-25})(-2 - \sqrt{-36})$

28. $(6 + \sqrt{-49})(6 - \sqrt{-49})$

29. $(\sqrt{-16} + 3)(2 + \sqrt{-9})$

30. $(12 - \sqrt{-4})(\sqrt{-25} + 7)$

31. $(2 + 3i)^2$

32. $(3 - 4i)^2$

33. $(2 - i)^3$

34. $(2 + 3i)^3$

35. $\dfrac{1}{2 + i}$

36. $\dfrac{-2}{3 - i}$

37. $\dfrac{3}{4i^2}$

38. $\dfrac{-11}{3i^3}$

39. $\dfrac{2i}{7 + i}$

40. $\dfrac{-3i}{2 + 5i}$

41. $\dfrac{2 + i}{3 - i}$

42. $\dfrac{4 - \sqrt{-25}}{2 + \sqrt{-9}}$

43. $\dfrac{-\sqrt{-16} + \sqrt{5}}{-8 + \sqrt{-4}}$

44. $\dfrac{34 + 2i}{\sqrt{2} - 4i}$

45. $\dfrac{2 + i\sqrt{3}}{3 + i}$

46. $\dfrac{3}{4 - i\sqrt{2}}$

47. Verify that $1 - i$ is a square root of $-2i$ by showing that $(1 - i)^2 = -2i$.

48. Verify that $2 - i$ is a square root of $3 - 4i$ by showing that $(2 - i)^2 = 3 - 4i$.

49. Verify that $1 + 2i$ is a square root of $-3 + 4i$ by showing that $(1 + 2i)^2 = -3 + 4i$.

50. Verify that $2 + i$ is a square root of $3 + 4i$ by showing that $(2 + i)^2 = 3 + 4i$.

51. Verify that $-\dfrac{1}{2} + \dfrac{\sqrt{3}}{2} i$ and $-\dfrac{1}{2} - \dfrac{\sqrt{3}}{2} i$ are roots of $x^2 + x + 1 = 0$.

52. Verify that $\dfrac{1}{2} + \dfrac{\sqrt{5}}{2} i$ and $\dfrac{1}{2} - \dfrac{\sqrt{5}}{2} i$ are not roots of $x^2 - x - 1 = 0$.

In Exercises 53–58, use the quadratic formula to solve each equation.

53. $x^2 + 2x + 2 = 0$ **54.** $x^2 + 4x + 8 = 0$

55. $x^2 + 4x + 5 = 0$ **56.** $x^2 + 2x + 5 = 0$

57. $9x^2 - 6x + 2 = 0$ **58.** $4x^2 - 4x + 5 = 0$

In Exercises 59–62, use a calculator to write each expression in $a + bi$ form.

59. $(2.3 + 4.5i)(8.9 - 3.2i)$ **60.** $(6.73 - 3.25i)^2 + (1.75 + 2.21i)$

61. $\dfrac{4.7 + 11.2i}{5.2 - 3.7i} - (6.5 - 7.4i)$ **62.** $\dfrac{29.8 - 45.3i}{-7.4 + 27.3i}$

63. Show that the product of the complex number $a + bi$ and its conjugate $a - bi$ is the real number $a^2 + b^2$.

64. Show that the quotient $\dfrac{a + bi}{c + di}$ is a complex number.

7.2 GRAPHING COMPLEX NUMBERS

Although the complex numbers obey most of the properties of real numbers, there is one property that does not carry over: there is no way of ordering the complex numbers that is consistent with the ordering established for the real numbers. It makes no sense to say that $3i$ is larger than 5, that 0 is less than $3 + i$, or that $5 + 5i$ is greater than $-5 + i^2$. Because the complex numbers are not linearly ordered, it is impossible to graph them on a number line in any way that preserves the ordering of the real numbers. This inability to graph complex numbers on the number line is not surprising because real numbers provide coordinates for all points on the number line. However, there *is* a method for graphing the complex numbers.

Graphing complex numbers   Construct two perpendicular axes. Consider the horizontal axis to be the axis of real numbers and the vertical axis to be the axis of imaginary numbers. These axes determine what is called the **complex plane.** Although these axes resemble the x-axis and y-axis encountered in previous chapters, instead of plotting x and y values, we plot ordered pairs of real numbers (a, b), where a and b are the parts of the complex number $a + bi$. The axis of reals is used for plotting a and the axis of imaginaries is used for plotting b.

Example 1   Graph the complex number $3 + 2i$.

Solution   To graph the complex number $3 + 2i$, plot the point $P(3, 2)$ as in Figure 7-1. The vector drawn from the origin, point O, to point P is considered to be the graph of the complex number $3 + 2i$.

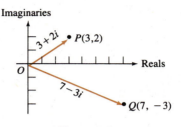

Figure 7-1

Example 2   Graph the complex number $7 - 3i$.

Solution   To graph the complex number $7 - 3i$, plot the point $Q(7, -3)$ as in Figure 7-1. The vector OQ is the graph of the complex number $7 - 3i$.

The length of the vector that represents a complex number is considered to be the **absolute value** of that complex number.

(7.6)   **Absolute Value of a Complex Number.**   If $a + bi$ is a complex number, then
$$|a + bi| = \sqrt{a^2 + b^2}$$

See Figure 7-2 and note that $\sqrt{a^2 + b^2}$ is the length of the hypotenuse of a right triangle with sides of length $|a|$ and $|b|$. On the complex plane, $|a + bi|$ is the distance between the point (a, b) and the origin $(0, 0)$. This is consistent with the definition of the absolute value of a real number. The expression $|x|$ represents the distance on the number line between the point with coordinate x and the origin O.

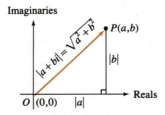

Figure 7-2

Example 3 Find the value of $|3 + 4i|$.

Solution $|3 + 4i| = \sqrt{3^2 + 4^2} = \sqrt{25} = 5$

Example 4 Evaluate $|2 - 5i|$.

Solution $|2 - 5i| = \sqrt{2^2 + (-5)^2} = \sqrt{29}$

Exercise 7.2

In Exercises 1–12, graph the given complex number.

1. $2 + 3i$ **2.** $2 - 3i$

3. $-2 + 3i$ **4.** $-2 - 3i$

5. $(4 + i)^2$ **6.** $(4 - i)^3$

7. $\dfrac{2 + i}{3 - i}$ **8.** $\dfrac{-2 + 2i}{-3 - i}$

9. 6 **10.** -8

11. $7i$ **12.** $-5i$

In Exercises 13–24, compute each absolute value.

13. $|2 + 3i|$ **14.** $|-5 - i|$

15. $|-7 + 7i|$ **16.** $\left| \dfrac{1}{2} - \dfrac{1}{4}i \right|$

17. $|6|$ **18.** $|-6|$

19. $|5i|$ **20.** $|-4i|$

21. $\left| \dfrac{-3i}{2 + i} \right|$ **22.** $\left| \dfrac{5i}{i - 2} \right|$

23. $\left| \dfrac{4 - i}{4 + i} \right|$ **24.** $|i^{365}|$

In Exercises 25–30, let $z = a + bi$. Assume that $\bar{z}$ represents the conjugate of z, that is, $\bar{z} = a - bi$.

25. Show that $|z| = |\bar{z}|$.   **26.** Show that $|z| + |\bar{z}| = 2|z|$.

27. Show that $|z|\,|\bar{z}| = |z|^2$.   **28.** Show that $|z + \bar{z}| = 2|a|$.

29. Show that $\sqrt{z\bar{z}} = |z|$.

30. Show that $z = \bar{z}$ if and only if z is a real number.

31. Show that the addition of two complex numbers is commutative. Do this by adding the complex numbers $a + bi$ and $c + di$ in both orders and observing that the sums are equal.

32. Show that the multiplication of two complex numbers is commutative. Do this by multiplying the complex numbers $a + bi$ and $c + di$ in both orders and observing that the products are equal.

33. Show that the addition of complex numbers is associative.

34. Find three examples of complex numbers that are reciprocals of their own conjugates.

In Exercises 35–36, let $x = a + bi$ and $y = c + di$. Assume that the bar over each symbol is read as "the conjugate of." For example, $\overline{x + y}$ means "the conjugate of the sum of x and y."

35. Show that $\overline{x + y} = \bar{x} + \bar{y}$.   **36.** Show that $\overline{xy} = \bar{x}\,\bar{y}$.

7.3   TRIGONOMETRIC FORM OF A COMPLEX NUMBER

If a and b are real numbers, then any complex number in the form $a + bi$ can be written in trigonometric form. Refer to Figure 7-3 and note that $\sin\theta = \frac{b}{r}$ or $b = r\sin\theta$. Similarly, $\cos\theta = \frac{a}{r}$ or $a = r\cos\theta$. Substitute these values for a and b in the complex number $a + bi$.

$$a + bi = r\cos\theta + r\sin\theta\, i$$
$$= r(\cos\theta + i\sin\theta)$$

Trigonometric form of a complex number   The expression $r(\cos\theta + i\sin\theta)$ is called the **trigonometric form** of the complex number $a + bi$. Refer to Figure 7-3 and note that $\tan\theta = \frac{b}{a}$.

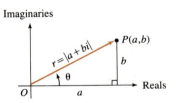

Figure 7-3

If $r = |a + bi|$ and θ is any angle for which $\sin\theta = \frac{b}{r}$ and $\cos\theta = \frac{a}{r}$, then

(7.7) $a + bi = r(\cos\theta + i\sin\theta)$

If a complex number is written in $r(\cos\theta + i\sin\theta)$ form, the number r is called the **modulus,** and the angle θ is called the **argument.** The notation $r(\cos\theta + i\sin\theta)$ is often abbreviated as r cis θ. Thus, we have

(7.8) $r\text{ cis }\theta = r(\cos\theta + i\sin\theta)$

Example 1   Write the complex number $3 + 4i$ in trigonometric form.

Solution   To write $3 + 4i$ in the form $r(\cos\theta + i\sin\theta)$, you must find the appropriate values for r and θ. To do this, graph the complex number as in Figure 7-4. Because triangle OAP is a right triangle, you have $r^2 = 3^2 + 4^2$ or $r = 5$. From the figure, it follows that $\tan\theta = \frac{4}{3}$, so $\theta \approx 53.1°$. Substituting these values for r and θ gives

$$3 + 4i \approx 5(\cos 53.1° + i\sin 53.1°)$$

Note that $5(\cos 53.1° + i\sin 53.1°)$ can be written more compactly as $5\ \text{cis}\ 53.1°$.

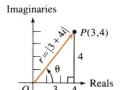

Figure 7-4

Example 2   Write the complex number $-2 - 7i$ in trigonometric form.

Solution   Again, to write $-2 - 7i$ in the form $r(\cos\theta + i\sin\theta)$, the appropriate values of r and θ must be found. Graph the number $-2 - 7i$ as in Figure 7-5.

$$r = |-2 - 7i| = \sqrt{(-2)^2 + (-7)^2} = \sqrt{4 + 49} = \sqrt{53}$$

Since $\tan\alpha = \frac{7}{2}$, it follows that $\alpha \approx 74.1°$. Find θ by noting that θ is a third-quadrant angle, so that

$$\theta = 180° + \alpha \approx 180° + 74.1° = 254.1°$$

Thus, you have

$$-2 - 7i \approx \sqrt{53}(\cos 254.1° + i\sin 254.1°)$$

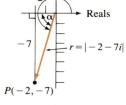

Figure 7-5

Example 3   Change the complex number $10(\cos 60° + i\sin 60°)$ to $a + bi$ form.

Solution $$10(\cos 60° + i\sin 60°) = 10\left(\frac{1}{2} + i\frac{\sqrt{3}}{2}\right)$$
$$= 5 + 5\sqrt{3}\,i$$

It is easy to add two complex numbers in $a + bi$ form—simply add their real and imaginary components. However, it is difficult to multiply and divide complex numbers if they are written in $a + bi$ form. Fortunately, it turns out to be easy to multiply and divide two complex numbers written in trigonometric form, and we now consider how.

Suppose that $N_1 = r_1(\cos\theta_1 + i\sin\theta_1)$ and $N_2 = r_2(\cos\theta_2 + i\sin\theta_2)$. The product N_1N_2 can be found as follows.

$$N_1N_2 = r_1(\cos\theta_1 + i\sin\theta_1)\cdot r_2(\cos\theta_2 + i\sin\theta_2)$$
$$= r_1r_2[\cos\theta_1\cos\theta_2 + i\cos\theta_1\sin\theta_2 + i\sin\theta_1\cos\theta_2 - \sin\theta_1\sin\theta_2]$$
$$= r_1r_2[(\cos\theta_1\cos\theta_2 - \sin\theta_1\sin\theta_2) + i(\cos\theta_1\sin\theta_2 + \sin\theta_1\cos\theta_2)]$$
$$= r_1r_2[\cos(\theta_1 + \theta_2) + i\sin(\theta_1 + \theta_2)]$$

This result can be stated more compactly by using $r\ \text{cis}\ \theta$ notation:

$$r_1\ \text{cis}\ \theta_1\cdot r_2\ \text{cis}\ \theta_2 = r_1r_2\ \text{cis}(\theta_1 + \theta_2)$$

Thus, to form a product of two complex numbers written in trigonometric form, multiply the moduli r_1 and r_2 and add the arguments θ_1 and θ_2. This property generalizes to products containing any number of factors.

(7.9)
$$r_1(\cos\theta_1 + i\sin\theta_1) \cdot r_2(\cos\theta_2 + i\sin\theta_2) \cdots \cdot r_n(\cos\theta_n + i\sin\theta_n)$$
$$= r_1 r_2 \cdots \cdot r_n[\cos(\theta_1 + \theta_2 + \cdots + \theta_n) + i\sin(\theta_1 + \theta_2 + \cdots + \theta_n)]$$

Equation (7.9) can be written more compactly as

(7.10)
$$r_1 \operatorname{cis}\theta_1 \cdot r_2 \operatorname{cis}\theta_2 \cdots \cdot r_n \operatorname{cis}\theta_n$$
$$= r_1 r_2 \cdots \cdot r_n \operatorname{cis}(\theta_1 + \theta_2 + \cdots + \theta_n)$$

Example 4   Find the product of $2(\cos 40° + i\sin 40°)$ and $3(\cos 30° + i\sin 30°)$.

Solution   To find the product, simply multiply the moduli and add the arguments.

$$2(\cos 40° + i\sin 40°) \cdot 3(\cos 30° + i\sin 30°)$$
$$= 2 \cdot 3\,[\cos(40° + 30°) + i\sin(40° + 30°)]$$
$$= 6(\cos 70° + i\sin 70°)$$
■

Example 5   Find the product of $2 \operatorname{cis} 10°$, $4 \operatorname{cis} 20°$, and $9 \operatorname{cis} 50°$.

Solution   To find the product, multiply the moduli and add the arguments.

$$(2 \operatorname{cis} 10°)(4 \operatorname{cis} 20°)(9 \operatorname{cis} 50°) = 2 \cdot 4 \cdot 9 \operatorname{cis}(10° + 20° + 50°)$$
$$= 72 \operatorname{cis} 80°$$
■

If two complex numbers are written in trigonometric form, their quotient can be found by dividing their moduli and subtracting their arguments. To show that this is true, we let $N_1 = r_1(\cos\theta_1 + i\sin\theta_1)$ and $N_2 = r_2(\cos\theta_2 + i\sin\theta_2)$. Then we have

$$\frac{N_1}{N_2} = \frac{r_1(\cos\theta_1 + i\sin\theta_1)}{r_2(\cos\theta_2 + i\sin\theta_2)}$$

$$= \frac{r_1}{r_2} \cdot \frac{\cos\theta_1 + i\sin\theta_1}{\cos\theta_2 + i\sin\theta_2} \cdot \frac{\cos\theta_2 - i\sin\theta_2}{\cos\theta_2 - i\sin\theta_2}$$

$$= \frac{r_1}{r_2} \cdot \frac{\cos\theta_1\cos\theta_2 - i\cos\theta_1\sin\theta_2 + i\sin\theta_1\cos\theta_2 - i^2\sin\theta_1\sin\theta_2}{\cos^2\theta_2 - i^2\sin^2\theta_2}$$

$$= \frac{r_1}{r_2} \cdot \frac{\cos\theta_1\cos\theta_2 + \sin\theta_1\sin\theta_2 + i(\sin\theta_1\cos\theta_2 - \cos\theta_1\sin\theta_2)}{\cos^2\theta_2 + \sin^2\theta_2}$$

$$= \frac{r_1}{r_2} \cdot \frac{\cos(\theta_1 - \theta_2) + i\sin(\theta_1 - \theta_2)}{1}$$

$$= \frac{r_1}{r_2}[\cos(\theta_1 - \theta_2) + i\sin(\theta_1 - \theta_2)]$$

Thus, to divide complex numbers written in trigonometric form, we use the following rule.

(7.11) $$\frac{r_1(\cos\theta_1 + i\sin\theta_1)}{r_2(\cos\theta_2 + i\sin\theta_2)} = \frac{r_1}{r_2}[\cos(\theta_1 - \theta_2) + i\sin(\theta_1 - \theta_2)]$$

Equation (7.11) can be written more compactly as

(7.12) $$\frac{r_1\,\text{cis}\,\theta_1}{r_2\,\text{cis}\,\theta_2} = \frac{r_1}{r_2}\,\text{cis}\,(\theta_1 - \theta_2)$$

Example 6   Divide 8 cis 110° by 4 cis 50°.

Solution   Use Equation (7.12).

$$\frac{8\,\text{cis}\,110°}{4\,\text{cis}\,50°} = \frac{8}{4}\,\text{cis}(110° - 50°)$$

$$= 2\,\text{cis}\,60°$$

Note that 2 cis 60° can be written in the form $2(\cos 60° + i\sin 60°)$.

Exercise 7.3

In Exercises 1–12, write the given complex number in trigonometric form.

1. $6 + 0i$ **2.** $-7 + 0i$

3. $0 - 3i$ **4.** $0 + 4i$

5. $-1 - i$ **6.** $-1 + i$

7. $3 + 3i\sqrt{3}$ **8.** $7 - 7i$

9. $-1 - \sqrt{3}\,i$ **10.** $-3\sqrt{3} + 3i$

11. $-\sqrt{3} - i$ **12.** $1 + i\sqrt{3}$

In Exercises 13–24, write the given complex number in $a + bi$ form.

13. $2(\cos 30° + i\sin 30°)$ **14.** $5(\cos 45° + i\sin 45°)$

15. $7(\cos 90° + i\sin 90°)$ **16.** $12(\cos 0° + i\sin 0°)$

17. $-2\left(\cos\dfrac{2\pi}{3} + i\sin\dfrac{2\pi}{3}\right)$ **18.** $3\left(\cos\dfrac{4\pi}{3} + i\sin\dfrac{4\pi}{3}\right)$

19. $\dfrac{1}{2}(\cos\pi + i\sin\pi)$ **20.** $-\dfrac{2}{3}(\cos 2\pi + i\sin 2\pi)$

21. $-3\,\text{cis}\,225°$ **22.** $3\,\text{cis}\,300°$

23. $11\,\text{cis}\,\dfrac{11\pi}{6}$ **24.** $9\,\text{cis}\,3$

In Exercises 25–36, find each product.

25. $[4(\cos 30° + i\sin 30°)][2(\cos 60° + i\sin 60°)]$

26. $[3(\cos 45° + i\sin 45°)][2(\cos 120° + i\sin 120°)]$

27. $(\cos 300° + i\sin 300°)(\cos 0° + i\sin 0°)$

28. $[5(\cos 85° + i \sin 85°)][2(\cos 65° + i \sin 65°)]$

29. $[2(\cos \pi + i \sin \pi)][3(\cos \pi + i \sin \pi)]$

30. $\left(\cos\frac{\pi}{2} + i \sin\frac{\pi}{2}\right)\left(\cos\frac{3\pi}{2} + i \sin\frac{3\pi}{2}\right)$

31. $\left[2\left(\cos\frac{\pi}{3} + i \sin\frac{\pi}{3}\right)\right]\left[3\left(\cos\frac{\pi}{6} + i \sin\frac{\pi}{6}\right)\right]$

32. $\left[3\left(\cos\frac{5\pi}{6} + i \sin\frac{5\pi}{6}\right)\right]\left[4\left(\cos\frac{7\pi}{6} + i \sin\frac{7\pi}{6}\right)\right]$

33. $(3 \text{ cis } 12°)(2 \text{ cis } 22°)(5 \text{ cis } 82°)$ **34.** $(2 \text{ cis } 50°)(3 \text{ cis } 100°)(6 \text{ cis } 2°)$

35. $\left(3 \text{ cis }\frac{\pi}{2}\right)\left(4 \text{ cis }\frac{\pi}{3}\right)\left(3 \text{ cis }\frac{\pi}{4}\right)$ **36.** $\left(4 \text{ cis }\frac{\pi}{6}\right)\left(2 \text{ cis }\frac{2\pi}{3}\right)\left(\text{cis }\frac{\pi}{4}\right)$

In Exercises 37–44, find each quotient.

37. $\dfrac{12(\cos 60° + i \sin 60°)}{2(\cos 30° + i \sin 30°)}$ **38.** $\dfrac{24(\cos 150° + i \sin 150°)}{48(\cos 50° + i \sin 50°)}$

39. $\dfrac{18(\cos \pi + i \sin \pi)}{12\left(\cos\frac{\pi}{2} + i \sin\frac{\pi}{2}\right)}$ **40.** $\dfrac{15(\cos 2\pi + i \sin 2\pi)}{45(\cos \pi + i \sin \pi)}$

41. $\dfrac{12 \text{ cis } 250°}{5 \text{ cis } 120°}$ **42.** $\dfrac{365 \text{ cis } 370°}{20 \text{ cis } 255°}$

43. $\dfrac{\text{cis }\frac{2\pi}{3}}{2 \text{ cis }\frac{\pi}{6}}$ **44.** $\dfrac{250 \text{ cis }\frac{7\pi}{16}}{50 \text{ cis }\frac{\pi}{3}}$

In Exercises 45–48, simplify each expression.

45. $\dfrac{(2 \text{ cis } 60°)(3 \text{ cis } 20°)}{6 \text{ cis } 40°}$ **46.** $\dfrac{36 \text{ cis } 200°}{(2 \text{ cis } 40°)(9 \text{ cis } 10°)}$

47. $\dfrac{48 \text{ cis }\frac{11\pi}{6}}{\left(3 \text{ cis }\frac{\pi}{3}\right)\left(4 \text{ cis }\frac{2\pi}{3}\right)}$ **48.** $\dfrac{(96 \text{ cis } \pi)(12 \text{ cis } 2\pi)}{\left(48 \text{ cis }\frac{\pi}{2}\right)\left(3 \text{ cis }\frac{3\pi}{2}\right)}$

7.4   DE MOIVRE'S THEOREM

We have shown that, if two complex numbers are written in trigonometric form, their product can be found by multiplying their moduli and adding their arguments. This property makes it easy to find powers of complex numbers that are expressed in trigonometric form. For example, we can find the cube of $3(\cos 40° + i \sin 40°)$ as follows.

$[3(\cos 40° + i \sin 40°)]^3$

$$= [3(\cos 40° + i \sin 40°)][3(\cos 40° + i \sin 40°)][3(\cos 40° + i \sin 40°)]$$
$$= 3^3[\cos(40° + 40° + 40°) + i \sin(40° + 40° + 40°)]$$
$$= 27[\cos 3(40°) + i \sin 3(40°)]$$
$$= 27(\cos 120° + i \sin 120°)$$

The generalization of the previous example is called **De Moivre's theorem.**

(7.13)

> **De Moivre's Theorem.**   If n is a real number and $r (\cos θ + i \sin θ)$ is a complex number in trigonometric form, then
>
> $$[r(\cos θ + i \sin θ)]^n = r^n[\cos nθ + i \sin nθ]$$
>
> or
>
> $$(r \operatorname{cis} θ)^n = r^n \operatorname{cis} nθ$$

This theorem was first developed about 1730 by the French mathematician Abraham De Moivre. It can be proved for all natural numbers n by using mathematical induction. However, the theorem is true for negative and fractional powers as well.

Example 1   Find $[2(\cos 15° + i \sin 15°)]^4$.

Solution   Use De Moivre's theorem.

$$[2(\cos 15° + i \sin 15°)]^4 = 2^4[\cos 4 \cdot 15° + i \sin 4 \cdot 15°]$$
$$= 16[\cos 60° + i \sin 60°]$$

This result could be changed easily to $a + bi$ form if desired:

$$16(\cos 60° + i \sin 60°) = 16\left(\frac{1}{2} + i\frac{\sqrt{3}}{2}\right) = 8 + 8\sqrt{3}\,i$$

Example 2   Find $[\sqrt{2}\,(\cos 10° + i \sin 10°)]^{10}$.

Solution   Use De Moivre's theorem.

$$[\sqrt{2}(\cos 10° + i \sin 10°)]^{10} = (\sqrt{2})^{10}[(\cos 10 \cdot 10° + i \sin 10 \cdot 10°)]$$
$$= 32(\cos 100° + i \sin 100°)$$

De Moivre's theorem can be used to find all of the nth roots of any number. Because both real and complex numbers can be written in the form $a + bi$, they can be written in trigonometric form as well. Thus, one nth root of $a + bi$ is

$$\sqrt[n]{a + bi} = (a + bi)^{1/n}$$
$$= [r(\cos θ + i \sin θ)]^{1/n}$$
$$= \sqrt[n]{r}\left(\cos\frac{θ}{n} + i \sin\frac{θ}{n}\right)$$

Recall from algebra that the equation $x^2 = 9$ has two distinct roots, 3 and -3, and each qualifies as a square root of 9. In like manner, the equation $x^n = a + bi$ has n distinct roots, and each qualifies as an nth root of the complex number $a + bi$. It follows that there are n distinct nth roots of any complex number, and De Moivre's theorem can be used to find them all. Because $\sin \theta = \sin(\theta + k \cdot 360°)$ and $\cos \theta = \cos(\theta + k \cdot 360°)$ for all integers k, De Moivre's theorem implies that

$$[r(\cos \theta + i \sin \theta)]^{1/n} = \{r[\cos(\theta + k \cdot 360°) + i \sin(\theta + k \cdot 360°)]\}^{1/n}$$

$$= r^{1/n}\left(\cos \frac{\theta + k \cdot 360°}{n} + i \sin \frac{\theta + k \cdot 360°}{n}\right)$$

Substituting the numbers 0, 1, 2, . . . , $(n - 1)$ for k yields the n nth roots of the given complex number.

Example 3 Find the three cube roots of 8.

Solution Because 8 can be expressed as $8 + 0i$, graph the complex number $8 + 0i$ as in Figure 7-6 to see that $r = 8$ and $\theta = 0°$. Write $8 + 0i$ in trigonometric form, and use the equation

$$[r(\cos \theta + i \sin \theta)]^{1/n} = \sqrt[n]{r}\left[\cos \frac{\theta + k \cdot 360°}{n} + i \sin \frac{\theta + k \cdot 360°}{n}\right]$$

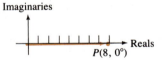

Imaginaries

Reals

$P(8, 0°)$

Figure 7-6

Substituting the values for n, r, and θ gives

$$[8(\cos 0° + i \sin 0°)]^{1/3} = 8^{1/3}\left[\cos \frac{0° + k \cdot 360°}{3} + i \sin \frac{0° + k \cdot 360°}{3}\right]$$

Substituting 0 for k and replacing $8^{1/3}$ with 2 gives

$$2\left(\cos \frac{0°}{3} + i \sin \frac{0°}{3}\right) = 2(\cos 0° + i \sin 0°)$$

$$= 2(1 + 0i)$$

$$= 2$$

Now substitute 1 for k.

$$2\left(\cos\frac{0° + 360°}{3} + i\sin\frac{0° + 360°}{3}\right) = 2(\cos 120° + i\sin 120°)$$

$$= 2\left(-\frac{1}{2} + i\frac{\sqrt{3}}{2}\right)$$

$$= -1 + i\sqrt{3}$$

Finally, substitute 2 for k.

$$2\left(\cos\frac{0° + 720°}{3} + i\sin\frac{0° + 720°}{3}\right) = 2(\cos 240° + i\sin 240°)$$

$$= 2\left(-\frac{1}{2} + i\frac{-\sqrt{3}}{2}\right)$$

$$= -1 - i\sqrt{3}$$

The values 2, $-1 + i\sqrt{3}$, and $-1 - i\sqrt{3}$ are the three cube roots of 8. If these three cube roots of 8 are graphed in the complex plane, they are equally spaced around a circle of radius 2; if the endpoints of these vectors are joined by straight line segments, an equilateral triangle is formed as in Figure 7-7.

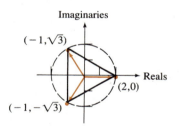

Figure 7-7

Example 4   Find the four fourth roots of $-16i$.

Solution   Express $-16i$ as $0 - 16i$ and graph it to determine that $r = 16$ and $\theta = 270°$. Hence,

$$\sqrt[4]{-16i} = [16(\cos 270° + i\sin 270°)]^{1/4}$$

$$= 16^{1/4}\left(\cos\frac{270° + k \cdot 360°}{4} + i\sin\frac{270° + k \cdot 360°}{4}\right)$$

Substitute 0, 1, 2, and 3 for k to get the four fourth roots of $-16i$.

$$2(\cos 67.5° + i\sin 67.5°) \approx 0.77 + 1.85i$$
$$2(\cos 157.5° + i\sin 157.5°) \approx -1.85 + 0.77i$$
$$2(\cos 247.5° + i\sin 247.5°) \approx -0.77 - 1.85i$$
$$2(\cos 337.5° + i\sin 337.5°) \approx 1.85 - 0.77i$$

Graphs of these four fourth roots are equally spaced about a circle of radius 2, and the endpoints of these vectors are the vertices of a square. See Figure 7-8.

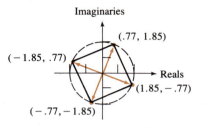

Figure 7-8

Example 5   Find the five fifth roots of $-4 - 4i$.

Solution   Express $-4 - 4i$ in trigonometric form by determining that $r = 4\sqrt{2}$ and $\theta = 225°$. Then, you have

$$[4\sqrt{2}(\cos 225° + i \sin 225°)]^{1/5}$$

$$= (4\sqrt{2})^{1/5}\left(\cos\frac{225° + k \cdot 360°}{5} + i \sin\frac{225° + k \cdot 360°}{5}\right)$$

$$= \sqrt{2}\left(\cos\frac{225° + k \cdot 360°}{5} + i \sin\frac{225° + k \cdot 360°}{5}\right)$$

Substituting 0, 1, 2, 3, and 4 for k generates the five fifth roots of $-4 - 4i$.

$\sqrt{2}(\cos 45° + i \sin 45°)$
$\sqrt{2}(\cos 117° + i \sin 117°)$
$\sqrt{2}(\cos 189° + i \sin 189°)$
$\sqrt{2}(\cos 261° + i \sin 261°)$
$\sqrt{2}(\cos 333° + i \sin 333°)$

If these five fifth roots are graphed, they are equally spaced about a circle with radius $\sqrt{2}$, and the endpoints of these vectors are the vertices of a regular pentagon. See Figure 7-9.

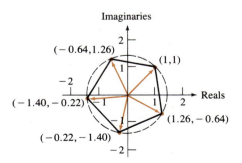

Figure 7-9

If $n > 2$ and the n nth roots of a complex number are graphed on the complex plane, the endpoints of the vectors that represent each root will always be at the vertices of a regular polygon. ■

Exercise 7.4

In Exercises 1–12, find the indicated power. Leave all answers in trigonometric form.

1. $[3(\cos 30° + i \sin 30°)]^3$

2. $[4(\cos 15° + i \sin 15°)]^6$

3. $(\cos 15° + i \sin 15°)^{12}$

4. $[2(\cos 120° + i \sin 120°)]^6$

5. $[5 \operatorname{cis} 2°]^5$

6. $[0.5 \operatorname{cis} 100°]^3$

7. $\left[3\left(\cos \dfrac{\pi}{4} + i \sin \dfrac{\pi}{4} \right) \right]^4$

8. $\left[2\left(\cos \dfrac{3\pi}{2} + i \sin \dfrac{3\pi}{2} \right) \right]^6$

9. $[4(\cos 3 + i \sin 3)]^4$

10. $[2(\cos 5 + i \sin 5)]^{20}$

11. $\left[\dfrac{1}{3} \operatorname{cis} \dfrac{\pi}{2} \right]^3$

12. $\left[\dfrac{1}{2} \operatorname{cis} \dfrac{\pi}{6} \right]^5$

In Exercises 13–18, find the indicated nth root of each expression in a + bi form.

13. A cube root of $8(\cos 180° + i \sin 180°)$.

14. A fifth root of $32(\cos 150° + i \sin 150°)$.

15. A fifth root of $(\cos 300° + i \sin 300°)$.

16. A fourth root of $64(\cos \pi + i \sin \pi)$.

17. A sixth root of $64(\cos 2\pi + i \sin 2\pi)$.

18. A sixth root of $3^6(\cos \pi + i \sin \pi)$.

In Exercises 19–26, find and graph the indicated roots of each complex number. Change your answers to a + bi form only if that answer would be exact. Otherwise, leave the answers in trigonometric form.

19. The three cube roots of -8.

20. The four fourth roots of 16.

21. The two square roots of i.

22. The three cube roots of i.

23. The five fifth roots of $-i$.

24. The three cube roots of $\dfrac{\sqrt{2}}{2} + \dfrac{\sqrt{2}}{2} i$.

25. The four fourth roots of $-8 + 8\sqrt{3}\, i$.

26. The six sixth roots of $-i$.

In Exercises 27–30, substitute the given value of n into De Moivre's theorem with $r = 1$, and raise the binomial on the left to the nth power. Follow the additional directions.

27. $n = 2$. Set the real parts of the complex numbers equal to each other and thereby show that $\cos 2\theta = \cos^2 \theta - \sin^2 \theta$.

28. $n = 2$. Set the imaginary parts of the complex numbers equal to each other and thereby show that $\sin 2\theta = 2 \cos \theta \sin \theta$.

29. $n = 3$. Set the imaginary parts of the complex numbers equal to each other and thereby show that $\sin 3\theta = 3 \cos^2 \theta \sin \theta - \sin^3 \theta$.

30. $n = 3$. Set the real parts of the complex numbers equal to each other and thereby show that $\cos 3\theta = \cos^3 \theta - 3 \cos \theta \sin^2 \theta$.

31. Use the right side of the identity in Exercise 29 to show that $\sin 3\theta = 3 \sin \theta - 4 \sin^3 \theta$.

32. Use the right side of the identity in Exercise 30 to show that $\cos 3\theta = 4 \cos^3 \theta - 3 \cos \theta$.

7.5 POLAR COORDINATES

Some equations such as $(x^2 + y^2)^{3/2} = x$ are difficult to graph using the x and y coordinates of the Cartesian rectangular coordinate system. However, these equations often can be written in a form using the variables r (a radius) and θ (an angle). These coordinates allow easy graphing of such equations in an alternative coordinate system called the **polar coordinate system.** The trigonometric functions will aid in the development of this new system.

Pole and polar axis

The polar coordinate system is based on a ray called the **polar axis** and its source called the **pole.** In Figure 7-10, ray OA is the polar axis and point O is the pole. Any point $P(r, \theta)$ in the plane can be located if the length of a radius and an angle in standard position are known. For example, the polar coordinates $(10, 30°)$ determine the position of point R in Figure 7-10. If θ is in radians, the coordinates $\left(5, \frac{2\pi}{3}\right)$ determine the point Q. In Figure 7-11, point L is determined by the coordinates $(7, 225°)$ and point M by the coordinates $(6, \frac{5\pi}{3})$. Note that point M is also determined by many other pairs of polar coordinates such as $(6, 660°)$ or $(6, -60°)$. Any pair of polar coordinates locates a single point. However, any point has infinitely many pairs of polar coordinates.

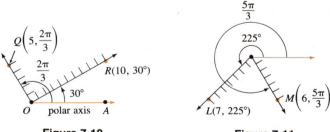

Figure 7-10   **Figure 7-11**

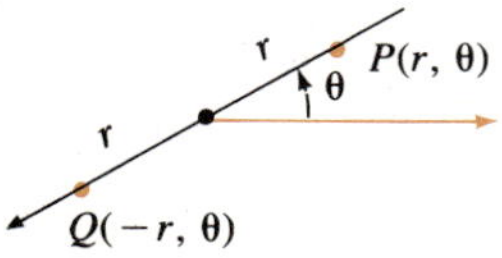

To plot point P with coordinates (r, θ) when r is positive, we draw angle θ in standard position and count r units along the terminal side of θ. This determines point P. See Figure 7-12. To plot point Q with coordinates (r, θ) when r is negative, we draw angle θ in standard position and count $|r|$ units along the extension of the terminal side of θ, but in the opposite direction. For example, to graph the point $P\left(-2, \frac{\pi}{6}\right)$, we first draw an angle of $\frac{\pi}{6}$ in standard position as in Figure 7-13. We then draw the extension of ray OC in the opposite direction

Figure 7-12

to obtain ray OB and count two units along ray OB to find point P. The graphs of the three points $R(5, \pi)$, $Q(-6, 100°)$, and $P(5, -30°)$ are shown in Figure 7-14.

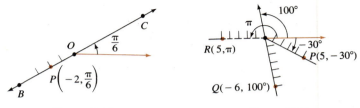

Figure 7-13   Figure 7-14

There is a relationship between the rectangular coordinates (x, y) and the polar coordinates (r, θ) of a point. Suppose point P in Figure 7-15 has rectangular coordinates (x, y) and polar coordinates (r, θ). Draw PA perpendicular to the x-axis to form right triangle OAP with $OA = x$, $AP = y$, and $OP = r$. Because angle θ is in standard position, the hypotenuse OP is the terminal side of angle θ. Thus, we have

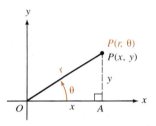

Figure 7-15

$$\cos \theta = \frac{x}{r} \qquad \text{and} \qquad \sin \theta = \frac{y}{r}$$

From these, we obtain the two relations

$$x = r \cos \theta \qquad \text{and} \qquad y = r \sin \theta$$

Hence, to change from polar to rectangular coordinates, we use the following equations.

(7.14)

> **Formulas to Convert from Polar to Rectangular Coordinates.**
>
> $x = r \cos \theta$
>
> $y = r \sin \theta$

Example 1   If the polar coordinates of point P are $(10, 150°)$, find the rectangular coordinates.

Solution $x = r \cos \theta = 10 \cos(150°) = 10 \left(\dfrac{-\sqrt{3}}{2} \right) = -5\sqrt{3}$

$$y = r \sin \theta = 10 \sin(150°) = 10\left(\frac{1}{2}\right) = 5$$

The rectangular coordinates of point P are $(-5\sqrt{3}, 5)$.   ■

To find the polar coordinates from the rectangular coordinates of point P, again refer to Figure 7-15. From the right triangle OAP, it follows that

$$r^2 = x^2 + y^2 \qquad \text{and} \qquad \tan \theta = \frac{y}{x}$$

To find possible polar coordinates for point P, let $r = \sqrt{x^2 + y^2}$. Then, find an angle θ $\left(\text{equal to } \tan^{-1} \frac{y}{x}\right)$ whose terminal side passes through the point (x, y). That is, if x is negative, for example, and y is positive, choose θ to be a second-quadrant angle. If x and y are both negative, choose θ to be a third-quadrant angle.

(7.15)

> **Formulas to Convert from Rectangular to Polar Coordinates.**
>
> $$r = \sqrt{x^2 + y^2}$$
>
> $$\theta = \tan^{-1} \frac{y}{x}$$
>
> where the terminal side of θ passes through the point (x, y). If $x = 0$, choose θ to be 90° (if $y > 0$) or 270° (if $y < 0$).

Example 2   If the rectangular coordinates of point P are $(4, 3)$, find a pair of polar coordinates for P.

Solution   Refer to Figure 7-16. First find the r-coordinate of point P, as follows.

$$\begin{aligned} r &= \sqrt{x^2 + y^2} \\ &= \sqrt{4^2 + 3^2} \\ &= 5 \end{aligned}$$

Secondly, determine the θ-coordinate of point P. Because x and y are both positive, θ is a first-quadrant angle.

$$\begin{aligned} \theta &= \tan^{-1} \frac{y}{x} \\ &= \tan^{-1} \frac{3}{4} \\ &\approx 36.9° \end{aligned}$$

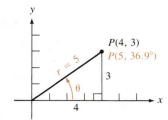

Figure 7-16

One possible choice of polar coordinates for point P is $(5, 36.9°)$, where θ is rounded to the nearest tenth of a degree.   ■

Example 3   Change the rectangular coordinates $(-1, -4)$ to polar coordinates.

Solution   First find the r-coordinate of point P, as follows.

$$
\begin{aligned}
r &= \sqrt{x^2 + y^2} \\
&= \sqrt{(-1)^2 + (-4)^2} \\
&= \sqrt{17}
\end{aligned}
$$

Secondly, determine the θ-coordinate of point P. Because x and y are both negative, θ is a third-quadrant angle.

$$
\begin{aligned}
\theta &= \tan^{-1}\frac{y}{x} \\
&= \tan^{-1}\frac{-4}{-1} \\
&= \tan^{-1}4 \\
&\approx 256.0°
\end{aligned}
$$
Note that $256.0° = 76.0° + 180°$.

One possible choice of polar coordinates for point P is $(\sqrt{17}, 256.0°)$, where θ is rounded to the nearest tenth of a degree.

There are other possible choices for the polar coordinates of P. One is $(-\sqrt{17}, 76.0°)$, which has a *negative* value of r. See Figure 7-17.

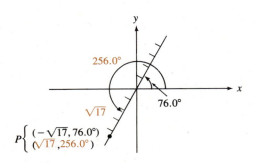

Figure 7-17   ■

Equations involving the variables r and θ can be graphed in a polar coordinate system. The next several examples discuss the graphing of polar equations.

Example 4   Graph the polar equation $r = \theta$.

Solution   Make a table of values, plot the points, and join them with a smooth curve as in Figure 7-18. As θ increases, r increases, and the graph is a spiral called an **Archimedean spiral.**

$r = \theta$

θ	r
0	0
$\frac{\pi}{6}$	0.52
$\frac{\pi}{3}$	1.05
$\frac{\pi}{2}$	1.57
$\frac{2\pi}{3}$	2.09
$\frac{5\pi}{6}$	2.62
π	3.14
$\frac{3\pi}{2}$	4.71

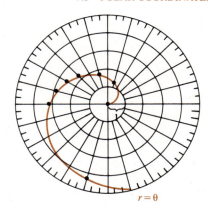

$r = \theta$

Figure 7-18

Example 5   Change the rectangular equation $x^2 + y^2 = y$ to polar coordinates and graph the curve.

Solution   Substitute r^2 for $x^2 + y^2$ and $r \sin \theta$ for y in the original equation.

$$x^2 + y^2 = y$$
$$r^2 = r \sin \theta$$

If $r = 0$, the graph is the point at the pole for all θ. If $r \neq 0$, you can divide both sides of the equation by r to obtain the equation

$$r = \sin \theta$$

Make a table of values, plot the points, and graph as in Figure 7-19. The graph is a circle with center at $\left(\frac{1}{2}, \frac{\pi}{2}\right)$ and with radius of $\frac{1}{2}$. You will be asked in an exercise to graph $x^2 + y^2 = y$ on a set of rectangular coordinate axes to verify that the graph is the described circle.

$r = \sin \theta$

θ	r
0	0
$\frac{\pi}{6}$	0.5
$\frac{\pi}{3}$	0.87
$\frac{\pi}{2}$	1
$\frac{2\pi}{3}$	0.87
$\frac{5\pi}{6}$	0.5
π	0
$\frac{7\pi}{6}$	-0.5
$\frac{3\pi}{2}$	-1

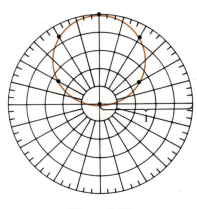

Figure 7-19

Example 6   Change the rectangular equation $(x^2 + y^2)^{3/2} = x$ to an equation having variables of r and θ. Then graph the curve using polar coordinates.

Solution   Because $x^2 + y^2 = r^2$ and $x = r \cos \theta$, you have

$$(x^2 + y^2)^{3/2} = x$$
$$r^3 = r \cos \theta$$

If $r = 0$, the graph is the pole for all θ. If $r \neq 0$, you can divide both sides by r and obtain

$$r^2 = \cos \theta$$
$$r = \pm \sqrt{\cos \theta}$$

Make a table of values, and plot the points as in Figure 7-20. Because r^2 is positive, $\cos \theta$ must be positive. Therefore, θ is in quadrant I or IV.

$r^3 = r \cos \theta$ or $r = \pm \sqrt{\cos \theta}$

θ	r
0	± 1
$\frac{\pi}{6}$	$\pm .93$
$\frac{\pi}{4}$	$\pm .84$
$\frac{\pi}{3}$	$\pm .7$
$\frac{\pi}{2}$	0
$\frac{5\pi}{3}$	$\pm .7$
$\frac{11\pi}{6}$	$\pm .93$

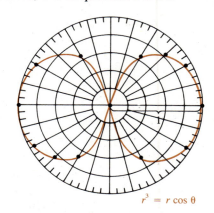

$r^3 = r \cos \theta$

Figure 7-20

Example 7   Change the polar equation $r(3 \cos \theta + 2 \sin \theta) = 7$ to an equation with rectangular coordinates.

Solution   Use the distributive law to remove parentheses.

$$r(3 \cos \theta + 2 \sin \theta) = 7$$
$$3r \cos \theta + 2r \sin \theta = 7$$

Because $r \cos \theta = x$ and $r \sin \theta = y$, you can write this equation as

$$3x + 2y = 7$$

This is the equation of a line that can easily be graphed.

Example 8   Change the polar equation $r^2 = 9 \cos 2\theta$ to an equation with rectangular coordinates.

Solution   Recall that $\cos 2\theta = \cos^2 \theta - \sin^2 \theta$, and substitute this in the original equation.

$$r^2 = 9 \cos 2\theta$$

$$r^2 = 9(\cos^2\theta - \sin^2\theta)$$
$$r^2 = 9\cos^2\theta - 9\sin^2\theta$$

Multiply both sides by r^2 to obtain

$$r^4 = 9r^2\cos^2\theta - 9r^2\sin^2\theta$$
$$r^4 = 9(r\cos\theta)^2 - 9(r\sin\theta)^2$$
$$(x^2 + y^2)^2 = 9x^2 - 9y^2$$

■

Example 9   Change the rectangular equation $(x^2 + y^2)^3 = 8x^2y^2$ to an equation with polar coordinates.

Solution   Substitute $r\cos\theta$ for x, $r\sin\theta$ for y, and r^2 for $x^2 + y^2$ in the rectangular equation and simplify.

$$(x^2 + y^2)^3 = 8x^2y^2$$
$$(r^2)^3 = 8r^2\cos^2\theta\, r^2\sin^2\theta$$
$$r^6 = 8r^4\cos^2\theta\sin^2\theta$$

If $r = 0$, the graph is the point at the pole for all θ. If $r \neq 0$, you can divide both sides by r^4 to obtain

$$r^2 = 8\cos^2\theta\sin^2\theta$$
$$r^2 = 2(2\cos\theta\sin\theta)(2\cos\theta\sin\theta)$$
$$r^2 = 2(\sin 2\theta)(\sin 2\theta)$$
$$r^2 = 2\sin^2 2\theta$$

■

Exercise 7.5

In Exercises 1–20, the polar coordinates of point P are given. Find the rectangular coordinates of point P.

1. $(2, 30°)$ **2.** $(5, 135°)$

3. $(7, 300°)$ **4.** $(20, 225°)$

5. $(-3, 60°)$ **6.** $(-7, 210°)$

7. $\left(2, \dfrac{\pi}{2}\right)$ **8.** $\left(4, \dfrac{3\pi}{2}\right)$

9. $\left(-2, \dfrac{13\pi}{6}\right)$ **10.** $(-5, 3\pi)$

11. $\left(5, \dfrac{13\pi}{4}\right)$ **12.** $\left(3, \dfrac{17\pi}{6}\right)$

13. $(2, -30°)$ **14.** $(6, -225°)$

15. $(-10, -90°)$ **16.** $(-15, -45°)$

17. $(0, 39°)$ **18.** $(0, 0°)$

19. $(6, 1230°)$ **20.** $(35, 11.5\pi)$

In Exercises 21–36, the rectangular coordinates of point P are given. Find a pair of polar coordinates for point P.

21. $(1, 1)$ **22.** $(1, \sqrt{3})$

23. $(2\sqrt{3}, -2)$ **24.** $(-2, -2\sqrt{3})$

25. $(-\sqrt{3}, -1)$ **26.** $(-1, \sqrt{3})$

27. $(-\sqrt{3}, 1)$ **28.** $(0, 3)$

29. $(0, 0)$ **30.** $(7, 0)$

31. $(-5, 0)$ **32.** $(7, 7)$

33. $(3, -3)$ **34.** (π, π)

35. $(7, 7\sqrt{3})$ **36.** $(-\sqrt{2}, -\sqrt{6})$

In Exercises 37–48, each equation contains rectangular coordinates. Change each equation to an equation containing polar coordinates.

37. $x = 3$ **38.** $y = -7$

39. $3x + 2y = 3$ **40.** $2x - y = 7$

41. $x^2 + y^2 = 9x$ **42.** $yx = 12$

43. $(x^2 + y^2)^3 = 4x^2y^2$ **44.** $x^2 + y^2 = 9$

45. $x^2 = 2x - x^2$ **46.** $(x^2 + y^2)^2 = x^2 - y^2$

47. $x^2 = 2y + 1$ **48.** $y^2 = 2x + 1$

In Exercises 49–60, each equation contains polar coordinates. Change each equation to an equation containing rectangular coordinates.

49. $r = 3$ **50.** $r \sin \theta = 4$

51. $\cos \theta = \dfrac{5}{r}$ **52.** $3r \cos \theta + 2r \sin \theta = 2$

53. $r = \dfrac{1}{1 + \sin \theta}$ **54.** $r = \dfrac{1}{1 - \cos \theta}$

55. $r^2 = \sin 2\theta$ **56.** $r^2 = \cos 2\theta$

57. $\theta = \pi$ **58.** $\theta = 90°$

59. $r(2 - \cos \theta) = 2$ **60.** $r = 3 \csc \theta + 2 \sec \theta$

In Exercises 61–72, graph the given equation.

61. $r = -\theta, \quad \theta \geq 0$ **62.** $r \sin \theta = 3$

63. $r \cos \theta = -3$ **64.** $r \cos \theta + r \sin \theta = 1$

65. $r = \cos 2\theta$ **66.** $r = \sin 2\theta$

67. $r = \sin 3\theta$ **68.** $r = 3 \cos 3\theta$

69. $r = 2(1 + \sin \theta)$ **70.** $r = \sqrt{2 \cos \theta}$

71. $r = 2 + \cos \theta$ **72.** $r = 3(1 - \cos \theta)$

73. Graph $x^2 + y^2 = y$ on a Cartesian coordinate system and show that its graph is identical to the graph in Figure 7-19.

7.6 MORE ON POLAR COORDINATES

In the previous section, you graphed several equations in polar coordinates by making an extensive table of values and plotting many points. A different approach will be used in this section. Although some specific points will be plotted, the behavior of functions of θ will be considered as θ increases from 0 to 2π.

Example 1   Graph $r = \sin 2\theta$.

Solution   Note that $\sin 2\theta$, and hence r as well, is zero when $\theta = 0$, $\frac{\pi}{2}$, π, and $\frac{3\pi}{2}$. For these values, the curve passes through the pole. When $\theta = \frac{\pi}{4}$ or $\frac{5\pi}{4}$, the value of $\sin 2\theta = 1$. When θ is $\frac{3\pi}{4}$ or $\frac{7\pi}{4}$, the value of $\sin 2\theta = -1$. Therefore, the curve passes through the points $(1, \frac{\pi}{4})$, $(-1, \frac{3\pi}{4})$, $(1, \frac{5\pi}{4})$, and $(-1, \frac{7\pi}{4})$ and several times through the pole: $(0, 0)$, $(0, \frac{\pi}{2})$, $(0, \pi)$, and $(0, \frac{3\pi}{2})$. Note that these last four ordered pairs represent four different pairs of polar coordinates for the pole. See Figure 7-21*i*. As θ increases from 0 to $\frac{\pi}{4}$, the value of r increases from 0 to 1; draw that portion of the curve as in Figure 7-21*ii*. As θ continues to increase from $\frac{\pi}{4}$ to $\frac{\pi}{2}$, r decreases, and the curve returns to the pole as in Figure 7-21*iii*. When θ continues to increase from $\frac{\pi}{2}$ to $\frac{3\pi}{4}$, the value of r goes from 0 to -1 and the points (r, θ) trace the additional portion of the curve shown in Figure 7-21*iv*. As θ increases from $\frac{3\pi}{4}$ to π, r goes from -1 back to 0 and the curve continues to develop as in Figure 7-21*v*. The complete curve is shown in Figure 7-21*vi*. It is called a **four-leaved rose.**

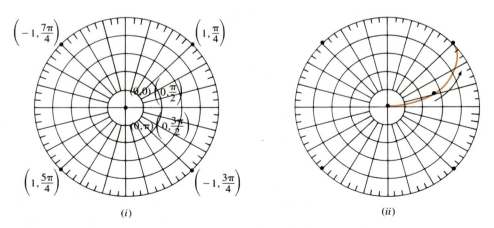

(*i*)   (*ii*)

Figure 7-21 (continues on page 278)

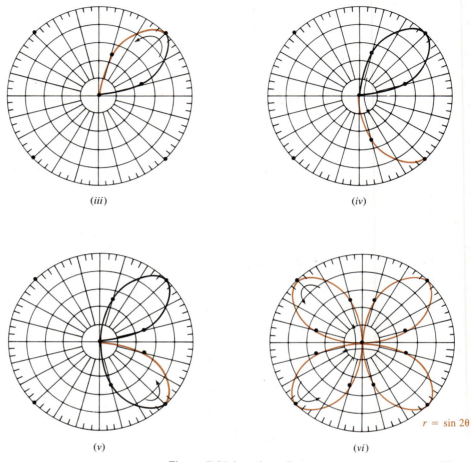

$r = \sin 2\theta$

Figure 7-21 (continued)

Example 2 Graph $r = \cos 3\theta$.

Solution When $\theta = \frac{\pi}{6}, \frac{\pi}{2}, \frac{5\pi}{6}, \frac{7\pi}{6}, \frac{3\pi}{2}$, or $\frac{11\pi}{6}$, the corresponding value of $\cos 3\theta$, and hence of r, is zero. When $\theta = 0, \frac{2\pi}{3}$, or $\frac{4\pi}{3}$, r reaches its maximum value of 1. When $\theta = \frac{\pi}{3}, \pi$, or $\frac{5\pi}{3}$, r reaches its minimum value of -1. Therefore, the curve passes through the points $(1, 0)$, $(1, \frac{2\pi}{3})$, $(1, \frac{4\pi}{3})$, $(-1, \frac{\pi}{3})$, $(-1, \pi)$, and $(-1, \frac{5\pi}{3})$. The curve also passes through the pole several times at $(0, \frac{\pi}{6})$, $(0, \frac{\pi}{2})$, $(0, \frac{5\pi}{6})$, $(0, \frac{7\pi}{6})$, $(0, \frac{3\pi}{2})$, and $(0, \frac{11\pi}{6})$. Note that these last six pairs of coordinates represent six different pairs of coordinates for the pole. Note also that the first six pairs of coordinates represent only three distinct points, as indicated in Figure 7-22i. As θ increases from 0 to $\frac{\pi}{6}$, $\cos 3\theta$ decreases from 1 to 0, and that portion of the curve is traced in Figure 7-22ii. As θ continues to increase from $\frac{\pi}{6}$ to $\frac{\pi}{3}$, the value of $\cos 3\theta$ continues to decrease to its minimum value of -1; the curve continues to develop as in Figure 7-22iii. The complete graph of $r = \cos 3\theta$ appears in Figure 7-22iv. It is called a **three-leaved rose**.

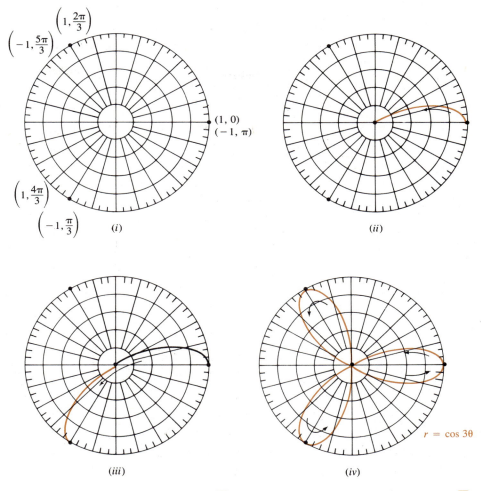

Figure 7-22

All of the rose curves fit into two categories.

(7.16)

> **Theorem.** If n is an odd integer, then $r = \cos n\theta$ or $r = \sin n\theta$ represents an n-leaved rose.
>
> If n is even, then $r = \cos n\theta$ or $r = \sin n\theta$ represents a $2n$-leaved rose.

Example 3 Graph the curve $r = 1 - \sin \theta$.

Solution The easiest values of r to compute are those associated with the quadrantal angle values of θ. When $\theta = 0$ or π, then $r = 1$. When $\theta = \frac{\pi}{2}$, then $r = 0$. When $\theta = \frac{3\pi}{2}$, then $r = 2$. These four points—$(1, 0)$, $(0, \frac{\pi}{2})$, $(1, \pi)$, and $(2, \frac{3\pi}{2})$—

are the intercepts of the curve with the polar axis and the perpendicular line to the polar axis at the pole. See Figure 7-23i. As θ increases from 0 to π, the value of $1 - \sin\theta$, and hence the value of r, decreases from 1 to 0 and then increases back to 1. This accounts for the two "bumps" in the curve in Figure 7-23ii. As θ increases from π to $\frac{3\pi}{2}$, the third-quadrant loop of Figure 7-23iii is formed. The complete curve, called a **cardioid,** is shown in Figure 7-23iv.

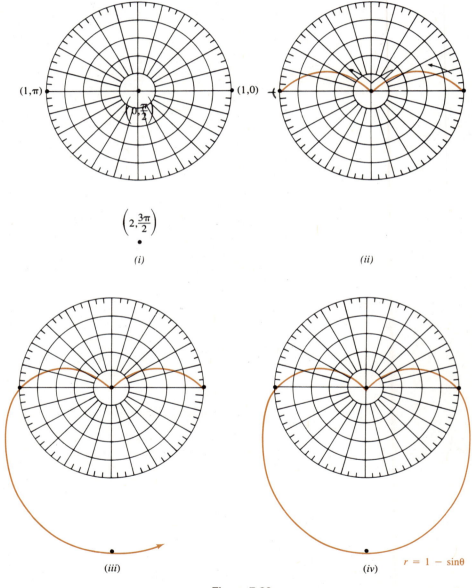

$r = 1 - \sin\theta$

Figure 7-23

Example 4   Graph the curve $r\theta = \pi$.

Solution   Because the product of variables r and θ is a constant, r and θ are inversely proportional—as θ increases, then r decreases. Write the equation as $r = \frac{\pi}{\theta}$, and calculate the intercepts of the curve with the polar axis and a line perpendicular to the axis through the pole.

$$\text{At } \theta = \frac{\pi}{2}, \quad r = \frac{\pi}{\frac{\pi}{2}} = 2, \quad \text{at} \quad \theta = \pi, \quad r = 1, \quad \text{and at} \quad \theta = \frac{3\pi}{2}, \quad r = \frac{2}{3}.$$

These points and a portion of the curve are plotted in Figure 7-24*i*; the curve is a spiral. The difficulty is in determining the shape of the curve as θ approaches 0 (and r approaches infinity). If θ is close to zero, and r consequently is very large, the distance PQ of Figure 7-24*ii* is very close to the length of arc PQ', centered at the pole. Because this arc length is $r\theta$ or π, the distance PQ is approximately π. The horizontal line that is π units above the polar axis is an asymptote of this curve. The graph of $r\theta = \pi$ is called a **hyperbolic spiral** and its graph appears in Figure 7-24*iii*. The "tail" heading off to the right approaches the horizontal line that is parallel to the polar axis and π units above it. Note that, unlike the Archimedean spiral of the previous section, the hyperbolic spiral does not pass through the pole.

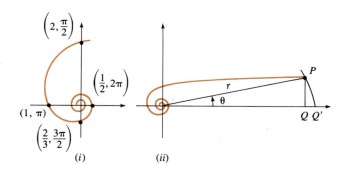

(i)   (ii)

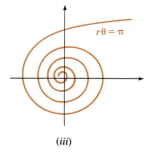

(iii)

Figure 7-24

The following chart shows the graphs and the general equations for the most important polar curves.

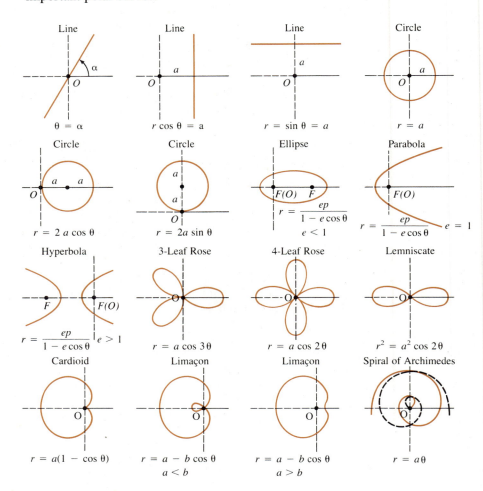

Exercise 7.6

Each of the following curves is important enough to receive a name. The equation and its title are given. Graph each polar equation.

1. $r = 2 \cos 2\theta$;   four-leaved rose

2. $r = 3 \sin 2\theta$;   four-leaved rose

3. $r = \sin 3\theta$;   three-leaved rose

4. $r = 2 \cos 3\theta$;   three-leaved rose

5. $r = 2(1 + \cos \theta)$;   cardioid

6. $r = 3(1 + \sin \theta)$;   cardioid

7. $r = 2 + \cos \theta$;   limaçon

8. $r = \dfrac{1}{2} + \cos \theta$;   limaçon

9. $r^2 = 2 \sin 2\theta$;   lemniscate

10. $r = \sin \theta \cos^2 \theta$;   bifolium

11. $r^2\theta = \pi, r > 0$;   lituus   (*Hint*: This curve has a horizontal asymptote.)

12. $r = \tan\theta$;   kappa curve ⎱
                                        ⎰ (*Hint*: These curves have vertical asymptotes.)
13. $r = \sin\theta\tan\theta$;   cissoid

14. $r = 2a\cos\theta$;   circle

15. $r = \dfrac{3}{2 - \cos\theta}$;   ellipse   **16.** $r = \dfrac{4}{2 - 3\cos\theta}$;   hyperbola

17. $r = \dfrac{4}{1 + \sin\theta}$;   parabola   **18.** $r = \dfrac{6}{4 - 3\cos\theta}$;   ellipse

7.7   APPLICATIONS OF COMPLEX NUMBERS AND POLAR COORDINATES

Because the complex number $a + bi$ can be represented as a vector from the origin to the point (a, b) in the complex plane, it is not surprising that the arithmetic of complex numbers can be used in the solution of problems involving vectors.

Example 1   Forces A and B, of 25 and 35 lb, respectively, make an angle of $40°$ with each other. What is the magnitude of the resultant force, and what angle does it make with the smaller force?

Solution   Represent the forces as complex numbers in trigonometric form as suggested by the diagram of Figure 7-25. Then, rewrite them in rectangular form.

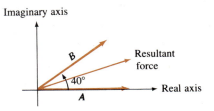

Figure 7-25

$$\text{Vector } \mathbf{A} = 25(\cos 0° + i\sin 0°)$$
$$= 25 + 0i$$
$$\text{Vector } \mathbf{B} = 35(\cos 40° + i\sin 40°)$$
$$\approx 35(0.7660 + 0.6428i)$$
$$\approx 26.8 + 22.5i$$

The sum of these two vectors is the sum of their complex representations.

$$\mathbf{A} + \mathbf{B} = (25 + 0i) + (26.8 + 22.5i)$$
$$= 51.8 + 22.5i$$

Converting this result to trigonometric form gives

$$\mathbf{A} + \mathbf{B} = k(\cos \phi + i \sin \phi)$$

where

$$k = \sqrt{51.8^2 + 22.5^2} \approx \sqrt{3189} \approx 56.5$$

and

$$\phi = \tan^{-1} \frac{22.5}{51.8} \approx 23.5°$$

Thus, you have

$$\mathbf{A} + \mathbf{B} = 56.5(\cos 23.5° + i \sin 23.5°)$$

Hence, the magnitude of the resultant force is 56.5 lb, and the angle that it makes with the weaker force is 23.5°. ■

The law of cosines can be used to derive a distance formula for points given in polar coordinates. If $P(r_1, \theta_1)$ and $Q(r_2, \theta_2)$ are two points in Figure 7-26, given in polar coordinates, then the distance d between them is found by applying the law of cosines to triangle OPQ.

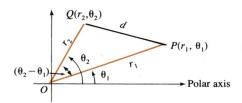

Figure 7-26

$$d^2 = r_1^2 + r_2^2 - 2r_1 r_2 \cos(\theta_2 - \theta_1)$$
$$d = \sqrt{r_1^2 + r_2^2 - 2r_1 r_2 \cos(\theta_2 - \theta_1)}$$

Example 2   What is the distance between $P\left(3, \frac{\pi}{6}\right)$ and $Q\left(2, \frac{\pi}{2}\right)$?

Solution   Let $r_1 = 3$, $\theta_1 = \frac{\pi}{6}$, $r_2 = 2$, and $\theta_2 = \frac{\pi}{2}$. Then, you have

$$d = \sqrt{r_1^2 + r_2^2 - 2r_1 r_2 \cos(\theta_2 - \theta_1)}$$

$$= \sqrt{3^2 + 2^2 - 2(3)(2) \cos\left(\frac{\pi}{2} - \frac{\pi}{6}\right)}$$

$$= \sqrt{9 + 4 - 12 \cos \frac{\pi}{3}}$$

$$= \sqrt{13 - 12\left(\frac{1}{2}\right)}$$
$$= \sqrt{13 - 6}$$
$$= \sqrt{7}$$

The distance PQ is $\sqrt{7}$ units. ∎

Exercise 7.7

In Exercises 1–2, use complex numbers to solve each problem.

1. Forces of 15.2 and 29.7 lb make an angle of 36.0° with each other. What is the magnitude of the resultant force, and what angle does it make with the weaker force?

2. Show that three equal forces, each acting at an angle of 120° from the other two, have a resultant force of zero.

3. Find the distance between points $P\left(5, \frac{5\pi}{6}\right)$ and $Q\left(3, \frac{-\pi}{6}\right)$.

4. Find the perimeter of the triangle formed by $P(2, 20°)$, $Q(5, 50°)$, and the pole.

CHAPTER SUMMARY

Key Words

absolute value of a complex number (7.2)
Archimedean spiral (7.5)
argument (7.3)
cardioid (7.6)
cis θ (7.3)
complex conjugates (7.1)
complex number (7.1)
complex plane (7.2)
De Moivre's theorem (7.4)
four-leaved rose (7.6)
hyperbolic spiral (7.6)
i (7.1)
imaginary number (7.1)

imaginary part of a complex number (7.1)
modulus (7.3)
polar axis (7.5)
polar coordinate system (7.5)
pole (7.5)
principal nth root (7.4)
rationalizing the denominator (7.1)
real part of a complex number (7.1)
three-leaved rose (7.6)
trigonometric form of a complex number (7.3)

Key Ideas

(7.1) $i^2 = -1$

Two complex numbers are equal if and only if their real parts are equal and their imaginary parts are equal:

$a + bi = c + di$ if and only if $a = c$ and $b = d$

Complex numbers are added as binomials:

$$(a + bi) + (c + di) = (a + c) + (b + d)i$$

Complex numbers are multiplied as binomials:

$$(a + bi)(c + di) = (ac - bd) + (ad + bc)i$$

(7.2) To graph the complex number $a + bi$, plot the point (a, b) in the complex plane. The vector drawn from the origin to the point (a, b) is the graph of the complex number $a + bi$.

$$|a + bi| = \sqrt{a^2 + b^2}$$

(7.3) Complex numbers can be written either in algebraic $(a + bi)$ or trigonometric $[r(\cos\theta + i\sin\theta)]$ form, where $r = |a + bi|$, and θ is an angle such that $\sin\theta = \frac{b}{r}$ and $\cos\theta = \frac{a}{r}$.

$$r \text{ cis } \theta = r(\cos\theta + i\sin\theta)$$

$$(r_1 \text{ cis } \theta_1)(r_2 \text{ cis } \theta_2) \bullet \cdots \bullet (r_n \text{ cis } \theta_n)$$
$$= r_1 r_2 \bullet \cdots \bullet r_n \text{ cis}(\theta_1 + \theta_2 + \cdots + \theta_n)$$

$$\frac{r_1 \text{ cis } \theta_1}{r_2 \text{ cis } \theta_2} = \frac{r_1}{r_2} \text{ cis}(\theta_1 - \theta_2)$$

(7.4) De Moivre's theorem

$$[r(\cos\theta + i\sin\theta)]^n = r^n[\cos n\theta + i\sin n\theta]$$

or

$$[r \text{ cis } \theta]^n = r^n \text{ cis } n\theta$$

If $n > 2$ and the n nth roots of a complex number are graphed on the complex plane, the end points of the vectors that represent each root lie at the vertices of a regular polygon.

(7.5) Formulas to convert from polar coordinates to rectangular coordinates.

$$\begin{cases} x = r\cos\theta \\ y = r\sin\theta \end{cases}$$

Formulas to convert from rectangular coordinates to polar coordinates.

$$\begin{cases} r = \sqrt{x^2 + y^2} \\ \theta = \tan^{-1}\dfrac{y}{x} \end{cases}$$

(7.6) If n is an odd integer, then $r = \cos n\theta$ and $r = \sin n\theta$ represent roses with n leaves. If n is even, then $r = \cos n\theta$ and $r = \sin n\theta$ represent roses with $2n$ leaves.

REVIEW EXERCISES

1. Simplify i^{11}.

2. Simplify i^{5003}.

3. Simplify i^{-33}.

4. Simplify i^{-1812}.

5. Solve $x + (x + y)i = 2y + 2i$ for x and y.

6. Solve $3x + (x - y)i = 2y + 3 + 7i$.

In Review Exercises 7–18, perform any indicated operations and express the final answer in a + bi form.

7. $(3 + 2i) + (-7 - i)$

8. $(-2 - i) - (3 - 2i)$

9. $(2 + i)(-3 - i)$

10. $(3 + 2i)(5 - 3i)$

11. $\dfrac{1}{5i}$

12. $\dfrac{13}{-6i}$

13. $\dfrac{2}{4 + i}$

14. $\dfrac{-5}{3 - i}$

15. $\dfrac{1 + \sqrt{-1}}{1 - \sqrt{-1}}$

16. $\dfrac{2 + \sqrt{-1}}{3 + \sqrt{-16}}$

17. $\dfrac{2 + 3i}{1 - \sqrt{2}\,i}$

18. $\dfrac{3 - i}{1 - \sqrt{3}\,i}$

In Review Exercises 19–22, graph each complex number.

19. $4 - 5i$

20. $-7 + 2i$

21. 6

22. $3i$

In Review Exercises 23–26, compute each absolute value.

23. $|8 + 3i|$

24. $|10 - 10i|$

25. $\left|\dfrac{3i}{i + 3}\right|$

26. $\left|\dfrac{4 - 3i}{4 + 3i}\right|$

In Review Exercises 27–30, write the given complex number in trigonometric form.

27. $-2 + 2i$

28. $5 - 5i$

29. $3 + 3i\sqrt{3}$

30. 4

In Review Exercises 31–34, write the given complex number in a + bi form.

31. $3(\cos 60° + i \sin 60°)$

32. $2(\cos 330° + i \sin 330°)$

33. $3\left(\cos\dfrac{4\pi}{3} + i \sin\dfrac{4\pi}{3}\right)$

34. $7\left(\cos\dfrac{5}{6}\pi + i \sin\dfrac{5}{6}\pi\right)$

In Review Exercises 35–40, perform the indicated operation. Simplify, but leave your answer in trigonometric form.

35. $(\text{cis } 60°)(\text{cis } 50°)$

36. $(2 \text{ cis } 330°)(3 \text{ cis } 240°)$

37. $\left[3\left(\cos\dfrac{\pi}{12} + i\sin\dfrac{\pi}{12}\right)\right]\left[2\left(\cos\dfrac{\pi}{6} + i\sin\dfrac{\pi}{6}\right)\right]$

38. $\left[7\left(\cos\dfrac{\pi}{5} + i\sin\dfrac{\pi}{5}\right)\right]\left[3\left(\cos\dfrac{4\pi}{5} + i\sin\dfrac{4\pi}{5}\right)\right]$

39. $\dfrac{10\operatorname{cis} 60°}{5\operatorname{cis} 10°}$

40. $\dfrac{20(\cos 50° + i\sin 50°)}{30(\cos 40° + i\sin 40°)}$

41. Find one cube root of $\cos 60° + i\sin 60°$.

42. Find one fourth root of $7\sqrt{2} + 7\sqrt{2}\,i$.

43. Find the three cube roots of 125.   **44.** Find the four fourth roots of 81.

In Review Exercises 45–48, change the polar coordinates to rectangular coordinates.

45. $(5, 60°)$ **46.** $(-2, 390°)$

47. $\left(-1, \dfrac{7\pi}{6}\right)$ **48.** $\left(10, -\dfrac{5\pi}{4}\right)$

In Review Exercises 49–52, change the rectangular coordinates to a pair of polar coordinates.

49. $(-\sqrt{2}, \sqrt{2})$ **50.** $(-\sqrt{3}, 1)$

51. $(1, 0)$ **52.** $(1, -\sqrt{3})$

In Review Exercises 53–56, change the rectangular equation to a polar equation.

53. $4xy = 4$ **54.** $x + 2y = 2$

55. $x^2 = 3y$ **56.** $(x^2 + y^2)^2 = 4xy$

In Review Exercises 57–60, change the polar equation to a rectangular equation.

57. $r^2 = 9\cos 2\theta$ **58.** $r = 5\sin\theta$

59. $r = \dfrac{1}{4 + \sin\theta}$ **60.** $r = \dfrac{2}{1 - \cos\theta}$

In Review Exercises 61–64, graph each polar equation.

61. $r = \dfrac{6}{1 + \sin\theta}$ **62.** $r = \dfrac{2}{1 - \sin\theta}$

63. $r = 4(1 + \cos\theta)$ **64.** $r = 8 - 4\cos\theta$

APPENDIXES

APPENDIX I
BASIC GEOMETRY

I.1 ANGLES, TRIANGLES, AND THE PYTHAGOREAN THEOREM

(I.1)

> **Definition.** An **angle** is a figure determined by two rays (not necessarily different) originating from a common point called the **vertex.** A line segment and a ray, or two line segments, with a common endpoint also determine an angle.

Ways of naming angles

In Figure I-1, the two rays *BA* and *BC* originate at point *B*, the vertex of the angle. There are various ways of naming this angle. One is to name the angle after its vertex: angle *B*. A second way is to use three letters: angle *ABC* or angle *CBA*. A third way is to use a Greek letter such as β: angle β.

In Figure I-2, angle *ABC* is the same angle as angle β. Angle *CBD* and angle α are two names for the same angle.

In geometry, angles are usually measured in **degrees.** One degree, denoted as 1°, is 1/360 of one complete revolution.

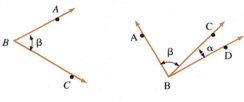

| Figure I-1 | Figure I-2 |

Types of angles

An **acute angle** is an angle greater than 0°, but less than 90°. See Figure I-3. An angle of 90° is called a **right angle.** An angle that is greater than 90° but less than 180° is called an **obtuse angle.** An angle of 180° is called a **straight angle.** An angle that is greater than 180° but less than 360° is called a **reflex angle.**

If the measures of two acute angles add to 90°, the angles are called **complementary.** If the measures of two angles add to 180°, they are called **supplementary.**

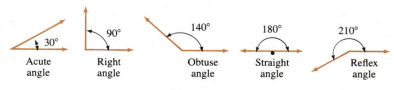

Figure I-3

Types of
triangles   A triangle with three acute angles is called an **acute triangle.** See Figure I-4. A triangle with a single obtuse angle is called an **obtuse triangle.** A triangle with a 90° angle (a right angle) is called a **right triangle,** and the side opposite the right angle is called the **hypotenuse.** The other two sides are often called **legs** of the triangle.

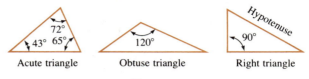

| Acute triangle | Obtuse triangle | Right triangle |

Figure I-4

If a triangle is not a right triangle, it is called an **oblique triangle.** The acute and obtuse triangles in Figure I-4 are oblique triangles. If a triangle has at least two sides of equal length, the triangle is called **isosceles.** In an isosceles triangle, the angles opposite the sides with equal lengths have equal measures also. A triangle with all three sides of equal length is called an **equilateral** triangle. Because an equilateral triangle has three equal angles, it is called **equiangular** also. See Figure 1-5.

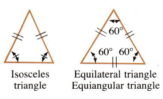

Isosceles   Equilateral triangle
triangle   Equiangular triangle

Figure I-5

There are two theorems from geometry that cannot be overemphasized. Their proofs can be found in any geometry book.

(I.2) | **Theorem.**   The sum of the three angles in any triangle is equal to 180°.

Example 1   If the vertex angle of an isosceles triangle is equal to 112°, find the measure of each of the base angles.

Solution   You know that the base angles of an isosceles triangle are equal, and that the vertex angle is 112°. Let x represent the measure of each base angle as in Figure I-6. Because the sum of the three angles in a triangle is 180°, it follows that

$$x + x + 112 = 180$$
$$2x + 112 = 180$$
$$2x = 68$$
$$x = 34$$

Each base angle measures 34°.

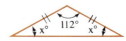

Figure I-6

(I.3)   **The Pythagorean Theorem.**   A triangle is a right triangle if and only if the square of the hypotenuse is equal to the sum of the squares of the two legs of the triangle.

Example 2   A triangle has sides of 5, 12, and 13 m. Is this triangle a right triangle?

Solution   By the Pythagorean Theorem, the triangle is a right triangle if the square of one side of the triangle equals the sum of the squares of the other two sides. Because

$$13^2 = 169 \quad \text{and} \quad 5^2 + 12^2 = 169$$

this triangle is a right triangle.   ∎

Example 3   If two legs of a right triangle measure 9 and 11 cm, find the length of the triangle's hypotenuse.

Solution   Let the length of the hypotenuse be h and use the Pythagorean Theorem.

$$h^2 = 9^2 + 11^2$$
$$h^2 = 81 + 121$$
$$h = 202$$
$$h = \sqrt{202}$$

The hypotenuse measures $\sqrt{202}$ cm.   ∎

There are infinitely many triples of numbers x, y, and z that have the property that $x^2 + y^2 = z^2$. Such triples of numbers are called **Pythagorean triples.** Some of the more common of these are

3, 4, 5   5, 12, 13   6, 8, 10
7, 24, 25   8, 15, 17   9, 40, 41

Principal   Recall that $\sqrt{202}$ represents the positive, or **principal square root,** of 202. The
square root   negative square root of 202 is denoted as $-\sqrt{202}$. Although the equation $h^2 = 202$ has two solutions, $h = \sqrt{202}$ and $h = -\sqrt{202}$, only the principal square root is considered in Example 3; the hypotenuse of a right triangle cannot be negative.

The following theorems, based on the Pythagorean Theorem, are important.

(I.4)   **Theorem.**   If a right triangle is isosceles, the length of the hypotenuse is $\sqrt{2}$ times one of the equal legs.

Proof   We consider the isosceles right triangle in Figure I-7 with equal legs of x units and hypotenuse of h units and use the Pythagorean Theorem.

Figure I-7

$$h^2 = x^2 + x^2$$
$$h^2 = 2x^2$$
$$h = x\sqrt{2}$$

Only the principal square root is considered because the sides of a triangle cannot be negative. The theorem is proved. □

Example 4 Find the length of the hypotenuse of an isosceles right triangle with one leg equal to 5 cm.

Solution By Theorem I.4, the hypotenuse measures $5\sqrt{2}$ cm. ■

Example 5 If the hypotenuse of an isosceles right triangle measures 9 m, how long is each leg of the triangle?

Solution Let each of the equal legs measure x m. Then by Theorem I.4, it follows that

$$\sqrt{2}x = 9$$
$$x = \frac{9}{\sqrt{2}}$$
$$x = \frac{9\sqrt{2}}{2}$$

Each leg of the triangle measures $\dfrac{9\sqrt{2}}{2}$ m. ■

The following theorem is presented without proof.

(I.5)

Theorem.   If a right triangle has acute angles that measure 30° and 60°, the hypotenuse is twice as long as the leg opposite the 30° angle. The leg opposite the 60° angle is $\sqrt{3}$ times as long as the leg opposite the 30° angle.

Example 6 The leg opposite the 30° angle in a 30°, 60° right triangle is 10 cm. Find the other two sides.

Solution The hypotenuse measures twice 10 cm, or 20 cm. The leg opposite the 60° angle is $\sqrt{3}$ times 10 cm, or $10\sqrt{3}$ cm. ■

Example 7 The leg opposite the 60° angle in a 30°, 60° right triangle is 12 decimeters. Find the length of the other two sides.

Solution Refer to Figure I-8. If DB measures x decimeters, side CD measures $\sqrt{3}x$ decimeters. Hence, you have

$$\sqrt{3}x = 12$$
$$x = \frac{12}{\sqrt{3}}$$
$$x = 4\sqrt{3}$$

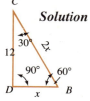

Figure I-8

Because the hypotenuse *CB* is twice as long as the leg opposite the 30° angle, it follows that

$$CB = 2(4\sqrt{3})$$
$$= 8\sqrt{3}$$

DB measures $4\sqrt{3}$ decimeters and *CB* measures $8\sqrt{3}$ decimeters.

Exercise I.1

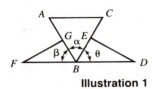

Illustration 1

Exercises 1–6 refer to Illustration 1. Give another name for the given angle.

1. angle α **2.** angle θ

3. angle *BAC*   **4.** angle *FBG*

5. angle *D*   **6.** angle *F*

In Exercises 7–16, tell if the given angle is acute, right, obtuse, straight, reflex, or is in none of these categories.

7. 30°   **8.** 89°

9. 135°   **10.** 180°

11. 90°   **12.** 146°

13. 0°   **14.** 212°

15. An angle whose measure equals the sum of the measures of two right angles.

16. An angle whose measure equals the sum of the measures of one acute and one right angle.

Exercises 17–26 refer to Illustration 2. Find the measure of angle BAC based on the given information.

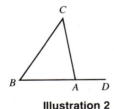

Illustration 2

17. angle *B* = 20° and angle *C* = 30°   **18.** angle *B* = 89° and angle *C* = 17°

19. *AB* = *AC* and angle *B* = 42°   **20.** *BC* = *AC* and angle *B* = 42°

21. *AB* = *BC* and angle *C* = 63°   **22.** *CB* = *CA* and angle *C* = 63°

23. Triangle *BCA* is equilateral.   **24.** Triangle *CAB* is equiangular.

25. Angle *CAD* measures 57° and points *B*, *A*, and *D* are on the same line.

26. angle *B* + angle *C* = 43°

In Exercises 27–32, tell if the triangles with the given sides are right triangles.

27. 3 cm, 4 cm, 5 cm   **28.** 6 ft, 8 ft, 10 ft

29. 4 cm, 5 cm, 6 cm   **30.** 7 cm, 9 cm, 11 cm

31. 15 ft, 20 ft, 25 ft   **32.** 10 cm, 26 cm, 24 cm

In Exercises 33–38, find the measure of the hypotenuse of a right triangle with the given legs.

33. 30 m and 40 m   **34.** 20 cm and 48 cm

Illustration 3

35. 1 cm and 2 cm

36. 11 in. and 17 in.

37. a ft and b ft

38. b cm and $(b + d)$ cm

Exercises 39–44 refer to Illustration 3. Give the measure of each of the remaining sides.

39. $AC = 2$ cm

40. $BC = 3$ ft

41. $AB = 10$ in.

42. $AB = 8$ cm

43. $AC = 3\sqrt{3}$ km

44. $BC = 4\sqrt{5}$ cm

Exercises 45–56 refer to Illustration 4. Give the measure of each of the remaining sides.

Illustration 4

45. $AC = 3$ cm

46. $BC = 3$ cm

47. $BC = 5$ mi

48. $AC = 5$ cm

49. $AB = 10$ cm

50. $AB = 15$ yd

51. $AC = 2\sqrt{3}$ cm

52. $AC = 2\sqrt{5}$ cm

53. $BC = 2\sqrt{3}$ ft

54. $BC = 2\sqrt{5}$ in.

55. $AB = 2\sqrt{3}$ cm

56. $AB = 2\sqrt{5}$ cm

Recall that $\sqrt{x^2}$ denotes the principal square root of x^2. If x can be negative, then absolute value notation is required to guarantee that $\sqrt{x^2}$ is positive. For this reason, $\sqrt{x^2} = |x|$. Absolute values are not required for cube roots, because all real numbers have a single real cube root; the cube root of a negative number is negative, the cube root of zero is zero, and the cube root of a positive number is positive.

In Exercises 57–64, simplify each radical using absolute value notation when necessary.

57. $\sqrt{16x^2}$

58. $\sqrt{25x^4}$

59. $\sqrt{625x^8}$

60. $\sqrt{64x^6}$

61. $\sqrt[3]{x^3}$

62. $\sqrt[3]{x^6}$

63. $\sqrt[n]{x^n}$ and n is even

64. $\sqrt[n]{x^n}$ and n is odd

I.2   SIMILAR TRIANGLES

Historically, much of the work of applied trigonometry has involved indirect measurement. For example, it would be inconvenient to measure the length of a flagpole by climbing up the pole with a tape measure. It would be much better to stay safely on the ground and measure the flagpole indirectly. We can do this by using similar triangles.

(I.6)

> **Definition.**   Two triangles are called **similar** if and only if the three angles of one triangle are equal, respectively, to the three angles of the second triangle, and all pairs of corresponding sides are in proportion.

This definition implies that similar triangles have the same shape. If similar triangles also have the same size (area), they are called **congruent.** We will accept the following theorem without proof.

(I.7)

> **Theorem.**   Two triangles are similar if and only if two angles of one triangle are equal to two angles of the other triangle.

Example 1   Find the values of x and y in Figure I-9 if angle A equals angle D and angle C equals angle F.

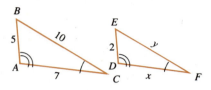

Figure I-9

Solution   Because two angles of the larger triangle are equal to two angles of the smaller triangle, the triangles are similar. Because they are similar, all corresponding sides are in proportion. So, you can set up the proportions

$$\frac{5}{2} = \frac{10}{y} \quad \text{and} \quad \frac{5}{2} = \frac{7}{x}$$

and solve for x and y.

$$5y = 20 \quad \text{and} \quad 5x = 14$$

$$y = 4 \qquad\qquad x = \frac{14}{5}$$

Example 2   A flagpole casts a shadow of 8 m at the same time a meter stick held perpendicular to the ground casts a shadow of 50 cm. Find the height of the flagpole.

Solution   Because the triangles in Figure I-10 are similar, all corresponding sides are in proportion. Note that 50 cm = 0.5 m. Form the proportion

$$\frac{h}{1} = \frac{8}{0.5}$$

and solve for h.

$$0.5h = 8$$

$$h = 16$$

Figure I-10   The height of the flagpole is 16 m.

Example 3   Two boy scouts wish to know the width of the river represented in Figure I-11. How can they find the width of the river without crossing to the other side?

Solution   The boys are lucky and notice a tree and a large rock on the other bank of the river. By using their line of sight, the boys put stakes at points A, B, and C.

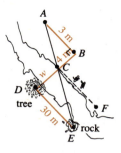

Figure I-11

The boys are able to measure sides AB and BC directly, and can estimate the measure of side DE indirectly by measuring the segment CF (F is on the bank straight across from E, and C is on the bank straight across from D). The boys make some measurements and find that $AB = 3$ m, $CB = 4$ m, and $CF = DE = 30$ m. Note that angles ACB and ECD are vertical angles and, therefore, must be equal. Also note that angles B and D are right angles and, therefore, must be equal. It follows that triangles ABC and EDC are similar triangles. Thus, the boys can form the following proportion and solve for w.

$$\frac{3}{30} = \frac{4}{w}$$
$$120 = 3w$$
$$40 = w$$

The river is 40 m wide.

In Example 3, the triangles ABC and EDC were shown to be similar. Because all corresponding pairs of angles in similar triangles are equal, it follows that angles A and E are equal. Note that line AE, called a **transversal,** intersects lines AB and DE to form angles A and E, which lie on opposite sides of the transversal AE and on the interior of lines AB and DE. Such angles are called **alternate interior angles.** Because transversal AE intersects lines AB and DE to form *equal alternate interior angles,* the lines AB and DE can never intersect. Such lines are called **parallel lines.** This important idea, along with its converse, is stated formally in the next theorem. Theorem (I.8) is stated without proof.

(I.8)

> **Theorem.**   If two lines are cut by a transversal so that alternate interior angles are equal, then the lines are parallel.
>
> If two parallel lines are cut by a transversal, then alternate interior angles are equal.

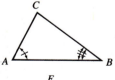

Illustration 1

Exercise I.2

Exercises 1–4 refer to the similar triangles in Illustration 1.

1. If $AC = 7$ cm, $DF = 4$ cm, and $AB = 15$ cm, find DE.
2. If $CB = 12$ m, $AB = 15$ m, and $DE = 7$ m, find FE.
3. If $AC = 21$ cm, $BC = 25$ cm, and $FE = 8$ cm, find DF.
4. If $DE = 6$ cm, $AC = 12$ cm, and $DF = 4$ cm, find AB.
5. A building casts a shadow of 50 ft at the same time a person casts a shadow of 8 ft. If the person is 6 ft tall, how tall is the building?

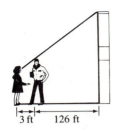

3 ft | 126 ft

Illustration 2

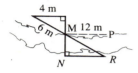

4 m

6 m M 12 m P

N   R

Illustration 3

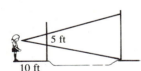

4 ft

6 ft   54 ft

Illustration 4

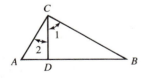

5 ft

10 ft

Illustration 5

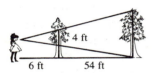

C

1

2

A D   B

Illustration 6

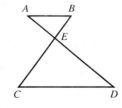

A   B

E

C   D

Illustration 7

6. A person $5\frac{1}{2}$ ft tall casts a 3-ft shadow. How long is the shadow of a 60-ft flagpole?

7. A straight road going up a constant grade rises 3 ft in the first 50 ft of roadway. If the car travels 1000 ft from the bottom of the hill to the top, how high is the hill?

8. A ski hill has a run $\frac{1}{2}$ mi long. If the hill drops 5 ft as the skier slides 9 ft, how high is the hill?

9. Sally is 5 ft 6 in. tall. To measure the height of a smokestack, she asks her 6-ft 6-in. boyfriend to help her. They stand as in Illustration 2 with her line of sight to the top of the stack being tangent to his head. If the boyfriend stands 126 ft from the smokestack and the girl is 3 ft from her friend, how tall is the smokestack?

10. Find the width of the river in Illustration 3. You may use the measurements in the figure, and you may assume that M and P are on the river bank, directly across from N and R, respectively.

11. An airplane ascends 100 ft as it travels 1000 ft. How much altitude will it gain if it flies one mile?

12. In a landing approach an airplane descends 45 ft as it travels 500 ft. How far does the plane travel as it drops 600 ft?

13. The two trees in Illustration 4 are 54 ft apart. To find the height of the large tree, a 5-ft-tall girl scout backs 6 ft away from the small tree and sights to both the top and bottom of the taller tree. If her lines of sight are separated by 4 ft on the smaller tree, how tall is the larger tree?

14. A 12-ft pole is placed on each side of a river as in Illustration 5. A trigonometry student steps back from one pole a distance of 10 ft. She sights to the top and bottom of the second pole. If her lines of sight are separated by 5 ft on the first pole, how wide is the river?

Exercises 15–20 refer to Illustration 6. Assume angle ACB = 90° and angle CDB = 90°.

15. If angle $A = 40°$, find angle 1.

16. If angle $2 = 60°$, find angle B.

17. Show that angle A is always equal to angle 1 and that angle B is always equal to angle 2.

18. Show that triangles ACD and ACB are similar.

19. Show that triangles BDC and BCA are similar.

20. Show that triangles ADC and CDB are similar.

Exercises 21–22 refer to Illustration 7.

21. If $AE = 4$, $DE = 6$, $BE = 3$, and AB is parallel to CD, find the length of CE.

22. If $AE = 4$, $DE = 6$, $BC = 8$, and AB is parallel to CD, find the length of BE.

I.3 CIRCLES

A knowledge of certain properties of circles is essential to understanding the principles of trigonometry. As a matter of fact, the trigonometric functions are often called the **circular functions.**

(I.9)

> **Definition.** A **circle** is a set of points all in the same plane that are equidistant from a fixed point. The fixed point is called the **center** of the circle.

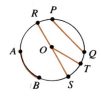

Figure I-12

Consider the circle with center at point O in Figure I-12. A segment, such as PQ, that joins two points on a circle is called a **chord.** Any chord that passes through the center of the circle is called a **diameter.** For example, RS is a diameter of circle O. A portion of the circle (such as $\overset{\frown}{AB}$) is called an **arc.** Any segment joining the center of the circle to a point on the circle (such as OT) is called a **radius.** Any angle determined by two radii (such as angle TOS) is called a **central angle.** The distance around the entire circle is called the **circumference** of the circle.

If the circumference of any circle is divided by the length of its diameter, the quotient is a constant that is independent of the size of the circle. This constant is the number pi ($\pi = 3.14159265...$).

$$\frac{C}{D} = \pi$$

Multiplying both sides by D gives the formula for the circumference of a circle.

(I.10)

> **Formula for Circumference of a Circle.** $C = \pi D$

Because $D = 2r$, where r is the length of a radius of the circle, this formula can be rewritten as

(I.11)

> **Formula for Circumference of a Circle.** $C = 2\pi r$

A rigorous proof for the formula for the area of a circle requires techniques from calculus. For this reason, we state without proof that the area of a circle is given by the formula

(I.12)

> **Formula for Area of a Circle.** $A = \pi r^2$

Example 1 Find the circumference and area of a circle with a diameter of 26 cm.

Solution Because the diameter is 26 cm, the radius is 13 cm. Use the formulas for the circumference and area of a circle.

$$C = \pi(26\,\text{cm}) = 26\pi\,\text{cm}$$

$$A = \pi(13\,\text{cm})^2 = 169\pi\,\text{sq cm}$$

■

(I.13)

> **Definition.**   A **sector** of a circle is the set of points bounded by two radii and the arc intercepted by those radii.

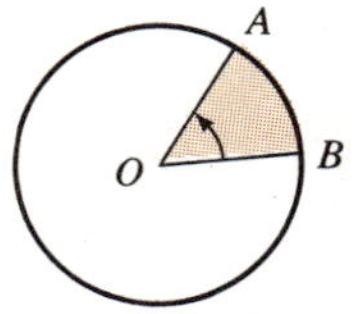

Figure I-13

The area shaded in Figure I-13 represents the area of sector *AOB* of circle *O*. Note from Figure I-14 that the central angle *AOB* is the same fractional part of one complete revolution that its intercepted arc is of the circle's circumference.

$$\frac{\text{angle}\,AOB}{360°} = \frac{\overset{\frown}{AB}}{C}$$

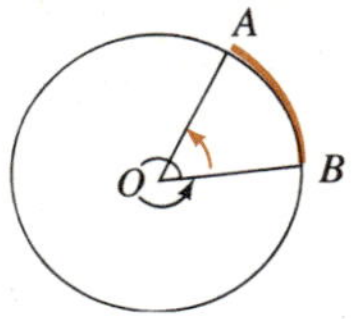

Figure I-14

The Greek mathematician Eratosthenes (275–195 B.C.) was aware of this relationship and used it to estimate the circumference of the earth. He knew that at noontime during the summer solstice, the sun was directly overhead in Syene, Egypt (now Aswan). At this time, the sundials in Syene produced no shadows. However, the sundials did cast a shadow in Alexandria, a city about 5000 stades (approximately 500 miles) north of Syene. See Figure I-15. In the figure, line *OA* passes from the center of the earth through the tip of a sundial in Alexandria. Line *OS* passes from the center of the earth through the tip of a sundial in Syene. Eratosthenes was able to determine that in Alexandria, the sun crossed the pointer of a sundial at an angle that was one-fiftieth of a complete revolution, or 7.2°. Because the rays of the sun are nearly parallel and because two parallel lines cut by a transversal determine equal corresponding angles, angle *AOS* is also 7.2°. Although Eratosthenes reasoned geometrically, elementary algebra can be used to solve a proportion for *C*.

$$\frac{\text{angle}\,AOS}{360°} = \frac{\overset{\frown}{AS}}{C}$$

$$\frac{7.2°}{360°} = \frac{500}{C}$$

$$C = \frac{500(360)}{7.2}$$

$$C = 25{,}000$$

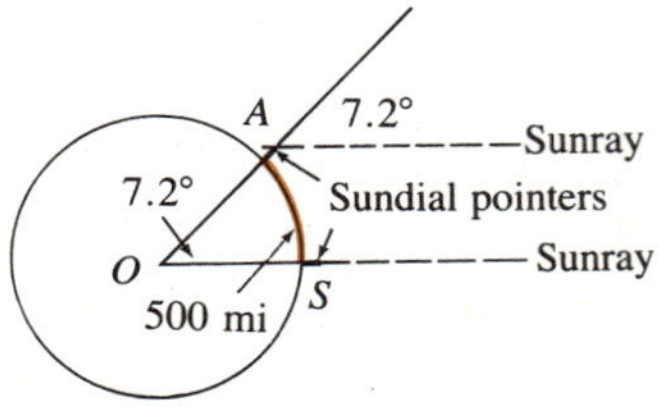

Figure I-15

Eratosthenes' estimate of the circumference of the earth was 25,000 mi. The actual circumference is closer to 24,900 mi.

Exercise I.3

Exercises 1–2 refer to Illustration 1. Circle O has AB as one of its diameters.

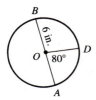

Illustration 1

1. Find the circumference of circle O.

2. Find the area of circle O.

3. Find the radius of the circle with an area of 81π sq units.

4. Find the radius of the circle with an area of 121π sq units.

5. Find the radius of the circle with a circumference of 81π units.

6. Find the radius of the circle with a circumference of 121π units.

7. Find the radius of the circle with an area of 63 sq units.

8. Find the radius of the circle with an area of 47 sq units.

9. Find the radius of the circle whose area has the same number of units as its circumference.

10. Find the radius of the circle whose area has twice the number of units as its circumference.

11. Find the diameter of the circle whose circumference has one-third the number of units as its area.

12. Find the diameter of the circle whose circumference has three times the number of units as its area.

Exercises 13–16 refer to Illustration 2.

Illustration 2

13. If $AB = 10$ cm, find the circumference of circle O.

14. If $AB = 12$ m, find the area of circle O.

15. If the area of circle O is 12 sq in., find AB.

16. If the circumference of circle O is 18 ft, find AB.

17. John wants to plant marigolds to border his circular-shaped garden. If he plants the marigolds 8 in. apart and the garden has a diameter of 20 ft, how many marigold plants does he need?

18. If canvas sells for $4 a square yard, how much will it cost for material to make a cover for a round swimming pool 25 ft in diameter? Ignore any waste or overlap.

19. If the cost of a belt is $4 per yard, how much would it cost to buy a belt to encircle the earth at its equator? Consider the radius of the earth to be 4000 mi.

20. If paint costs $10 a gallon and if one gallon covers 500 sq ft, how much would it cost to paint a circular helicopter landing pad 60 ft in diameter?

21. A bicycle with 27-in. diameter wheels is ridden for 1 mi. How many times will each of the wheels rotate during the trip?

22. A vehicle with 30-in. diameter wheels travels down the road. How many miles does the vehicle go if the tires rotate 1000 times?

23. If a college has a dining room with a circular floor plan with a diameter of 40 ft, how much will it cost for material to carpet the floor? Assume no waste and that carpet can be purchased for $15 per square yard.

Illustration 3

24. Find the area of the shaded portion of Illustration 3. The large circle has a radius of R and the small circle has a radius of r.

25. If a central angle of a circle measures 15° and its intercepted arc measures 1 m, what is the circumference of the circle?

26. If the intercepted arc of a central angle in a circle measures 12 cm and the circumference of the circle is 31 cm, what is the measure of the central angle to the nearest tenth of a degree?

27. A central angle of a circle measures 125.5° and intercepts an arc of x cm. Find x if the diameter of the circle is 50 cm.

28. If the radius of a circle is 10 m and an arc of the circle is 60 cm, find, to the nearest tenth of a degree, the measure of the central angle that intercepts the given arc.

29. Use Eratosthenes' estimate of the earth's circumference to estimate the length of the earth's radius.

30. Use Eratosthenes' estimate of the earth's circumference to estimate the earth's volume. ($V = \frac{4}{3}\pi r^3$)

I.4   DEGREES–MINUTES–SECONDS

One unit of angular measure is the **degree.** One degree is $\frac{1}{360}$ of one complete rotation. In this book, we indicate fractional parts of a degree by using decimals. For example, $\frac{1}{2}^\circ = 0.5°$, or $33\frac{3}{4}^\circ = 33.75°$. **Minutes** and **seconds** provide another way of expressing fractional parts of a degree.

(I.14)

> **Definition.**   1 minute is $\frac{1}{60}$ of 1 degree.
>
> 1 second is $\frac{1}{60}$ of 1 minute.

This definition implies that

60 minutes = 1 degree
60 seconds = 1 minute
3600 seconds = 1 degree

The symbol for minute is ′ and for second is ″. Thus, the angle 37° 24′ 18″ is read 37 degrees 24 minutes 18 seconds.

A time may come when you may need to convert from degrees–minutes–seconds to decimal degrees, or vice versa.

Example 1   Change 37° 24′ 18″ to decimal degrees.

Solution   Note that 24′ is $\frac{24}{60}$ of 1 degree. Change $\frac{24}{60}$ to a decimal to find that 24′ = 0.4°. 18″ is $\frac{18}{3600}$ of 1 degree. Because $\frac{18}{3600} = 0.005$, 18″ = 0.005°. Thus,

$$37° \ 24′ \ 18″ = 37° + 0.4° + 0.005°$$
$$= 37.405°$$

Example 2   Change 83.41° to degrees–minutes–seconds.

Solution   The 83 to the left of the decimal point represents 83°. The task is to change 0.41° to minutes and seconds. First change the 0.41° to seconds using the proportion

$$\frac{41}{100} = \frac{x}{3600}$$

$$1476 = x$$

Thus, 0.41° = 1476″. Divide 1476″ by 60 to find the number of minutes in 1476 seconds.

```
        24
   60)1476
      120
      276
      240
       36
```

Thus, 0.41° = 24′ 36″ and 83.41° = 83° 24′ 36″. ■

Exercise I.4

In Exercises 1–6, change each angle to decimal degrees. Give answers to the nearest thousandth.

1. 84° 36′ 24″   **2.** 67° 12′ 48″

3. 25° 48′ 6″   **4.** 11° 30′ 15″

5. 5° 25′ 20″   **6.** 43° 43′ 51″

In Exercises 7–12, change each angle to degree–minute–second notation.

7. 23.14°   **8.** 51.45°

9. 73.87°   **10.** 81.45°

11. 123.29°   **12.** 312.57°

APPENDIX II

EXPONENTIAL AND LOGARITHMIC FUNCTIONS

In the late 16th century, **logarithms** were invented to simplify computations in astronomy. In fact, they proved to be so successful that they were soon used to simplify calculations in many other fields. Today, the widespread use of calculators has eliminated the need for this computational aid. However, because logarithms have maintained their importance in mathematics, science, engineering, and economics in other practical ways, we discuss their properties in the next section.

II.1 EXPONENTIAL AND LOGARITHMIC FUNCTIONS

If b is a positive real number, and x is any real number, then the exponential expression b^x represents a real number also. The function defined by the equation $y = f(x) = b^x$ is called an **exponential function.**

(II.1)

> **Definition.** An **exponential function** with base b is defined by the equation
>
> $$y = b^x$$
>
> where $b > 0$, $b \neq 1$, and x is any real number.

Because the exponential function $y = f(x) = b^x$ has the real number set for its domain and the set of positive real numbers for its range, it is possible to graph the function $y = b^x$. Figure II-1 shows the graph of two exponential functions: $y = f(x) = 2^x$ and $y = f(x) = \left(\frac{1}{2}\right)^x = 2^{-x}$.

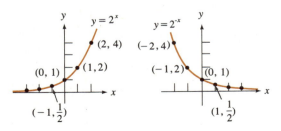

Figure II-1

We now consider the graphs of the exponential function $y = f(x) = 3^x$ (see Figure II-2) and the exponential equation $x = 3^y$ (see Figure II-3).

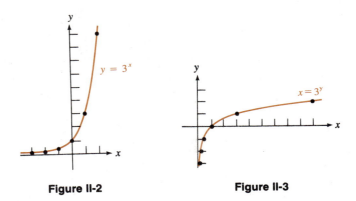

Figure II-2   **Figure II-3**

Because the roles of x and y have been interchanged, each of these functions is the inverse of the other. Merging the graphs of the two functions onto a single set of coordinate axes reveals that the graphs are symmetric to each other with the line $y = x$ as the axis of symmetry. See Figure II-4.

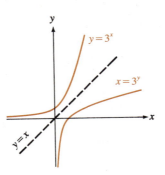

Figure II-4

To express the function $x = 3^y$ in a form that is solved for y, we give the following definition.

(II.2)   **Definition.** The **logarithmic function** $y = \log_b x$, where $b > 0$ and $b \neq 1$, is equivalent to the exponential equation $x = b^y$: any pair of values (x, y) that satisfies $x = b^y$ also satisfies $y = \log_b x$. In either notation, b is called the **base**.

From this definition, it follows that

$$\log_3 x = y \quad \text{means} \quad x = 3^y$$
$$\log_{10} x = y \quad \text{means} \quad x = 10^y$$
$$\log_7 1 = 0 \quad \text{means} \quad 1 = 7^0$$
$$\log_4 16 = 2 \quad \text{means} \quad 16 = 4^2$$
$$\log_2 32 = 5 \quad \text{means} \quad 32 = 2^5$$

Note that $y = \log_b x$ and $y = b^x$ are inverse functions.

Because logarithms can be written as exponential expressions, it is not surprising that the laws of exponents have counterparts in logarithmic notation. We will use this fact to develop the important properties of logarithms.

(II.3)

> **Property 1.** $\log_b MN = \log_b M + \log_b N$

Proof   Let $x = \log_b M$ and $y = \log_b N$. Using the definition of logarithms, these equations can be written in the form

$$M = b^x \quad \text{and} \quad N = b^y$$

Multiplying these equations together and using properties of exponents gives

$$MN = b^x b^y$$
$$MN = b^{x+y}$$

Using the definition of logarithms in reverse gives

$$\log_b MN = x + y$$

Substituting the values of x and y completes the proof.

$$\log_b MN = \log_b M + \log_b N \qquad \square$$

(II.4)

> **Property 2.** $\log_b \dfrac{M}{N} = \log_b M - \log_b N$

Proof   Again, let $x = \log_b M$ and $y = \log_b N$. These equations can be written as

$$M = b^x \quad \text{and} \quad N = b^y$$

This time, we divide one equation by the other and use the properties of exponents.

$$\frac{M}{N} = \frac{b^x}{b^y}$$
$$\frac{M}{N} = b^{x-y}$$

Using the definition of logarithms in reverse gives

$$\log_b \frac{M}{N} = x - y$$

Substituting the values for x and y completes the proof.

$$\log_b \frac{M}{N} = \log_b M - \log_b N$$

□

(II.5) | **Property 3.** $\log_b M^p = p \log_b M$

Proof   Again let $x = \log_b M$. Write this expression in exponential form and raise both sides to the power of p.

$$M = b^x$$
$$(M)^p = (b^x)^p$$
$$M^p = b^{px}$$

Using the definition of logarithms in reverse gives

$$\log_b M^p = px$$

Substituting the value for x completes the proof.

$$\log_b M^p = p \log_b M$$

□

Two other important properties follow directly from the definition.

(II.6) | **Property 4.** $\log_b 1 = 0$ since $b^0 = 1$.
Property 5. $\log_b b = 1$ since $b^1 = b$.

Example 1   Find the value of x in the equations **a.** $\log_3 81 = x$, **b.** $\log_x 125 = 3$, and **c.** $\log_4 x = 3$.

Solution   **a.** $\log_3 81 = x$ is equivalent to $3^x = 81$. Note that $3^4 = 81$. Hence, $x = 4$.
b. $\log_x 125 = 3$ is equivalent to $x^3 = 125$. Because $5^3 = 125$, it follows that $x = 5$.
c. $\log_4 x = 3$ is equivalent to $4^3 = x$. Because $4^3 = 64$, it follows that $x = 64$.

■

Example 2   Find the value of x in the equations **a.** $\log_{1/3} x = 2$, **b.** $\log_{1/3} x = -2$, and **c.** $\log_{1/3} \frac{1}{27} = x$.

Solution   **a.** $\log_{1/3} x = 2$ is equivalent to $\left(\frac{1}{3}\right)^2 = x$. Because $\left(\frac{1}{3}\right)^2 = \frac{1}{9}$, it follows that $x = \frac{1}{9}$.

b. $\log_{1/3} x = -2$ is equivalent to $\left(\frac{1}{3}\right)^{-2} = x$. Because

$$\left(\frac{1}{3}\right)^{-2} = \frac{1}{\left(\frac{1}{3}\right)^2} = \frac{1}{\frac{1}{9}} = 9$$

it follows that $x = 9$.

c. $\log_{1/3} \frac{1}{27} = x$ is equivalent to $\left(\frac{1}{3}\right)^x = \frac{1}{27}$. Because $\left(\frac{1}{3}\right)^3 = \frac{1}{27}$, it follows that $x = 3$.   ■

Example 3   If $\log_b A = a$, $\log_b C = c$, and $\log_b D = d$, find **a.** $\log_b(ACD)$, **b.** $\log_b A^3$, **c.** $\log_b(A^2D^3)$, and **d.** $\log_b \left(\dfrac{A + C}{D^2}\right)$.

Solution   **a.** $\log_b (ACD) = \log_b A + \log_b C + \log_b D$

$$= a + c + d$$

b. $\log_b A^3 = 3 \log_b A$

$$= 3a$$

c. $\log_b(A^2D^3) = \log_b A^2 + \log_b D^3$

$$= 2 \log_b A + 3 \log_b D$$

$$= 2a + 3d$$

d. $\log_b \left(\dfrac{A + C}{D^2}\right) = \log_b(A + C) - \log_b D^2$

$$= \log_b(A + C) - 2d$$

Notice that $\log_b(A + C)$ is not equal to $\log_b A + \log_b C$.   ■

Finding Logarithms by Using Calculators

It is easy to find logarithms of certain numbers. For example, $\log_{10} 10 = 1$, $\log_{10} 100 = 2$, and $\log_{10} 1000 = 3$. However, finding the value of $\log_{10} 2$ is not so easy. The best way to find this value is to use a calculator. To find $\log_{10} 2$, enter the number 2 in your calculator and press the $\boxed{\text{LOG}}$ key. (*Caution:* You may need to press a $\boxed{\text{2ND}}$ function key first.) The display on your calculator should read .30103000. Hence, $\log_{10} 2 \approx 0.3010$. Remember that $\log_{10} 2 \approx 0.3010$ is equivalent to $10^{0.3010} \approx 2$. If this procedure does not work on your calculator, consult your owner's manual.

Example 4   Use a calculator to find the values of x in the equations **a.** $\log_{10} 8.75 = x$, **b.** $\log_{10} 379 = x$, and **c.** $\log_{10} x = -2.1180$.

Solution   **a.** Enter the number 8.75. Press the $\boxed{\text{LOG}}$ key. The display reads 0.94200805. Hence, $\log_{10} 8.75 \approx 0.9420$.

b. Enter the number 379. Press the ☐LOG☐ key. The display reads 2.57863921. Hence, $\log_{10} 379 \approx 2.5786$.

c. This problem is worked differently. To find x in the equation $\log_{10} x = -2.1180$, change the equation to its exponential form: $10^{-2.1180} = x$. To evaluate $10^{-2.1180}$, enter the number 10, press the ☐y^x☐ key, and enter the number 2.1180. Then, press the ☐+/−☐ key and the ☐=☐ key. The display reads .00762079. Hence, $x = 10^{-2.1180} \approx 0.0076$.   ■

Note that it is impossible to find a logarithm for a negative number. For example, $\log_3(-9) = x$ is a meaningless statement because there is no number x such that $3^x = -9$. The graph of $y = 3^x$, shown in Figure II-2, shows that y can never be negative; y can never be -9.

Finding Logarithms by Using Tables

Before the widespread use of calculators, mathematicians used extensive tables to find logarithms of numbers. The following examples show how. In these examples, if the base b is not written, assume that it is 10; log A means $\log_{10} A$. Base-10 logarithms are called **common logarithms.**

Example 5 Use Table B in Appendix IV to find the logarithm of 2.71 to base 10.

Solution Because log 1 = 0 and log 10 = 1 (by Properties 4 and 5 of logarithms), log 2.71 will be between 0 and 1 (2.71 lies between 1 and 10). Run your finger down the left column of Table B until you reach 27. Then slide your finger along that row until you reach the entry in the column headed 1. The value listed there is the common logarithm of 2.71. Thus, log 2.71 $\approx$ 0.4330 ($10^{0.4330} \approx 2.71$). Notice that all decimal points are missing from the table; the 27 in the left column is understood to be 2.7, and the table's entry, 4330, is understood to be 0.4330.   ■

Example 6 Find the common logarithm of 2,710,000.0.

Solution Although 2,710,000.0 does not appear in the left column of Table B, its logarithm can be found by using properties of logarithms. Because 2,710,000.0 can be written in scientific notation as 2.71×10^6,

$$
\begin{aligned}
\log 2{,}710{,}000.0 &= \log(2.71 \times 10^6) \\
&= \log 2.71 + \log 10^6 & \text{by Property 1} \\
&= \log 2.71 + 6 \log 10 & \text{by Property 3} \\
&= \log 2.71 + 6 & \text{by Property 5} \\
&\approx 0.4330 + 6 & \text{by Example 5} \\
&\approx 6.4330
\end{aligned}
$$

Notice that 2,710,000.0 and 2.71 differ only in the position of the decimal point and that their base-10 logarithms differ only by an integer.   ■

If a common logarithm is written as a sum of an integer (6, in the previous example) and a decimal between 0 and 1 (0.4330 in the previous example), the decimal part is called the **mantissa** and the integer is called the **characteristic.** The characteristic of the logarithm of a number is determined solely by the position of the decimal point in that number.

Mantissa and characteristic

Example 7   Find log 0.000271.

Solution   This problem is similar to the previous one. Write 0.000271 in scientific notation and use logarithmic properties to simplify it.

$$\log 0.000271 = \log\ (2.71 \times 10^{-4})$$
$$= \log\ 2.71 + \log 10^{-4}$$
$$= \log\ 2.71 - 4 \log 10$$
$$= \log\ 2.71 - 4$$
$$\approx 0.4330 - 4$$
$$\approx -3.5670$$

In this example, the characteristic is -4 and the mantissa is 0.4330 because a mantissa must be a number between 0 and 1.   ◼

Example 8   Find N if $\log N = -2.1180$.

Solution   A logarithm can be negative; any number between 0 and 1 has a negative common logarithm, as in Example 7. Table B contains logarithms of numbers between 1 and 10, but these logarithms are positive. Here is how a negative logarithm can be found in a table of positive values.

Add the number 3 to the negative logarithm (-2.1180), making it a positive number. Then subtract 3 so that the value does not change.

$$\log N = -2.1180 = (-2.1180\ \mathbf{+\ 3)\ -\ 3}$$
$$= 0.8820 - 3$$

The positive decimal (or mantissa), 0.8820, can be found in the body of the table: $0.8820 \approx \log 7.62$. In exponential form, $10^{0.8820} \approx 7.62$. The characteristic -3 determines the location of the decimal point.

$$\log N = 0.8820 - 3$$

means that

$$N = 10^{(0.8820 - 3)}$$
$$= 10^{0.8820} \cdot 10^{-3}$$
$$\approx 7.62 \cdot 10^{-3}$$
$$\approx 0.00762$$ ◼

Finding the characteristic

An easy way to determine the characteristic of a number is to begin just to the right of the first nonzero digit of the number and count the number of digits to the decimal point. Counting to the left means that the characteristic is negative; counting to the right means that it is positive.

Example 9   Find the characteristic of the logarithm of the number 210.

Solution   Begin to the right of the 2 and count toward the decimal point, which is *two* places to the *right*. Hence, the characteristic is $+2$. ■

Example 10   Find the characteristic of the logarithm of the number 0.0034.

Solution   Begin to the right of the 3. Count to the decimal point, which is *three* places to the *left*. The characteristic is -3. ■

If the logarithm of a number N is known, use the characteristic to place the decimal point in the digits of N. If the digits of N are 3762, for example, and the characteristic of the logarithm of N is 2, the decimal point is determined by counting two places to the right of the first nonzero digit. Thus,

$$N = 3 \quad 7 \quad 6. \quad 2$$

Example 11   Use logarithms to calculate $\sqrt[3]{25,000}$.

Solution   First calculate the logarithm of $\sqrt[3]{25,000}$.

$$\log \sqrt[3]{25,000} = \log 25,000^{1/3}$$
$$= \frac{1}{3} \log 25,000$$
$$= \frac{1}{3} \log(2.5 \cdot 10^4)$$
$$\approx \frac{1}{3}(4.3979)$$
$$\approx 1.4660$$

The characteristic is 1 and the mantissa is 0.4660. Look up 0.4660 in the body of Table B and find that $0.4660 \approx \log 2.92$. The characteristic of 1 fixes the location of the decimal point.

$$\sqrt[3]{25,000} \approx 2.92 \cdot 10^1$$
$$\approx 29.2$$ ■

Exercise II.1

In Exercises 1–24, find the value of x. A calculator is of no value.

1. $\log_2 8 = x$

2. $\log_{1/2} \frac{1}{8} = x$

3. $\log_{1/2} 8 = x$

4. $\log_{25} 5 = x$

5. $\log_5 25 = x$

6. $\log_8 x = 2$

7. $\log_x 8 = 3$

8. $\log_7 x = 0$

9. $\log_7 x = 1$

10. $\log_4 x = \frac{1}{2}$

11. $\log_x \dfrac{1}{16} = -2$

12. $\log_{125} x = \dfrac{2}{3}$

13. $\log_{100} \dfrac{1}{1000} = x$

14. $\log_{5/2} \dfrac{4}{25} = x$

15. $\log_{27} 9 = x$

16. $\log_{12} x = 0$

17. $\log_x 5^3 = 3$

18. $\log_x 5 = 1$

19. $\log_x \dfrac{9}{4} = 2$

20. $\log_x \dfrac{\sqrt{3}}{3} = \dfrac{1}{2}$

21. $\log_{\sqrt{3}} x = -4$

22. $\log_\pi x = 3$

23. $\log_{\sqrt{8}} x = 2$

24. $\log_4 8 = x$

25. On one set of coordinate axes, graph $y = 2^x$ and $y = \log_2 x$. What is the line of symmetry?

26. On one set of coordinate axes, graph $y = 4^x$ and $y = \log_4 x$. What is the line of symmetry?

27. On one set of coordinate axes, graph $y = \left(\frac{1}{2}\right)^x$ and $y = \log_{1/2} x$. What is the line of symmetry?

28. On one set of coordinate axes, graph $y = \left(\frac{4}{3}\right)^x$ and $y = \log_{4/3} x$. What is the line of symmetry?

In Exercises 29–34, find the value of b, if any, that would cause the graph of $y = b^x$ to look like the graph indicated.

29.

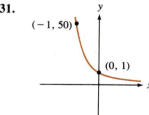

30.

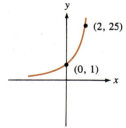

31.

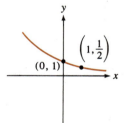

32.

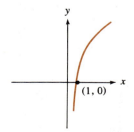

33.

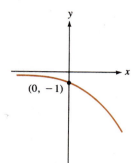

(0, −1)

34.

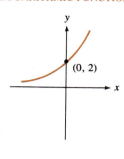

(0, 2)

In Exercises 35–40, find the value of b, if any, that would cause the graph of y = $\log_b x$ *to look like the graph indicated.*

35.

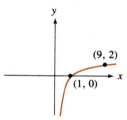

(2, 0)

36.

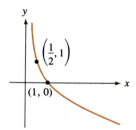

(−1, 0)

37.

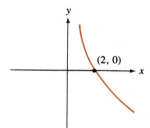

(9, 2)
(1, 0)

38.

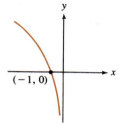

$\left(\frac{1}{2}, 1\right)$
(1, 0)

39.

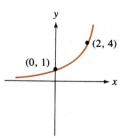

(2, 4)
(0, 1)

40.

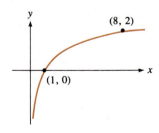

(8, 2)
(1, 0)

In Exercises 41–66, tell if the given statement is always true. If not, answer false.

41. $\log_b ab = \log_b a + 1$

42. $\log_b \dfrac{1}{a} = -\log_b a$

43. $\log_b 0 = 1$

44. $\log_b 2 = \log_2 b$

45. $\log_b(x + y) = \log_b x + \log_b y$

46. $\log_b xy = (\log_b x)(\log_b y)$

47. If $\log_a b = c$, then $\log_b a = c$.

48. If $\log_a b = c$, then $\log_b a = \dfrac{1}{c}$.

49. $\log_7 7^7 = 7$

50. $7^{\log_7 7} = 7$

51. $\log_b(-x) = -\log_b x$

52. If $\log_b a = c$, then $\log_b a^p = pc$.

53. $\dfrac{\log A}{\log B} = \log A - \log B$

54. $\log(A - B) = \dfrac{\log A}{\log B}$

55. The base of a system of logarithms cannot be negative.

56. $\log_b 1 = 0$

57. $\log_b 1 = 1$

58. $\log_b b = 1$

59. $\dfrac{1}{3} \log_b \sqrt[3]{a} = \log_b a$

60. $\dfrac{1}{3} \log_b a^3 = \log_b a$

61. A logarithm cannot be negative.

62. The logarithm of a negative number is negative.

63. $\log_b 0 = 0$

64. If x lies between 0 and 1, $\log_b x$ is negative.

65. $\log_{4/3} y = -\log_{3/4} y$

66. $\log_b y + \log_{1/b} y = 0$

In Exercises 67–76, use Table B in Appendix IV to calculate the following values.

67. $(2.3)^{4/3}$

68. $\sqrt[3]{0.007}$

69. $(0.012)^{-0.03}$

70. $(1.05)^{25}$

71. $10^{5.4942}$

72. $(1.73)^{1.73}$

73. $\sqrt{0.071}$

74. $4.3^{-5.2} + 3.1^{1.3}$

75. $(\log 4.1)^{2.4}$

76. $(2.3 + 1.79)^{-0.157}$

II.2   EXPONENTIAL AND LOGARITHMIC EQUATIONS

An **exponential equation** is one in which the variable appears in an exponent, and a **logarithmic equation** involves the logarithm of an expression containing the variable. Some exponential and logarithmic equations are exceedingly difficult to solve, but others (the ones considered here) yield to the usual equation-solving techniques, supplemented by the properties of logarithms.

Example 1   Solve the exponential equation $3^x = 5$.

Solution   You may be uncomfortable with the variable x as an exponent. Luckily, it can be moved from that position and made more accessible. Property 3 provides the means for moving this variable. Because the logarithms of equal numbers must be equal, take the common logarithm of each side of the equation. Then, apply Property 3 to the expression $\log 3^x$.

$$3^x = 5$$
$$\log 3^x = \log 5$$
$$x \log 3 = \log 5$$

Divide both sides of the equation by log 3, substitute values, and simplify.

1. $x = \dfrac{\log 5}{\log 3}$

$x \approx \dfrac{0.6990}{0.4771}$

$x \approx 1.465$

A careless reading of Equation 1 can lead to a common error. Because $\log \frac{A}{B} = \log A - \log B$, you may think that the expression $\frac{\log 5}{\log 3}$ also involves subtraction. It does not. The expression $\frac{\log 5}{\log 3}$ calls for division. ■

Example 2   Solve the exponential equation $6^{x-3} = 2$.

Solution   As in the previous example, the exponent is made more accessible by taking the common logarithm of both sides, and applying Property 3. Then, solve for x.

$$6^{x-3} = 2$$

$\log 6^{x-3} = \log 2$ \qquad Take the logarithm of both sides.

$(x - 3) \log 6 = \log 2$ \qquad Use Property 3.

$x - 3 = \dfrac{\log 2}{\log 6}$ \qquad Divide both sides by log 6.

$x = \dfrac{\log 2}{\log 6} + 3$ \qquad Add 3 to both sides.

$x \approx \dfrac{0.3010}{0.7782} + 3$ \qquad Substitute values.

$x \approx 0.3868 + 3$ \qquad Simplify.

$x \approx 3.3868$ \qquad Simplify. ■

Example 3   Solve the logarithmic equation $\log_{10} x + \log_{10}(x - 3) = 1$.

Solution

$\log_{10} x + \log_{10}(x - 3) = 1$

$\log_{10} x(x - 3) = 1$ \qquad Use Property 1.

$x(x - 3) = 10^1$ \qquad Change to exponential form.

$x^2 - 3x - 10 = 0$ \qquad Remove parentheses and add -10 to both sides.

$(x + 2)(x - 5) = 0$ \qquad Factor.

$x + 2 = 0$ \quad or \quad $x - 5 = 0$ \qquad Set each factor equal to 0.

$x = -2$ \qquad\qquad $x = 5$ \qquad Solve each possibility.

The value -2 is not a member of the replacement set of the original equation because the logarithm of a negative number is not defined. The remaining solution, $x = 5$, satisfies the original equation.

$$\log 5 + \log(5 - 3) = 1$$
$$\log 5 + \log 2 = 1$$
$$\log 5 \cdot 2 = 1$$
$$\log 10 = 1$$
$$1 = 1$$

Example 4   Solve the logarithmic equation $\log_b(3x + 2) - \log_b(2x - 3) = 0$.

Solution   Property 2 yields

$$\log_b\left(\frac{3x + 2}{2x - 3}\right) = 0$$

which can be written in equivalent exponential form as

$$\frac{3x + 2}{2x - 3} = b^0$$

$$\frac{3x + 2}{2x - 3} = 1 \qquad \text{Recall that } b^0 = 1.$$

$$3x + 2 = 2x - 3 \qquad \text{Multiply both sides by } 2x - 3.$$

$$x = -5 \qquad \text{Add } -2x - 2 \text{ to both sides.}$$

After substituting this value into the original logarithmic equation, you get $\log_b[3(-5)+2] - \log_b[2(-5)-3] = 0$. A logarithm of a negative number results, $\log_b(-13) - \log_b(-13) = 0$, and this is not defined. Thus, the original equation has no solution; its solution set is $\emptyset$.

A useful formula for finding logarithms to different bases can be found by solving the exponential equation $b^x = y$ for the variable x. This equation is very similar to that of Example 1 and it can be solved by taking the common logarithm of both sides of the equation. However, in order to produce the desired result, we will take the base-a logarithm of both sides of the equation and proceed as follows.

$$b^x = y$$
$$\log_a b^x = \log_a y$$
$$x \log_a b = \log_a y \qquad \text{Use Property 3.}$$
$$x = \frac{\log_a y}{\log_a b} \qquad \text{Divide both sides by } \log_a b.$$

Notice that the original equation, $b^x = y$, is equivalent to the logarithmic expression $x = \log_b y$. Combining this logarithmic expression with the expression $x = (\log_a y)/(\log_a b)$ results in the formula for converting logarithms from one base to another:

(II.7)   **Change-of-Base Formula.**

$$\log_b y = \frac{\log_a y}{\log_a b}$$

If you know logarithms to base a (for example, $a = 10$), we can find the logarithm of y to a new base, b. We divide the base-a logarithm of y by the base-a logarithm of the new base, b.

Example 5   Find $\log_2 8$. (Comment: You know that $\log_2 8 = 3$ because $2^3 = 8$. However, doing an easy problem first will help you understand the new ideas.)

Solution 1   To find the value of x that satisfies the equation $x = \log_2 8$ (in exponential form, $2^x = 8$), take the base-10 logarithm of both sides of $2^x = 8$.

$$\log_{10} 2^x = \log_{10} 8$$

$$x \log_{10} 2 = \log_{10} 8$$

$$x = \frac{\log_{10} 8}{\log_{10} 2}$$

$$x = \frac{\log_{10} 2^3}{\log_{10} 2}$$

$$x = \frac{3 \log_{10} 2}{\log_{10} 2}$$

$$x = 3$$

Solution 2   Use the change-of-base formula.

$$\log_b y = \frac{\log_a y}{\log_a b}$$

with $b = 2$, $y = 8$, and $a = 10$:

$$\log_2 8 = \frac{\log_{10} 8}{\log_{10} 2}$$

$$= 3 \qquad\qquad \text{See Solution 1.} \qquad\qquad ■$$

Example 6   Find $\log_3 5$.

Solution 1   Set x equal to $\log_3 5$ and write the equation in exponential form.

$$x = \log_3 5$$

$$3^x = 5$$

Take the common logarithm of both sides of this equation, apply some properties of logarithms, and solve for x.

$$\log_{10} 3^x = \log_{10} 5$$

$$x \cdot \log_{10} 3 = \log_{10} 5$$

$$x = \frac{\log_{10} 5}{\log_{10} 3}$$

$$x \approx \frac{0.6990}{0.4771}$$

$$x \approx 1.465$$

Solution 2   Use the change-of-base formula.

$$\log_b y = \frac{\log_a y}{\log_a b}$$

with $b = 3$, $y = 5$, and $a = 10$:

$$\log_3 5 = \frac{\log_{10} 5}{\log_{10} 3}$$

$$\approx 1.465 \qquad\qquad \text{See Solution 1.}$$

Exercise II.2

Solve each of the following exponential equations for x.

1. $4^x = 5$
2. $7^x = 12$
3. $2^{x-1} = 10$
4. $3^{x+1} = 100$
5. $2^{x+1} = 2^{2x+3}$
6. $3^{3x+2} = 3^{2x+7}$
7. $10^{x^2} = 100$
8. $10^{x^2} = 1000$
9. $3^x = 4^x$
10. $3^{2x} = 4^x$
11. $3^{x+1} = 4^x$
12. $3^x - 4^{x+1} = 0$
13. $10^{\sqrt{x}} = 1000$
14. $100^{\sqrt{x}} = 10$
15. $10^{\sqrt[3]{x}} - 100 = 0$
16. $1000^{\sqrt[3]{x}} - 10 = 0$

Solve each of the following logarithmic equations for x.

17. $\log 2x = \log 4$
18. $\log 3x = \log 9$
19. $\log(3x + 1) = \log(x + 7)$
20. $\log(x^2 + 4x) = \log(x^2 + 16)$
21. $\log(2x - 3) - \log(x + 4) = 0$
22. $\log(3x + 5) - \log(2x + 6) = 0$
23. $\log \dfrac{4x + 1}{2x + 9} = 0$
24. $\log \dfrac{5x + 2}{2(x + 7)} = 0$
25. $\log x^2 = 2$
26. $\log x^3 = 3$
27. $\log x + \log(x - 48) = 2$
28. $\log x + \log(x + 9) = 1$

29. $\log x + \log(x - 15) = 2$ **30.** $\log x + \log(x + 21) = 2$

31. $\log(x + 90) = 3 - \log x$ **32.** $\log(x - 90) = 3 - \log x$

33. $\log(x - 6) - \log(x - 2) = \log \dfrac{5}{x}$

34. $\log(x - 1) - \log 6 = \log(x - 2) - \log x$

35. $\log_7(2x - 3) - \log_7(x - 1) = 0$ **36.** $\log_{10} x^2 = (\log_{10} x)^2$

37. $\log_{10}(\log_{10} x) = 1$ **38.** $\log_3 x = \log_3 \left(\dfrac{1}{x}\right) + 4$

Find the following logarithms to the bases indicated.

39. $\log_3 7$ **40.** $\log_7 3$

41. $\log_{1/3} 3$ **42.** $\log_{2.3} 5.8$

43. $\log_8 \sqrt{2}$ **44.** $\log_9 3$

45. $\log_\pi \sqrt{2}$ **46.** $\log_{\sqrt{2}} \pi$

47. $\log_5 \sqrt{7}$ **48.** $\log_{\sqrt{2}} 3$

49. $\log_{2.5} \sqrt{5}$ **50.** $\log_7 14$

51. $\log_{3.9} 9.3$ **52.** $\log_{\sqrt{2}} \sqrt{3}$

53. $\log_{\sqrt{3}} \sqrt{2}$ **54.** $\log_{\sqrt{5}} 1.3$

55. Prove that $\log_{b^2} x = \dfrac{1}{2} \log_b x$.   **56.** Prove that $\log_b (b^x) = x$.

57. Prove that $b^{(\log_b x)} = x$.

58. Would logarithms to base 1 be useful? Why or why not?

II.3   APPLICATIONS OF LOGARITHMS AND EXPONENTS

Logarithms were welcomed by 17th-century mathematicians for much the same reason that a homemaker enjoys a dishwasher—logarithms and dishwashers are both labor-saving devices. Recognizing the importance of logarithms in his work, the French mathematician and astronomer Pierre-Simon Laplace (1749–1827) remarked that "the invention of logarithms has, by shortening the labors, doubled the life of the astronomer."

Base-10 logarithms have been well suited to simplify certain arithmetic calculations. However, the widespread availability of inexpensive calculators makes the use of logarithms for this purpose unnecessary. But, logarithms remain important for other reasons. Here are several examples that illustrate some applications of logarithms and exponential equations.

Example 1   In chemistry, logarithms are used to express the acidity of solutions. The more acidic a solution, the greater the concentration of hydrogen ions. This concentration is indicated indirectly by the *pH scale*, or *hydrogen-ion index*. The pH of a solution is defined by the equation pH $= -\log_{10}[H^+]$, where H^+ is the hydrogen-ion concentration in gram-ions per liter.

Pure water has a few free hydrogen ions—approximately 10^{-7} gram-ions of H^+ per liter. The pH of pure water is

$$\text{pH} = -\log_{10} 10^{-7}$$
$$= -(-7)$$
$$= +7$$

Seawater has a pH of approximately 8.5, and its hydrogen-ion concentration is found by solving the logarithmic equation $8.5 = -\log_{10}[H^+]$.

$$8.5 = -\log_{10}[H^+]$$
$$-8.5 = \log_{10}[H^+]$$
$$H^+ = 10^{-8.5}$$

Use a calculator to find that

$$H^+ \approx 3.2 \times 10^{-9} \text{ gram-ions per liter}$$

Example 2   In electrical engineering, logarithms are used to express the voltage gain or loss of an electronic device, such as an amplifier or a length of transmission line. The unit of a gain or loss, called the *decibel,* is defined by a logarithmic relation. If E_O is the output voltage of a device and E_I is the input voltage, the decibel gain is defined as

$$\text{db gain} = 20 \log_{10} \frac{E_O}{E_I}$$

If, for example, the voltage input to an amplifier is 0.5 volts and the output is 40 volts, the db gain is calculated by substituting into the formula.

$$\text{db gain} = 20 \log_{10} \frac{E_O}{E_I}$$
$$= 20 \log_{10} \frac{40}{0.5}$$
$$= 20 \log_{10} 80$$
$$\approx 20(1.9031)$$
$$\approx 38$$

The gain is approximately 38 decibels.

Example 3   The intensity of earthquakes is measured on the Richter Scale. This scale is based on a logarithmic relation, $R = \log_{10} \frac{A}{P}$, where A is the amplitude of the tremor (measured in microns) and P is the period of the tremor (measured in seconds) or the time necessary for the completion of one oscillation of the earth's surface. An earthquake with an amplitude of 10,000 microns (one centimeter) and a period of 0.1 seconds would measure

$$\log_{10}\frac{10,000}{0.1} = \log_{10} 100,000 = 5$$

on the Richter Scale. ■

Example 4   Radioactive materials emit radiation in the form of alpha particles (helium nuclei), beta particles (electrons), and gamma rays (high-energy X rays). The loss of these particles and rays changes the atomic structure of the radioactive material involved. Uranium, for example, decays into thorium, then into radium, and eventually into lead. The rate at which radioactive decay occurs is not constant; it depends on the amount of radioactive material still present. Theoretically, a block of radioactive material could not decay completely, because the more it decays, the less is present and the slower the rate of decay becomes.

Experiments have shown, however, that the time it takes *half* of a given radioactive material to decompose can be determined. This time, called the element's *half-life,* is constant for any given substance.

The amount A of any radioactive material present at any time t is given by the equation

$$A = A_0 2^{-t/h}$$

when A_0 is the amount present at $t = 0$ and h is the half-life of the material.

How long will it take 1 gram of radium, with a half-life of 1600 years, to decompose to 0.75 grams? To answer this question, evaluate t in the formula,

$$A = A_0 2^{-t/h}$$

where $A_0 = 1$, $A = 0.75$, and $h = 1600$:

$$0.75 = 2^{-t/1600}$$

Taking the common logarithm of both sides of the equation gives

$$\log 0.75 = \log 2^{-t/1600}$$

$$\log 0.75 = -\frac{t}{1600} \log 2$$

$$t = -1600 \cdot \frac{\log 0.75}{\log 2}$$

$$t \approx 664$$

It will take approximately 664 years. ■

Example 5   When a living organism dies, the oxygen/carbon-dioxide cycle common to all living things ceases and carbon-14, a radioactive isotope of carbon that is present in the atmosphere, is no longer absorbed. A block of wood taken from the tomb of an Egyptian pharaoh, for example, might have half the radioactivity of the wood of a living tree. Because the half-life of carbon-14 is about 5700 years, the wood from the tomb would have to have been carved from a tree that grew

5700 years ago. Carbon-14 dating is a useful tool for archeologists, who have used it to date objects as old as 25,000 years. ■

Example 6   Exponential equations enter into the compound-interest computations of any financial institution. If, for example, $1000 is deposited in a savings account that earns 8% per year, at the end of the first year the account will contain the original $1000 plus $80 in interest. This is $1000(1 + .08) = 1000(1.08) = $1080. At the end of the second year, the entire $1080 has earned interest and amounts to $1080(1.08) = $1000(1.08)^2 \approx 1166.40.

In general, an initial amount A_0 will grow to $A = A_0(1 + r)^t$ when compounded annually at a rate of r (expressed as a decimal) for t years.

This $1000, growing at 8% per year for 20 years, would become

$$A = 1000(1.08)^{20}$$
$$\approx 1000(4.66)$$
$$\approx \$4660$$

■

Example 7   At most financial institutions interest is compounded more frequently than once a year. A bank that computes interest quarterly, for example, would credit an account with one-fourth of the quoted annual interest after each quarter. For one year, the $1000 compounded quarterly at 8% per year would receive four increases at 2% each. Instead of $1000(1.08) = $1080, the account would contain $1000(1.02)^4 \approx $1000(1.08243) \approx 1082.43 after one year. In general, an initial amount A_0 compounded at an annual rate r, k times a year for n years, would become

$$A = A_0\left(1 + \frac{r}{k}\right)^{kn}$$

If $1000 on deposit at 8% for 20 years were compounded four times each year (quarterly interest), it would amount to

$$A = 1000\left(1 + \frac{0.08}{4}\right)^{4 \cdot 20}$$
$$= 1000(1.02)^{80}$$
$$\approx 1000(4.87544)$$
$$\approx \$4875.44$$

For the saver, quarterly compounding of interest is a substantial improvement over annual compounding.

If the interest is computed 365 times a year (daily interest), the $1000 becomes

$$A = 1000\left(1 + \frac{0.08}{365}\right)^{365 \cdot 20}$$
$$\approx 1000(1.00021918)^{7300}$$
$$\approx \$4952.16$$

This is $76.72 greater than the interest earned by compounding quarterly. ■

Exercise II.3

1. Find the pH of a solution with a hydrogen-ion concentration of 1.7×10^{-5} gram-ions/liter.

2. What is the H^+ concentration of a saturated solution of calcium hydroxide, whose pH is 13.2?

3. The decibel voltage gain of an amplifier is 29. If the output is 20 volts, what is the input voltage?

4. The power output (or input) of an amplifier is directly proportional to the square of the voltage output (or input). Show that the formula for decibel voltage gain is

$$\text{db voltage gain} = 10 \log_{10} \frac{P_O}{P_I}$$

where P_O is the power output and P_I is the power input.

5. An amplifier produces an output of 30 watts when driven by an input signal of 0.1 watt. What is the amplifier's voltage gain?

6. The decibel voltage gain of an amplifier is 35. If the input signal is 0.05 volts, what is the output voltage?

7. After studying exponents, a shrewd mathematics student agrees to work for 30 days under an unusual financial arrangement: Her salary for the first day is only 1¢, but for the second day it will be 2¢. The third day brings in 4¢ and so on, each day doubling the salary of the day before. What did the student earn on that last day?

8. If $1 had been invested in 1776 at 5% interest, compounded annually, what would it be worth in 2076?

9. A bacteria culture doubles in size every 24 hours. By how much will it have increased in 36 hours?

10. The population of South Waving Grass, Kansas, is presently 140 and is expected to grow exponentially, tripling every 15 years. When might the city planners expect the population to double the present census?

11. D dollars are added once a year to an account that is earning interest at a rate of r, compounded annually. Find an expression that will give the value of the account after n years.

12. The half-life of tritium is 12.4 years. How long will it take for 25% of a sample of tritium to decompose?

13. Twenty percent of a newly discovered radioactive element, mathematicium, decays in 2 years. What is its half-life?

II.4 NATURAL LOGARITHMS

For the saver, quarterly compounding of interest is a substantial improvement over annual compounding, but daily interest is not as big an improvement over quarterly interest. Could you get rich if you could convince a bank to compound interest every hour? Every second? Every millionth of a second? What happens to the amount A when the value of k in the formula

$$A = A_0 \left(1 + \frac{r}{k} \right)^{kn}$$

is allowed to become very large? Using the laws of exponents and some algebraic manipulations, rewrite the right side of the equation, first making a substitution: let $p = \frac{k}{r}$ so that $k = rp$.

$$A = A_0 \left(1 + \frac{r}{k} \right)^{kn}$$

$$= A_0 \left(1 + \frac{r}{rp} \right)^{rpn}$$

$$= A_0 \left(1 + \frac{1}{p} \right)^{rpn}$$

$$A = A_0 \left[\left(1 + \frac{1}{p} \right)^{p} \right]^{rn}$$

What happens to the amount A as k becomes very large? Because of the substitution $p = \frac{k}{r}$, p also becomes very large as k increases. The question of what happens to A, therefore, becomes tied in with another question: what is the value of $(1 + \frac{1}{p})^p$ for large values of p?

The number e   Some results calculated for the increasing values of p are tabulated in Table II-1. As p increases, the value of $(1 + \frac{1}{p})^p$ approaches more and more closely the nonterminating, nonrepeating decimal 2.71828182845904. . . . . This is known as the number e, named after Leonhard Euler (1707–1783), who was a dominant figure in 18th-century mathematics.

Table II-1

p	$[1 + (1/p)]^p$
1	2
10	2.5937
100	2.7048
1000	2.7169
.	.
.	.
.	.
1,000,000	2.7182805

If $1000 could be compounded not every minute or even every second but, rather, *continuously* at 8% a year for 20 years, the account would contain

$$A = A_0 e^{rn}$$

$$= 1000 \, e^{(.08)(20)}$$

$$\approx 1000(4.95303)$$

$$\approx \$4953.03$$

This is a mere 87¢ greater than the result of daily compounding.

The exponential function

The number e is so important in mathematics that the equation $y = e^x$ is called simply *the exponential function* to distinguish it from all the other exponential functions, such as $y = 10^x$ and $y = 2^x$. Logarithms to base e are called **natural logarithms,** or **Naperian logarithms,** after John Napier (1550–1617), and are usually denoted by the symbol ln x, rather than $\log_e x$. Natural logarithms can be found by using the $\boxed{\text{ln X}}$ key on a calculator or by using a table of natural logarithms, such as Table C in Appendix IV.

Natural, or Naperian, logarithms

Example 1

The exponential function appears in many problems involving growth and decay. An automobile battery, for example, charges at a rate that depends on how close it is to being fully charged; it charges fastest when it is most discharged. The charge C at any instant t is given by the formula

$$C = M(1 - e^{-kt})$$

where M is the theoretical maximum charge that the battery can contain and k is a positive constant that depends on the battery and the charger. Plotting the variable C against t, the curve is typical of many encountered in electronics (see Figure II-5). Notice that the full charge M is never reached; the actual charge can come very close to M, however, if the charger is left connected to the battery for a suitably long time.

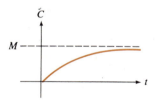

Figure II-5

Example 2

The loudness of a sound is not directly proportional to the intensity of the sound; that is, two mathematics teachers shouting at a class do not sound twice as loud as one teacher. Experiments in psychology suggest that the relationship of loudness and intensity is a logarithmic one known as the **Weber–Fechner law:** the apparent loudness L of a sound is proportional to the natural logarithm of its actual intensity I. In symbols, $L = k \ln I$.

For example, what actual increase in intensity will cause a doubling of the apparent loudness? If the original loudness is L_0, caused by an actual intensity I_0, then $L_0 = k \ln I_0$. To double the apparent loudness, multiply both sides of the equation by 2 and apply a law of logarithms:

$$2L_0 = 2k \ln I_0$$
$$= k \ln(I_0)^2$$

To double the apparent volume of a sound, then, the actual intensity must be squared. There is a moral in this—it's more difficult to shout twice as loud than it is to whisper twice as loud.

Exercise II.4

1. If the amount A_0 is invested at rate r and the interest is compounded continuously, how long will it take to double the capital? (*Hint*:   Solve $2A_0 = A_0 e^{rt}$ for t.)

2. A quick rule of thumb for determining how long it takes an investment to double is known as the "rule of seventy": divide 70 by the interest (as a percent); the quotient gives, approximately, the time it takes an initial amount compounded at that rate to double. At 7%, for example, it takes $70/7 = 10$ years to double the capital. At 5% it requires $70/5 = 14$ years. Why does this formula work?

3. If the intensity of a sound is doubled, what is the apparent change in loudness?

4. Prove that $e^{x \ln a} = a^x$.

5. The value of e can be calculated to any degree of accuracy by adding the first several terms of the following list.

$$1, 1, \frac{1}{2}, \frac{1}{2 \cdot 3}, \frac{1}{2 \cdot 3 \cdot 4}, \frac{1}{2 \cdot 3 \cdot 4 \cdot 5}, \frac{1}{2 \cdot 3 \cdot 4 \cdot 5 \cdot 6}, \cdots \cdots, \frac{1}{2 \cdot 3 \cdot \cdots \cdot n}, \cdots$$

The more terms that are added, the closer the sum is to the actual value of e. Calculate an approximation of the value of e by adding the first ten values in the list above. (A calculator would help!) To how many decimal places is your sum accurate?

6. (A calculator will be handy for this problem.) On a sheet of graph paper, graph the function $y = \frac{1}{2}(e^x + e^{-x})$ for values of x between -2 and $+2$. You'll need to calculate and plot about a dozen points before joining them with a smooth curve. The final graph looks like a parabola, but it is not. It is called a *catenary* and is important in the design of power-distribution networks because it represents the shape of a cable hanging between its supporting poles.

APPENDIX III

LINEAR INTERPOLATION

Before the widespread use of pocket calculators, people had to rely on published tables such as Table A in Appendix IV to find the values of the trigonometric functions. However, Table A includes angles that are measured to the nearest tenth of a degree only. Fortunately, there is a method, called **linear interpolation,** that lets us calculate good estimates of the values of the functions for angles measured to the nearest hundredth.

The method of linear interpolation is based on the fact that the graph of any continuous curve appears straight when we look at only a very small part of it. For example, the sine curve appears linear between points A and B in Figure III-1. If AB is assumed to *be* straight, we can set up proportions involving points A and B. We illustrate with an example.

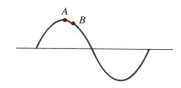

Figure III-1

Example 1   Use Table A in Appendix IV to find sin 33.23°.

Solution   As expected, Table A does not give the value of sin 33.23°. It does, however, list values for sin 33.20° and sin 33.30°.

$$10\left(\ 3\left(\begin{array}{l}\sin 33.20° \ \dots\dots\dots\dots\dots \ 0.5476\\ \sin 33.23° \ \dots\dots\dots\dots\dots \ ?\end{array}\right)x\\ \sin 33.30° \ \dots\dots\dots\dots\dots \ 0.5490\right)14$$

Note that the difference between 33.20 and 33.30 is 10 hundredths and that the difference between 33.20 and 33.23 is 3 hundredths. Note also that the difference between 0.5476 and 0.5490 is 14 ten-thousandths. Make use of the assumption of linearity and note that because 33.23 is $\frac{3}{10}$ of the way from 33.20 to 33.30, sin 33.23° must be $\frac{3}{10}$ of the way from 0.5476 to 0.5490. This can be expressed with the equation

$$x = \frac{3}{10}(14) \qquad \text{or with the proportion} \qquad \frac{x}{14} = \frac{3}{10}$$

Solve for x.

$$10x = 42$$
$$x = 4.2$$

To get a good estimate of sin 33.23°, you must add 4 ten-thousandths to 0.5476.

$$\sin 33.23° \approx 0.5476 + 0.0004 = 0.5480$$

Verify this result with a calculator. ■

Example 2   Find cos 15.28°.

Solution   Look up cos 15.20° and cos 15.30° in Table A, compute the differences, and set up a proportion.

$$10\left(\ 8\left(\begin{array}{l}\cos 15.20° \ \cdots\cdots\cdots\cdots\cdots\ 0.9650 \\ \cos 15.28° \ \cdots\cdots\cdots\cdots\cdots\ ? \\ \cos 15.30° \ \cdots\cdots\cdots\cdots\cdots\ 0.9646\end{array}\right)x\ \right)4$$

$$\frac{8}{10} = \frac{x}{4}$$
$$10x = 32$$
$$x = 3.2$$

This time, 3 ten-thousandths must be subtracted from 0.9650.

$$\cos 15.28° \approx 0.9650 - 0.0003 = 0.9647$$

Note that your calculator gives a value of 0.9646. Oh well, we said only that the method of linear interpolation gave good estimates. ■

Example 3   If tan θ = 4.598, find θ.

Solution   Look in Table A to find 4.598 in the Tan column. Although it does not appear, the numbers 4.586 and 4.625 do.

$$39\left(\ 12\left(\begin{array}{l}4.586 \ \cdots\cdots\cdots\cdots\cdots\ \tan 77.70° \\ 4.598 \ \cdots\cdots\cdots\cdots\cdots\ \tan θ \\ 4.625 \ \cdots\cdots\cdots\cdots\cdots\ \tan 77.80°\end{array}\right)x\ \right)10$$

Once again, set up a proportion.

$$\frac{12}{39} = \frac{x}{10}$$
$$39x = 120$$
$$x = 3.07$$

To find θ, add 0.03 to 77.70.

$$θ \approx 77.70° + 0.03° = 77.73°$$ ■

Exercise III.1

In Exercises 1–6, use linear interpolation to estimate the required values. Give answers to the nearest thousandth.

1. sin 23.35°

2. sin 83.17°

3. cos 81.13°

4. cos 13.21°

5. tan 48.14°

6. tan 88.23°

In Exercises 7–12, use linear interpolation to estimate angle θ, 0° ≤ θ < 90°. Give answers to the nearest hundredth of a degree.

7. sin θ = 0.4444

8. sin θ = 0.1234

9. cos θ = 0.9876

10. cos θ = 0.8642

11. tan θ = 0.2342

12. tan θ = 7.555

13. Jeff wanted to find sin 15°. So he used linear interpolation.

$$\sin \theta° = 0$$
$$\sin 15° = ?$$
$$\sin 30° = 0.5$$

He concluded that sin 15° = 0.25. However, a calculator shows that sin 15° ≈ 0.2588. What went wrong?

14. Martha wanted to find cos 45°. So she used linear interpolation.

$$\cos 0° = 1$$
$$\cos 45° = ?$$
$$\cos 90° = 0$$

She concluded that cos 45° = 0.5. However, a calculator shows that cos 45° ≈ 0.7071. What went wrong?

APPENDIX IV
TABLES

π radians = 180°

Table A   Values of the Trigonometric Functions

Radians	Degrees	Sin	Cos	Tan	Cot		
.0000	.0°	.0000	1.0000	.0000	—	90.0°	1.5708
.0017	.1°	.0017	1.0000	.0017	573.0	89.9°	1.5691
.0035	.2°	.0035	1.0000	.0035	286.5	89.8°	1.5673
.0052	.3°	.0052	1.0000	.0052	191.0	89.7°	1.5656
.0070	.4°	.0070	1.0000	.0070	143.2	89.6°	1.5638
.0087	.5°	.0087	1.0000	.0087	114.6	89.5°	1.5621
.0105	.6°	.0105	.9999	.0105	95.49	89.4°	1.5603
.0122	.7°	.0122	.9999	.0122	81.85	89.3°	1.5586
.0140	.8°	.0140	.9999	.0140	71.62	89.2°	1.5568
.0157	.9°	.0157	.9999	.0157	63.66	89.1°	1.5551
.0175	1.0°	.0175	.9998	.0175	57.29	89.0°	1.5533
.0192	1.1°	.0192	.9998	.0192	52.08	88.9°	1.5516
.0209	1.2°	.0209	.9998	.0209	47.74	88.8°	1.5499
.0227	1.3°	.0227	.9997	.0227	44.07	88.7°	1.5481
.0244	1.4°	.0244	.9997	.0244	40.92	88.6°	1.5464
.0262	1.5°	.0262	.9997	.0262	38.19	88.5°	1.5446
.0279	1.6°	.0279	.9996	.0279	35.80	88.4°	1.5429
.0297	1.7°	.0297	.9996	.0297	33.69	88.3°	1.5411
.0314	1.8°	.0314	.9995	.0314	31.82	88.2°	1.5394
.0332	1.9°	.0332	.9995	.0332	30.14	88.1°	1.5376
.0349	2.0°	.0349	.9994	.0349	28.64	88.0°	1.5359
.0367	2.1°	.0366	.9993	.0367	27.27	87.9°	1.5341
.0384	2.2°	.0384	.9993	.0384	26.03	87.8°	1.5324
.0401	2.3°	.0401	.9992	.0402	24.90	87.7°	1.5307
.0419	2.4°	.0419	.9991	.0419	23.86	87.6°	1.5289
.0436	2.5°	.0436	.9990	.0437	22.90	87.5°	1.5272
.0454	2.6°	.0454	.9990	.0454	22.02	87.4°	1.5254
.0471	2.7°	.0471	.9989	.0472	21.20	87.3°	1.5237
.0489	2.8°	.0488	.9988	.0489	20.45	87.2°	1.5219
.0506	2.9°	.0506	.9987	.0507	19.74	87.1°	1.5202
.0524	3.0°	.0523	.9986	.0524	19.08	87.0°	1.5184
.0541	3.1°	.0541	.9985	.0542	18.46	86.9°	1.5167
.0559	3.2°	.0558	.9984	.0559	17.89	86.8°	1.5149
.0576	3.3°	.0576	.9983	.0577	17.34	86.7°	1.5132
.0593	3.4°	.0593	.9982	.0594	16.83	86.6°	1.5115
.0611	3.5°	.0610	.9981	.0612	16.35	86.5°	1.5097
.0628	3.6°	.0628	.9980	.0629	15.89	86.4°	1.5080
.0646	3.7°	.0645	.9979	.0647	15.46	86.3°	1.5062
.0663	3.8°	.0663	.9978	.0664	15.06	86.2°	1.5045
.0681	3.9°	.0680	.9977	.0682	14.67	86.1°	1.5027
		Cos	Sin	Cot	Tan	Degrees	Radians

Table A *(Continued)*

Radians	Degrees	Sin	Cos	Tan	Cot		
.0698	4.0°	.0698	.9976	.0699	14.30	86.0°	1.5010
.0716	4.1°	.0715	.9974	.0717	13.95	85.9°	1.4992
.0733	4.2°	.0732	.9973	.0734	13.62	85.8°	1.4975
.0750	4.3°	.0750	.9972	.0752	13.30	85.7°	1.4957
.0768	4.4°	.0767	.9971	.0769	13.00	85.6°	1.4940
.0785	4.5°	.0785	.9969	.0787	12.71	85.5°	1.4923
.0803	4.6°	.0802	.9968	.0805	12.43	85.4°	1.4905
.0820	4.7°	.0819	.9966	.0822	12.16	85.3°	1.4888
.0838	4.8°	.0837	.9965	.0840	11.91	85.2°	1.4870
.0855	4.9°	.0854	.9963	.0857	11.66	85.1°	1.4853
.0873	5.0°	.0872	.9962	.0875	11.43	85.0°	1.4835
.0890	5.1°	.0889	.9960	.0892	11.20	84.9°	1.4818
.0908	5.2°	.0906	.9959	.0910	10.99	84.8°	1.4800
.0925	5.3°	.0924	.9957	.0928	10.78	84.7°	1.4783
.0942	5.4°	.0941	.9956	.0945	10.58	84.6°	1.4765
.0960	5.5°	.0958	.9954	.0963	10.39	84.5°	1.4748
.0977	5.6°	.0976	.9952	.0981	10.20	84.4°	1.4731
.0995	5.7°	.0993	.9951	.0998	10.02	84.3°	1.4713
.1012	5.8°	.1011	.9949	.1016	9.845	84.2°	1.4696
.1030	5.9°	.1028	.9947	.1033	9.677	84.1°	1.4678
.1047	6.0°	.1045	.9945	.1051	9.514	84.0°	1.4661
.1065	6.1°	.1063	.9943	.1069	9.357	83.9°	1.4643
.1082	6.2°	.1080	.9942	.1086	9.205	83.8°	1.4626
.1100	6.3°	.1097	.9940	.1104	9.058	83.7°	1.4608
.1117	6.4°	.1115	.9938	.1122	8.915	83.6°	1.4591
.1134	6.5°	.1132	.9936	.1139	8.777	83.5°	1.4573
.1152	6.6°	.1149	.9934	.1157	8.643	83.4°	1.4556
.1169	6.7°	.1167	.9932	.1175	8.513	83.3°	1.4539
.1187	6.8°	.1184	.9930	.1192	8.386	83.2°	1.4521
.1204	6.9°	.1201	.9928	.1210	8.264	83.1°	1.4504
.1222	7.0°	.1219	.9925	.1228	8.144	83.0°	1.4486
.1239	7.1°	.1236	.9923	.1246	8.028	82.9°	1.4469
.1257	7.2°	.1253	.9921	.1263	7.916	82.8°	1.4451
.1274	7.3°	.1271	.9919	.1281	7.806	82.7°	1.4434
.1292	7.4°	.1288	.9917	.1299	7.700	82.6°	1.4416
.1309	7.5°	.1305	.9914	.1317	7.596	82.5°	1.4399
.1326	7.6°	.1323	.9912	.1334	7.495	82.4°	1.4382
.1344	7.7°	.1340	.9910	.1352	7.396	82.3°	1.4364
.1361	7.8°	.1357	.9907	.1370	7.300	82.2°	1.4347
.1379	7.9°	.1374	.9905	.1388	7.207	82.1°	1.4329
.1396	8.0°	.1392	.9903	.1405	7.115	82.0°	1.4312
.1414	8.1°	.1409	.9900	.1423	7.026	81.9°	1.4294
.1431	8.2°	.1426	.9898	.1441	6.940	81.8°	1.4277
.1449	8.3°	.1444	.9895	.1459	6.855	81.7°	1.4259
.1466	8.4°	.1461	.9893	.1477	6.772	81.6°	1.4242
.1484	8.5°	.1478	.9890	.1495	6.691	81.5°	1.4224
.1501	8.6°	.1495	.9888	.1512	6.612	81.4°	1.4207
.1518	8.7°	.1513	.9885	.1530	6.535	81.3°	1.4190
.1536	8.8°	.1530	.9882	.1548	6.460	81.2°	1.4172
.1553	8.9°	.1547	.9880	.1566	6.386	81.1°	1.4155
		Cos	Sin	Cot	Tan	Degrees	Radians

Table A *(Continued)*

Radians	Degrees	Sin	Cos	Tan	Cot		
.1571	9.0°	.1564	.9877	.1584	6.314	81.0°	1.4137
.1588	9.1°	.1582	.9874	.1602	6.243	80.9°	1.4120
.1606	9.2°	.1599	.9871	.1620	6.174	80.8°	1.4102
.1623	9.3°	.1616	.9869	.1638	6.107	80.7°	1.4085
.1641	9.4°	.1633	.9866	.1655	6.041	80.6°	1.4067
.1658	9.5°	.1650	.9863	.1673	5.976	80.5°	1.4050
.1676	9.6°	.1668	.9860	.1691	5.912	80.4°	1.4032
.1693	9.7°	.1685	.9857	.1709	5.850	80.3°	1.4015
.1710	9.8°	.1702	.9854	.1727	5.789	80.2°	1.3998
.1728	9.9°	.1719	.9851	.1745	5.730	80.1°	1.3980
.1745	10.0°	.1736	.9848	.1763	5.671	80.0°	1.3963
.1763	10.1°	.1754	.9845	.1781	5.614	79.9°	1.3945
.1780	10.2°	.1771	.9842	.1799	5.558	79.8°	1.3928
.1798	10.3°	.1788	.9839	.1817	5.503	79.7°	1.3910
.1815	10.4°	.1805	.9836	.1835	5.449	79.6°	1.3893
.1833	10.5°	.1822	.9833	.1853	5.396	79.5°	1.3875
.1850	10.6°	.1840	.9829	.1871	5.343	79.4°	1.3858
.1868	10.7°	.1857	.9826	.1890	5.292	79.3°	1.3840
.1885	10.8°	.1874	.9823	.1908	5.242	79.2°	1.3823
.1902	10.9°	.1891	.9820	.1926	5.193	79.1°	1.3806
.1920	11.0°	.1908	.9816	.1944	5.145	79.0°	1.3788
.1937	11.1°	.1925	.9813	.1962	5.097	78.9°	1.3771
.1955	11.2°	.1942	.9810	.1980	5.050	78.8°	1.3753
.1972	11.3°	.1959	.9806	.1998	5.005	78.7°	1.3736
.1990	11.4°	.1977	.9803	.2016	4.959	78.6°	1.3718
.2007	11.5°	.1994	.9799	.2035	4.915	78.5°	1.3701
.2025	11.6°	.2011	.9796	.2053	4.872	78.4°	1.3683
.2042	11.7°	.2028	.9792	.2071	4.829	78.3°	1.3666
.2059	11.8°	.2045	.9789	.2089	4.787	78.2°	1.3648
.2077	11.9°	.2062	.9785	.2107	4.745	78.1°	1.3631
.2094	12.0°	.2079	.9781	.2126	4.705	78.0°	1.3614
.2112	12.1°	.2096	.9778	.2144	4.665	77.9°	1.3596
.2129	12.2°	.2113	.9774	.2162	4.625	77.8°	1.3579
.2147	12.3°	.2130	.9770	.2180	4.586	77.7°	1.3561
.2164	12.4°	.2147	.9767	.2199	4.548	77.6°	1.3544
.2182	12.5°	.2164	.9763	.2217	4.511	77.5°	1.3526
.2199	12.6°	.2181	.9759	.2235	4.474	77.4°	1.3509
.2217	12.7°	.2198	.9755	.2254	4.437	77.3°	1.3491
.2234	12.8°	.2215	.9751	.2272	4.402	77.2°	1.3474
.2251	12.9°	.2233	.9748	.2290	4.366	77.1°	1.3456
.2269	13.0°	.2250	.9744	.2309	4.331	77.0°	1.3439
.2286	13.1°	.2267	.9740	.2327	4.297	76.9°	1.3422
.2304	13.2°	.2284	.9736	.2345	4.264	76.8°	1.3404
.2321	13.3°	.2300	.9732	.2364	4.230	76.7°	1.3387
.2339	13.4°	.2317	.9728	.2382	4.198	76.6°	1.3369
.2356	13.5°	.2334	.9724	.2401	4.165	76.5°	1.3352
.2374	13.6°	.2351	.9720	.2419	4.134	76.4°	1.3334
.2391	13.7°	.2368	.9715	.2438	4.102	76.3°	1.3317
.2409	13.8°	.2385	.9711	.2456	4.071	76.2°	1.3299
.2426	13.9°	.2402	.9707	.2475	4.041	76.1°	1.3282
		Cos	Sin	Cot	Tan	Degrees	Radians

Table A *(Continued)*

Radians	Degrees	Sin	Cos	Tan	Cot		
.2443	14.0°	.2419	.9703	.2493	4.011	76.0°	1.3265
.2461	14.1°	.2436	.9699	.2512	3.981	75.9°	1.3247
.2478	14.2°	.2453	.9694	.2530	3.952	75.8°	1.3230
.2496	14.3°	.2470	.9690	.2549	3.923	75.7°	1.3212
.2513	14.4°	.2487	.9686	.2568	3.895	75.6°	1.3195
.2531	14.5°	.2504	.9681	.2586	3.867	75.5°	1.3177
.2548	14.6°	.2521	.9677	.2605	3.839	75.4°	1.3160
.2566	14.7°	.2538	.9673	.2623	3.812	75.3°	1.3142
.2583	14.8°	.2554	.9668	.2642	3.785	75.2°	1.3125
.2601	14.9°	.2571	.9664	.2661	3.758	75.1°	1.3107
.2618	15.0°	.2588	.9659	.2679	3.732	75.0°	1.3090
.2635	15.1°	.2605	.9655	.2698	3.706	74.9°	1.3073
.2653	15.2°	.2622	.9650	.2717	3.681	74.8°	1.3055
.2670	15.3°	.2639	.9646	.2736	3.655	74.7°	1.3038
.2688	15.4°	.2656	.9641	.2754	3.630	74.6°	1.3020
.2705	15.5°	.2672	.9636	.2773	3.606	74.5°	1.3003
.2723	15.6°	.2689	.9632	.2792	3.582	74.4°	1.2985
.2740	15.7°	.2706	.9627	.2811	3.558	74.3°	1.2968
.2758	15.8°	.2723	.9622	.2830	3.534	74.2°	1.2950
.2775	15.9°	.2740	.9617	.2849	3.511	74.1°	1.2933
.2793	16.0°	.2756	.9613	.2867	3.487	74.0°	1.2915
.2810	16.1°	.2773	.9608	.2886	3.465	73.9°	1.2898
.2827	16.2°	.2790	.9603	.2905	3.442	73.8°	1.2881
.2845	16.3°	.2807	.9598	.2924	3.420	73.7°	1.2863
.2862	16.4°	.2823	.9593	.2943	3.398	73.6°	1.2846
.2880	16.5°	.2840	.9588	.2962	3.376	73.5°	1.2828
.2897	16.6°	.2857	.9583	.2981	3.354	73.4°	1.2811
.2915	16.7°	.2874	.9578	.3000	3.333	73.3°	1.2793
.2932	16.8°	.2890	.9573	.3019	3.312	73.2°	1.2776
.2950	16.9°	.2907	.9568	.3038	3.291	73.1°	1.2758
.2967	17.0°	.2924	.9563	.3057	3.271	73.0°	1.2741
.2985	17.1°	.2940	.9558	.3076	3.251	72.9°	1.2723
.3002	17.2°	.2957	.9553	.3096	3.230	72.8°	1.2706
.3019	17.3°	.2974	.9548	.3115	3.211	72.7°	1.2689
.3037	17.4°	.2990	.9542	.3134	3.191	72.6°	1.2671
.3054	17.5°	.3007	.9537	.3153	3.172	72.5°	1.2654
.3072	17.6°	.3024	.9532	.3172	3.152	72.4°	1.2636
.3089	17.7°	.3040	.9527	.3191	3.133	72.3°	1.2619
.3107	17.8°	.3057	.9521	.3211	3.115	72.2°	1.2601
.3124	17.9°	.3074	.9516	.3230	3.096	72.1°	1.2584
.3142	18.0°	.3090	.9511	.3249	3.078	72.0°	1.2566
.3159	18.1°	.3107	.9505	.3269	3.060	71.9°	1.2549
.3176	18.2°	.3123	.9500	.3288	3.042	71.8°	1.2531
.3194	18.3°	.3140	.9494	.3307	3.024	71.7°	1.2514
.3211	18.4°	.3156	.9489	.3327	3.006	71.6°	1.2497
.3229	18.5°	.3173	.9483	.3346	2.989	71.5°	1.2479
.3246	18.6°	.3190	.9478	.3365	2.971	71.4°	1.2462
.3264	18.7°	.3206	.9472	.3385	2.954	71.3°	1.2444
.3281	18.8°	.3223	.9466	.3404	2.937	71.2°	1.2427
.3299	18.9°	.3239	.9461	.3424	2.921	71.1°	1.2409
		Cos	Sin	Cot	Tan	Degrees	Radians

Table A *(Continued)*

Radians	Degrees	Sin	Cos	Tan	Cot		
.3316	19.0°	.3256	.9455	.3443	2.904	71.0°	1.2392
.3334	19.1°	.3272	.9449	.3463	2.888	70.9°	1.2374
.3351	19.2°	.3289	.9444	.3482	2.872	70.8°	1.2357
.3368	19.3°	.3305	.9438	.3502	2.856	70.7°	1.2339
.3386	19.4°	.3322	.9432	.3522	2.840	70.6°	1.2322
.3403	19.5°	.3338	.9426	.3541	2.824	70.5°	1.2305
.3421	19.6°	.3355	.9421	.3561	2.808	70.4°	1.2287
.3438	19.7°	.3371	.9415	.3581	2.793	70.3°	1.2270
.3456	19.8°	.3387	.9409	.3600	2.778	70.2°	1.2252
.3473	19.9°	.3404	.9403	.3620	2.762	70.1°	1.2235
.3491	20.0°	.3420	.9397	.3640	2.747	70.0°	1.2217
.3508	20.1°	.3437	.9391	.3659	2.733	69.9°	1.2200
.3526	20.2°	.3453	.9385	.3679	2.718	69.8°	1.2182
.3543	20.3°	.3469	.9379	.3699	2.703	69.7°	1.2165
.3560	20.4°	.3486	.9373	.3719	2.689	69.6°	1.2147
.3578	20.5°	.3502	.9367	.3739	2.675	69.5°	1.2130
.3595	20.6°	.3518	.9361	.3759	2.660	69.4°	1.2113
.3613	20.7°	.3535	.9354	.3779	2.646	69.3°	1.2095
.3630	20.8°	.3551	.9348	.3799	2.633	69.2°	1.2078
.3648	20.9°	.3567	.9342	.3819	2.619	69.1°	1.2060
.3665	21.0°	.3584	.9336	.3839	2.605	69.0°	1.2043
.3683	21.1°	.3600	.9330	.3859	2.592	68.9°	1.2025
.3700	21.2°	.3616	.9323	.3879	2.578	68.8°	1.2008
.3718	21.3°	.3633	.9317	.3899	2.565	68.7°	1.1990
.3735	21.4°	.3649	.9311	.3919	2.552	68.6°	1.1973
.3752	21.5°	.3665	.9304	.3939	2.539	68.5°	1.1956
.3770	21.6°	.3681	.9298	.3959	2.526	68.4°	1.1938
.3787	21.7°	.3697	.9291	.3979	2.513	68.3°	1.1921
.3805	21.8°	.3714	.9285	.4000	2.500	68.2°	1.1903
.3822	21.9°	.3730	.9278	.4020	2.488	68.1°	1.1886
.3840	22.0°	.3746	.9272	.4040	2.475	68.0°	1.1868
.3857	22.1°	.3762	.9265	.4061	2.463	67.9°	1.1851
.3875	22.2°	.3778	.9259	.4081	2.450	67.8°	1.1833
.3892	22.3°	.3795	.9252	.4101	2.438	67.7°	1.1816
.3910	22.4°	.3811	.9245	.4122	2.426	67.6°	1.1798
.3927	22.5°	.3827	.9239	.4142	2.414	67.5°	1.1781
.3944	22.6°	.3843	.9232	.4163	2.402	67.4°	1.1764
.3962	22.7°	.3859	.9225	.4183	2.391	67.3°	1.1746
.3979	22.8°	.3875	.9219	.4204	2.379	67.2°	1.1729
.3997	22.9°	.3891	.9212	.4224	2.367	67.1°	1.1711
.4014	23.0°	.3907	.9205	.4245	2.356	67.0°	1.1694
.4032	23.1°	.3923	.9198	.4265	2.344	66.9°	1.1676
.4049	23.2°	.3939	.9191	.4286	2.333	66.8°	1.1659
.4067	23.3°	.3955	.9184	.4307	2.322	66.7°	1.1641
.4084	23.4°	.3971	.9178	.4327	2.311	66.6°	1.1624
.4102	23.5°	.3987	.9171	.4348	2.300	66.5°	1.1606
.4119	23.6°	.4003	.9164	.4369	2.289	66.4°	1.1589
.4136	23.7°	.4019	.9157	.4390	2.278	66.3°	1.1572
.4154	23.8°	.4035	.9150	.4411	2.267	66.2°	1.1554
.4171	23.9°	.4051	.9143	.4431	2.257	66.1°	1.1537
		Cos	Sin	Cot	Tan	Degrees	Radians

Table A *(Continued)*

Radians	Degrees	Sin	Cos	Tan	Cot		
.4189	24.0°	.4067	.9135	.4452	2.246	66.0°	1.1519
.4206	24.1°	.4083	.9128	.4473	2.236	65.9°	1.1502
.4224	24.2°	.4099	.9121	.4494	2.225	65.8°	1.1484
.4241	24.3°	.4115	.9114	.4515	2.215	65.7°	1.1467
.4259	24.4°	.4131	.9107	.4536	2.204	65.6°	1.1449
.4276	24.5°	.4147	.9100	.4557	2.194	65.5°	1.1432
.4294	24.6°	.4163	.9092	.4578	2.184	65.4°	1.1414
.4311	24.7°	.4179	.9085	.4599	2.174	65.3°	1.1397
.4328	24.8°	.4195	.9078	.4621	2.164	65.2°	1.1380
.4346	24.9°	.4210	.9070	.4642	2.154	65.1°	1.1362
.4363	25.0°	.4226	.9063	.4663	2.145	65.0°	1.1345
.4381	25.1°	.4242	.9056	.4684	2.135	64.9°	1.1327
.4398	25.2°	.4258	.9048	.4706	2.125	64.8°	1.1310
.4416	25.3°	.4274	.9041	.4727	2.116	64.7°	1.1292
.4433	25.4°	.4289	.9033	.4748	2.106	64.6°	1.1275
.4451	25.5°	.4305	.9026	.4770	2.097	64.5°	1.1257
.4468	25.6°	.4321	.9018	.4791	2.087	64.4°	1.1240
.4485	25.7°	.4337	.9011	.4813	2.078	64.3°	1.1222
.4503	25.8°	.4352	.9003	.4834	2.069	64.2°	1.1205
.4520	25.9°	.4368	.8996	.4856	2.059	64.1°	1.1188
.4538	26.0°	.4384	.8988	.4877	2.050	64.0°	1.1170
.4555	26.1°	.4399	.8980	.4899	2.041	63.9°	1.1153
.4573	26.2°	.4415	.8973	.4921	2.032	63.8°	1.1135
.4590	26.3°	.4431	.8965	.4942	2.023	63.7°	1.1118
.4608	26.4°	.4446	.8957	.4964	2.014	63.6°	1.1100
.4625	26.5°	.4462	.8949	.4986	2.006	63.5°	1.1083
.4643	26.6°	.4478	.8942	.5008	1.997	63.4°	1.1065
.4660	26.7°	.4493	.8934	.5029	1.988	63.3°	1.1048
.4677	26.8°	.4509	.8926	.5051	1.980	63.2°	1.1030
.4695	26.9°	.4524	.8918	.5073	1.971	63.1°	1.1013
.4712	27.0°	.4540	.8910	.5095	1.963	63.0°	1.0996
.4730	27.1°	.4555	.8902	.5117	1.954	62.9°	1.0978
.4747	27.2°	.4571	.8894	.5139	1.946	62.8°	1.0961
.4765	27.3°	.4586	.8886	.5161	1.937	62.7°	1.0943
.4782	27.4°	.4602	.8878	.5184	1.929	62.6°	1.0926
.4800	27.5°	.4617	.8870	.5206	1.921	62.5°	1.0908
.4817	27.6°	.4633	.8862	.5228	1.913	62.4°	1.0891
.4835	27.7°	.4648	.8854	.5250	1.905	62.3°	1.0873
.4852	27.8°	.4664	.8846	.5272	1.897	62.2°	1.0856
.4869	27.9°	.4679	.8838	.5295	1.889	62.1°	1.0838
.4887	28.0°	.4695	.8829	.5317	1.881	62.0°	1.0821
.4904	28.1°	.4710	.8821	.5340	1.873	61.9°	1.0804
.4922	28.2°	.4726	.8813	.5362	1.865	61.8°	1.0786
.4939	28.3°	.4741	.8805	.5384	1.857	61.7°	1.0769
.4957	28.4°	.4756	.8796	.5407	1.849	61.6°	1.0751
.4974	28.5°	.4772	.8788	.5430	1.842	61.5°	1.0734
.4992	28.6°	.4787	.8780	.5452	1.834	61.4°	1.0716
.5009	28.7°	.4802	.8771	.5475	1.827	61.3°	1.0699
.5027	28.8°	.4818	.8763	.5498	1.819	61.2°	1.0681
.5044	28.9°	.4833	.8755	.5520	1.811	61.1°	1.0664
		Cos	Sin	Cot	Tan	Degrees	Radians

Table A *(Continued)*

Radians	Degrees	Sin	Cos	Tan	Cot		
.5061	29.0°	.4848	.8746	.5543	1.804	61.0°	1.0647
.5079	29.1°	.4863	.8738	.5566	1.797	60.9°	1.0629
.5096	29.2°	.4879	.8729	.5589	1.789	60.8°	1.0612
.5114	29.3°	.4894	.8721	.5612	1.782	60.7°	1.0594
.5131	29.4°	.4909	.8712	.5635	1.775	60.6°	1.0577
.5149	29.5°	.4924	.8704	.5658	1.767	60.5°	1.0559
.5166	29.6°	.4939	.8695	.5681	1.760	60.4°	1.0542
.5184	29.7°	.4955	.8686	.5704	1.753	60.3°	1.0524
.5201	29.8°	.4970	.8678	.5727	1.746	60.2°	1.0507
.5219	29.9°	.4985	.8669	.5750	1.739	60.1°	1.0489
.5236	30.0°	.5000	.8660	.5774	1.732	60.0°	1.0472
.5253	30.1°	.5015	.8652	.5797	1.725	59.9°	1.0455
.5271	30.2°	.5030	.8643	.5820	1.718	59.8°	1.0437
.5288	30.3°	.5045	.8634	.5844	1.711	59.7°	1.0420
.5306	30.4°	.5060	.8625	.5867	1.704	59.6°	1.0402
.5323	30.5°	.5075	.8616	.5890	1.698	59.5°	1.0385
.5341	30.6°	.5090	.8607	.5914	1.691	59.4°	1.0367
.5358	30.7°	.5105	.8599	.5938	1.684	59.3°	1.0350
.5376	30.8°	.5120	.8590	.5961	1.678	59.2°	1.0332
.5393	30.9°	.5135	.8581	.5985	1.671	59.1°	1.0315
.5411	31.0°	.5150	.8572	.6009	1.664	59.0°	1.0297
.5428	31.1°	.5165	.8563	.6032	1.658	58.9°	1.0280
.5445	31.2°	.5180	.8554	.6056	1.651	58.8°	1.0263
.5463	31.3°	.5195	.8545	.6080	1.645	58.7°	1.0245
.5480	31.4°	.5210	.8536	.6104	1.638	58.6°	1.0228
.5498	31.5°	.5225	.8526	.6128	1.632	58.5°	1.0210
.5515	31.6°	.5240	.8517	.6152	1.625	58.4°	1.0193
.5533	31.7°	.5255	.8508	.6176	1.619	58.3°	1.0175
.5550	31.8°	.5270	.8499	.6200	1.613	58.2°	1.0158
.5568	31.9°	.5284	.8490	.6224	1.607	58.1°	1.0140
.5585	32.0°	.5299	.8480	.6249	1.600	58.0°	1.0123
.5603	32.1°	.5314	.8471	.6273	1.594	57.9°	1.0105
.5620	32.2°	.5329	.8462	.6297	1.588	57.8°	1.0088
.5637	32.3°	.5344	.8453	.6322	1.582	57.7°	1.0071
.5655	32.4°	.5358	.8443	.6346	1.576	57.6°	1.0053
.5672	32.5°	.5373	.8434	.6371	1.570	57.5°	1.0036
.5690	32.6°	.5388	.8425	.6395	1.564	57.4°	1.0018
.5707	32.7°	.5402	.8415	.6420	1.558	57.3°	1.0001
.5725	32.8°	.5417	.8406	.6445	1.552	57.2°	.9983
.5742	32.9°	.5432	.8396	.6469	1.546	57.1°	.9966
.5760	33.0°	.5446	.8387	.6494	1.540	57.0°	.9948
.5777	33.1°	.5461	.8377	.6519	1.534	56.9°	.9931
.5794	33.2°	.5476	.8368	.6544	1.528	56.8°	.9913
.5812	33.3°	.5490	.8358	.6569	1.522	56.7°	.9896
.5829	33.4°	.5505	.8348	.6594	1.517	56.6°	.9879
.5847	33.5°	.5519	.8339	.6619	1.511	56.5°	.9861
.5864	33.6°	.5534	.8329	.6644	1.505	56.4°	.9844
.5882	33.7°	.5548	.8320	.6669	1.499	56.3°	.9826
.5899	33.8°	.5563	.8310	.6694	1.494	56.2°	.9809
.5917	33.9°	.5577	.8300	.6720	1.488	56.1°	.9791
		Cos	Sin	Cot	Tan	Degrees	Radians

Table A *(Continued)*

Radians	Degrees	Sin	Cos	Tan	Cot		
.5934	34.0°	.5592	.8290	.6745	1.483	56.0°	.9774
.5952	34.1°	.5606	.8281	.6771	1.477	55.9°	.9756
.5969	34.2°	.5621	.8271	.6796	1.471	55.8°	.9739
.5986	34.3°	.5635	.8261	.6822	1.466	55.7°	.9721
.6004	34.4°	.5650	.8251	.6847	1.460	55.6°	.9704
.6021	34.5°	.5664	.8241	.6873	1.455	55.5°	.9687
.6039	34.6°	.5678	.8231	.6899	1.450	55.4°	.9669
.6056	34.7°	.5693	.8221	.6924	1.444	55.3°	.9652
.6074	34.8°	.5707	.8211	.6950	1.439	55.2°	.9634
.6091	34.9°	.5721	.8202	.6976	1.433	55.1°	.9617
.6109	35.0°	.5736	.8192	.7002	1.428	55.0°	.9599
.6126	35.1°	.5750	.8181	.7028	1.423	54.9°	.9582
.6144	35.2°	.5764	.8171	.7054	1.418	54.8°	.9564
.6161	35.3°	.5779	.8161	.7080	1.412	54.7°	.9547
.6178	35.4°	.5793	.8151	.7107	1.407	54.6°	.9530
.6196	35.5°	.5807	.8141	.7133	1.402	54.5°	.9512
.6213	35.6°	.5821	.8131	.7159	1.397	54.4°	.9495
.6231	35.7°	.5835	.8121	.7186	1.392	54.3°	.9477
.6248	35.8°	.5850	.8111	.7212	1.387	54.2°	.9460
.6266	35.9°	.5864	.8100	.7239	1.381	54.1°	.9442
.6283	36.0°	.5878	.8090	.7265	1.376	54.0°	.9425
.6301	36.1°	.5892	.8080	.7292	1.371	53.9°	.9407
.6318	36.2°	.5906	.8070	.7319	1.366	53.8°	.9390
.6336	36.3°	.5920	.8059	.7346	1.361	53.7°	.9372
.6353	36.4°	.5934	.8049	.7373	1.356	53.6°	.9355
.6370	36.5°	.5948	.8039	.7400	1.351	53.5°	.9338
.6388	36.6°	.5962	.8028	.7427	1.347	53.4°	.9320
.6405	36.7°	.5976	.8018	.7454	1.342	53.3°	.9303
.6423	36.8°	.5990	.8007	.7481	1.337	53.2°	.9285
.6440	36.9°	.6004	.7997	.7508	1.332	53.1°	.9268
.6458	37.0°	.6018	.7986	.7536	1.327	53.0°	.9250
.6475	37.1°	.6032	.7976	.7563	1.322	52.9°	.9233
.6493	37.2°	.6046	.7965	.7590	1.317	52.8°	.9215
.6510	37.3°	.6060	.7955	.7618	1.313	52.7°	.9198
.6528	37.4°	.6074	.7944	.7646	1.308	52.6°	.9180
.6545	37.5°	.6088	.7934	.7673	1.303	52.5°	.9163
.6562	37.6°	.6101	.7923	.7701	1.299	52.4°	.9146
.6580	37.7°	.6115	.7912	.7729	1.294	52.3°	.9128
.6597	37.8°	.6129	.7902	.7757	1.289	52.2°	.9111
.6615	37.9°	.6143	.7891	.7785	1.285	52.1°	.9093
.6632	38.0°	.6157	.7880	.7813	1.280	52.0°	.9076
.6650	38.1°	.6170	.7869	.7841	1.275	51.9°	.9058
.6667	38.2°	.6184	.7859	.7869	1.271	51.8°	.9041
.6685	38.3°	.6198	.7848	.7898	1.266	51.7°	.9023
.6702	38.4°	.6211	.7837	.7926	1.262	51.6°	.9006
.6720	38.5°	.6225	.7826	.7954	1.257	51.5°	.8988
.6737	38.6°	.6239	.7815	.7983	1.253	51.4°	.8971
.6754	38.7°	.6252	.7804	.8012	1.248	51.3°	.8954
.6772	38.8°	.6266	.7793	.8040	1.244	51.2°	.8936
.6789	38.9°	.6280	.7782	.8069	1.239	51.1°	.8919
		Cos	Sin	Cot	Tan	Degrees	Radians

Table A *(Continued)*

Radians	Degrees	Sin	Cos	Tan	Cot		
.6807	39.0°	.6293	.7771	.8098	1.235	51.0°	.8901
.6824	39.1°	.6307	.7760	.8127	1.230	50.9°	.8884
.6842	39.2°	.6320	.7749	.8156	1.226	50.8°	.8866
.6859	39.3°	.6334	.7738	.8185	1.222	50.7°	.8849
.6877	39.4°	.6347	.7727	.8214	1.217	50.6°	.8831
.6894	39.5°	.6361	.7716	.8243	1.213	50.5°	.8814
.6912	39.6°	.6374	.7705	.8273	1.209	50.4°	.8796
.6929	39.7°	.6388	.7694	.8302	1.205	50.3°	.8779
.6946	39.8°	.6401	.7683	.8332	1.200	50.2°	.8762
.6964	39.9°	.6414	.7672	.8361	1.196	50.1°	.8744
.6981	40.0°	.6428	.7660	.8391	1.192	50.0°	.8727
.6999	40.1°	.6441	.7649	.8421	1.188	49.9°	.8709
.7016	40.2°	.6455	.7638	.8451	1.183	49.8°	.8692
.7034	40.3°	.6468	.7627	.8481	1.179	49.7°	.8674
.7051	40.4°	.6481	.7615	.8511	1.175	49.6°	.8657
.7069	40.5°	.6494	.7604	.8541	1.171	49.5°	.8639
.7086	40.6°	.6508	.7593	.8571	1.167	49.4°	.8622
.7103	40.7°	.6521	.7581	.8601	1.163	49.3°	.8604
.7121	40.8°	.6534	.7570	.8632	1.159	49.2°	.8587
.7138	40.9°	.6547	.7559	.8662	1.154	49.1°	.8570
.7156	41.0°	.6561	.7547	.8693	1.150	49.0°	.8552
.7173	41.1°	.6574	.7536	.8724	1.146	48.9°	.8535
.7191	41.2°	.6587	.7524	.8754	1.142	48.8°	.8517
.7208	41.3°	.6600	.7513	.8785	1.138	48.7°	.8500
.7226	41.4°	.6613	.7501	.8816	1.134	48.6°	.8482
.7243	41.5°	.6626	.7490	.8847	1.130	48.5°	.8465
.7261	41.6°	.6639	.7478	.8878	1.126	48.4°	.8447
.7278	41.7°	.6652	.7466	.8910	1.122	48.3°	.8430
.7295	41.8°	.6665	.7455	.8941	1.118	48.2°	.8412
.7313	41.9°	.6678	.7443	.8972	1.115	48.1°	.8395
.7330	42.0°	.6691	.7431	.9004	1.111	48.0°	.8378
.7348	42.1°	.6704	.7420	.9036	1.107	47.9°	.8360
.7365	42.2°	.6717	.7408	.9067	1.103	47.8°	.8343
.7383	42.3°	.6730	.7396	.9099	1.099	47.7°	.8325
.7400	42.4°	.6743	.7385	.9131	1.095	47.6°	.8308
.7418	42.5°	.6756	.7373	.9163	1.091	47.5°	.8290
.7435	42.6°	.6769	.7361	.9195	1.087	47.4°	.8273
.7453	42.7°	.6782	.7349	.9228	1.084	47.3°	.8255
.7470	42.8°	.6794	.7337	.9260	1.080	47.2°	.8238
.7487	42.9°	.6807	.7325	.9293	1.076	47.1°	.8221
.7505	43.0°	.6820	.7314	.9325	1.072	47.0°	.8203
.7522	43.1°	.6833	.7302	.9358	1.069	46.9°	.8186
.7540	43.2°	.6845	.7290	.9391	1.065	46.8°	.8168
.7557	43.3°	.6858	.7278	.9424	1.061	46.7°	.8151
.7575	43.4°	.6871	.7266	.9457	1.057	46.6°	.8133
.7592	43.5°	.6884	.7254	.9490	1.054	46.5°	.8116
.7610	43.6°	.6896	.7242	.9523	1.050	46.4°	.8098
.7627	43.7°	.6909	.7230	.9556	1.046	46.3°	.8081
.7645	43.8°	.6921	.7218	.9590	1.043	46.2°	.8063
.7662	43.9°	.6934	.7206	.9623	1.039	46.1°	.8046
		Cos	Sin	Cot	Tan	Degrees	Radians

Table A *(Continued)*

Radians	Degrees	Sin	Cos	Tan	Cot		
.7679	44.0°	.6947	.7193	.9657	1.036	46.0°	.8029
.7697	44.1°	.6959	.7181	.9691	1.032	45.9°	.8011
.7714	44.2°	.6972	.7169	.9725	1.028	45.8°	.7994
.7732	44.3°	.6984	.7157	.9759	1.025	45.7°	.7976
.7749	44.4°	.6997	.7145	.9793	1.021	45.6°	.7959
.7767	44.5°	.7009	.7133	.9827	1.018	45.5°	.7941
.7784	44.6°	.7022	.7120	.9861	1.014	45.4°	.7924
.7802	44.7°	.7034	.7108	.9896	1.011	45.3°	.7906
.7819	44.8°	.7046	.7096	.9930	1.007	45.2°	.7889
.7837	44.9°	.7059	.7083	.9965	1.003	45.1°	.7871
.7854	45.0°	.7071	.7071	1.0000	1.000	45.0°	.7854
		Cos	Sin	Cot	Tan	Degrees	Radians

Table B   Base 10 Logarithms

N	0	1	2	3	4	5	6	7	8	9
10	0000	0043	0086	0128	0170	0212	0253	0294	0334	0374
11	0414	0453	0492	0531	0569	0607	0645	0682	0719	0755
12	0792	0828	0864	0899	0934	0969	1004	1038	1072	1106
13	1139	1173	1206	1239	1271	1303	1335	1367	1399	1430
14	1461	1492	1523	1553	1584	1614	1644	1673	1703	1732
15	1761	1790	1818	1847	1875	1903	1931	1959	1987	2014
16	2041	2068	2095	2122	2148	2175	2201	2227	2253	2279
17	2304	2330	2355	2380	2405	2430	2455	2480	2504	2529
18	2553	2577	2601	2625	2648	2672	2695	2718	2742	2765
19	2788	2810	2833	2856	2878	2900	2923	2945	2967	2989
20	3010	3032	3054	3075	3096	3118	3139	3160	3181	3201
21	3222	3243	3263	3284	3304	3324	3345	3365	3385	3404
22	3424	3444	3464	3483	3502	3522	3541	3560	3579	3598
23	3617	3636	3655	3674	3692	3711	3729	3747	3766	3784
24	3802	3820	3838	3856	3874	3892	3909	3927	3945	3962
25	3979	3997	4014	4031	4048	4065	4082	4099	4116	4133
26	4150	4166	4183	4200	4216	4232	4249	4265	4281	4298
27	4314	4330	4346	4362	4378	4393	4409	4425	4440	4456
28	4472	4487	4502	4518	4533	4548	4564	4579	4594	4609
29	4624	4639	4654	4669	4683	4698	4713	4728	4742	4757
30	4771	4786	4800	4814	4829	4843	4857	4871	4886	4900
31	4914	4928	4942	4955	4969	4983	4997	5011	5024	5038
32	5051	5065	5079	5092	5105	5119	5132	5145	5159	5172
33	5185	5198	5211	5224	5237	5250	5263	5276	5289	5302
34	5315	5328	5340	5353	5366	5378	5391	5403	5416	5428
35	5441	5453	5465	5478	5490	5502	5514	5527	5539	5551
36	5563	5575	5587	5599	5611	5623	5635	5647	5658	5670
37	5682	5694	5705	5717	5729	5740	5752	5763	5775	5786
38	5798	5809	5821	5832	5843	5855	5866	5877	5888	5899
39	5911	5922	5933	5944	5955	5966	5977	5988	5999	6010
40	6021	6031	6042	6053	6064	6075	6085	6096	6107	6117
41	6128	6138	6149	6160	6170	6180	6191	6201	6212	6222
42	6232	6243	6253	6263	6274	6284	6294	6304	6314	6325
43	6335	6345	6355	6365	6375	6385	6395	6405	6415	6425
44	6435	6444	6454	6464	6474	6484	6493	6503	6513	6522
45	6532	6542	6551	6561	6571	6580	6590	6599	6609	6618
46	6628	6637	6646	6656	6665	6675	6684	6693	6702	6712
47	6721	6730	6739	6749	6758	6767	6776	6785	6794	6803
48	6812	6821	6830	6839	6848	6857	6866	6875	6884	6893
49	6902	6911	6920	6928	6937	6946	6955	6964	6972	6981
50	6990	6998	7007	7016	7024	7033	7042	7050	7059	7067
51	7076	7084	7093	7101	7110	7118	7126	7135	7143	7152
52	7160	7168	7177	7185	7193	7202	7210	7218	7226	7235
53	7243	7251	7259	7267	7275	7284	7292	7300	7308	7316
54	7324	7332	7340	7348	7356	7364	7372	7380	7388	7396

Table B (*Continued*)

N	0	1	2	3	4	5	6	7	8	9
55	7404	7412	7419	7427	7435	7443	7451	7459	7466	7474
56	7482	7490	7497	7505	7513	7520	7528	7536	7543	7551
57	7559	7566	7574	7582	7589	7597	7604	7612	7619	7627
58	7634	7642	7649	7657	7664	7672	7679	7686	7694	7701
59	7709	7716	7723	7731	7738	7745	7752	7760	7767	7774
60	7782	7789	7796	7803	7810	7818	7825	7832	7839	7846
61	7853	7860	7868	7875	7882	7889	7896	7903	7910	7917
62	7924	7931	7938	7945	7952	7959	7966	7973	7980	7987
63	7993	8000	8007	8014	8021	8028	8035	8041	8048	8055
64	8062	8069	8075	8082	8089	8096	8102	8109	8116	8122
65	8129	8136	8142	8149	8156	8162	8169	8176	8182	8189
66	8195	8202	8209	8215	8222	8228	8235	8241	8248	8254
67	8261	8267	8274	8280	8287	8293	8299	8306	8312	8319
68	8325	8331	8338	8344	8351	8357	8363	8370	8376	8382
69	8388	8395	8401	8407	8414	8420	8426	8432	8439	8445
70	8451	8457	8463	8470	8476	8482	8488	8494	8500	8506
71	8513	8519	8525	8531	8537	8543	8549	8555	8561	8567
72	8573	8579	8585	8591	8597	8603	8609	8615	8621	8627
73	8633	8639	8645	8651	8657	8663	8669	8675	8681	8686
74	8692	8698	8704	8710	8716	8722	8727	8733	8739	8745
75	8751	8756	8762	8768	8774	8779	8785	8791	8797	8802
76	8808	8814	8820	8825	8831	8837	8842	8848	8854	8859
77	8865	8871	8876	8882	8887	8893	8899	8904	8910	8915
78	8921	8927	8932	8938	8943	8949	8954	8960	8965	8971
79	8976	8982	8987	8993	8998	9004	9009	9015	9020	9025
80	9031	9036	9042	9047	9053	9058	9063	9069	9074	9079
81	9085	9090	9096	9101	9106	9112	9117	9122	9128	9133
82	9138	9143	9149	9154	9159	9165	9170	9175	9180	9186
83	9191	9196	9201	9206	9212	9217	9222	9227	9232	9238
84	9243	9248	9253	9258	9263	9269	9274	9279	9284	9289
85	9294	9299	9304	9309	9315	9320	9325	9330	9335	9340
86	9345	9350	9355	9360	9365	9370	9375	9380	9385	9390
87	9395	9400	9405	9410	9415	9420	9425	9430	9435	9440
88	9445	9450	9455	9460	9465	9469	9474	9479	9484	9489
89	9494	9499	9504	9509	9513	9518	9523	9528	9533	9538
90	9542	9547	9552	9557	9562	9566	9571	9576	9581	9586
91	9590	9595	9600	9605	9609	9614	9619	9624	9628	9633
92	9638	9643	9647	9652	9657	9661	9666	9671	9675	9680
93	9685	9689	9694	9699	9703	9708	9713	9717	9722	9727
94	9731	9736	9741	9745	9750	9754	9759	9763	9768	9773
95	9777	9782	9786	9791	9795	9800	9805	9809	9814	9818
96	9823	9827	9832	9836	9841	9845	9850	9854	9859	9863
97	9868	9872	9877	9881	9886	9890	9894	9899	9903	9908
98	9912	9917	9921	9926	9930	9934	9939	9943	9948	9952
99	9956	9961	9965	9969	9974	9978	9983	9987	9991	9996

Table C   Base *e* Logarithms

N	0	1	2	3	4	5	6	7	8	9
1.0	.0000	.0100	.0198	.0296	.0392	.0488	.0583	.0677	.0770	.0862
1.1	.0953	.1044	.1133	.1222	.1310	.1398	.1484	.1570	.1655	.1740
1.2	.1823	.1906	.1989	.2070	.2151	.2231	.2311	.2390	.2469	.2546
1.3	.2624	.2700	.2776	.2852	.2927	.3001	.3075	.3148	.3221	.3293
1.4	.3365	.3436	.3507	.3577	.3646	.3716	.3784	.3853	.3920	.3988
1.5	.4055	.4121	.4187	.4253	.4318	.4383	.4447	.4511	.4574	.4637
1.6	.4700	.4762	.4824	.4886	.4947	.5008	.5068	.5128	.5188	.5247
1.7	.5306	.5365	.5423	.5481	.5539	.5596	.5653	.5710	.5766	.5822
1.8	.5878	.5933	.5988	.6043	.6098	.6152	.6206	.6259	.6313	.6366
1.9	.6419	.6471	.6523	.6575	.6627	.6678	.6729	.6780	.6831	.6881
2.0	.6931	.6981	.7031	.7080	.7129	.7178	.7227	.7275	.7324	.7372
2.1	.7419	.7467	.7514	.7561	.7608	.7655	.7701	.7747	.7793	.7839
2.2	.7885	.7930	.7975	.8020	.8065	.8109	.8154	.8198	.8242	.8286
2.3	.8329	.8372	.8416	.8459	.8502	.8544	.8587	.8629	.8671	.8713
2.4	.8755	.8796	.8838	.8879	.8920	.8961	.9002	.9042	.9083	.9123
2.5	.9163	.9203	.9243	.9282	.9322	.9361	.9400	.9439	.9478	.9517
2.6	.9555	.9594	.9632	.9670	.9708	.9746	.9783	.9821	.9858	.9895
2.7	.9933	.9969	1.0006	.0043	.0080	.0116	.0152	.0188	.0225	.0260
2.8	1.0296	.0332	.0367	.0403	.0438	.0473	.0508	.0543	.0578	.0613
2.9	.0647	.0682	.0716	.0750	.0784	.0818	.0852	.0886	.0919	.0953
3.0	1.0986	.1019	.1053	.1086	.1119	.1151	.1184	.1217	.1249	.1282
3.1	.1314	.1346	.1378	.1410	.1442	.1474	.1506	.1537	.1569	.1600
3.2	.1632	.1663	.1694	.1725	.1756	.1787	.1817	.1848	.1878	.1909
3.3	.1939	.1969	.2000	.2030	.2060	.2090	.2119	.2149	.2179	.2208
3.4	.2238	.2267	.2296	.2326	.2355	.2384	.2413	.2442	.2470	.2499
3.5	1.2528	.2556	.2585	.2613	.2641	.2669	.2698	.2726	.2754	.2782
3.6	.2809	.2837	.2865	.2892	.2920	.2947	.2975	.3002	.3029	.3056
3.7	.3083	.3110	.3137	.3164	.3191	.3218	.3244	.3271	.3297	.3324
3.8	.3350	.3376	.3403	.3429	.3455	.3481	.3507	.3533	.3558	.3584
3.9	.3610	.3635	.3661	.3686	.3712	.3737	.3762	.3788	.3813	.3838
4.0	1.3863	.3888	.3913	.3938	.3962	.3987	.4012	.4036	.4061	.4085
4.1	.4110	.4134	.4159	.4183	.4207	.4231	.4255	.4279	.4303	.4327
4.2	.4351	.4375	.4398	.4422	.4446	.4469	.4493	.4516	.4540	.4563
4.3	.4586	.4609	.4633	.4656	.4679	.4702	.4725	.4748	.4770	.4793
4.4	.4816	.4839	.4861	.4884	.4907	.4929	.4951	.4974	.4996	.5019
4.5	1.5041	.5063	.5085	.5107	.5129	.5151	.5173	.5195	.5217	.5239
4.6	.5261	.5282	.5304	.5326	.5347	.5369	.5390	.5412	.5433	.5454
4.7	.5476	.5497	.5518	.5539	.5560	.5581	.5602	.5623	.5644	.5665
4.8	.5686	.5707	.5728	.5748	.5769	.5790	.5810	.5831	.5851	.5872
4.9	.5892	.5913	.5933	.5953	.5974	.5994	.6014	.6034	.6054	.6074
5.0	1.6094	.6114	.6134	.6154	.6174	.6194	.6214	.6233	.6253	.6273
5.1	.6292	.6312	.6332	.6351	.6371	.6390	.6409	.6429	.6448	.6467
5.2	.6487	.6506	.6525	.6544	.6563	.6582	.6601	.6620	.6639	.6658
5.3	.6677	.6696	.6715	.6734	.6752	.6771	.6790	.6808	.6827	.6845
5.4	.6864	.6882	.6901	.6919	.6938	.6956	.6974	.6993	.7011	.7029

TABLE C BASE *e* LOGARITHMS **A-55**

Table C *(Continued)*

N	0	1	2	3	4	5	6	7	8	9
5.5	1.7047	.7066	.7084	.7102	.7120	.7138	.7156	.7174	.7192	.7210
5.6	.7228	.7246	.7263	.7281	.7299	.7317	.7334	.7352	.7370	.7387
5.7	.7405	.7422	.7440	.7457	.7475	.7492	.7509	.7527	.7544	.7561
5.8	.7579	.7596	.7613	.7630	.7647	.7664	.7681	.7699	.7716	.7733
5.9	.7750	.7766	.7783	.7800	.7817	.7834	.7851	.7867	.7884	.7901
6.0	1.7918	.7934	.7951	.7967	.7984	.8001	.8017	.8034	.8050	.8066
6.1	.8083	.8099	.8116	.8132	.8148	.8165	.8181	.8197	.8213	.8229
6.2	.8245	.8262	.8278	.8294	.8310	.8326	.8342	.8358	.8374	.8390
6.3	.8405	.8421	.8437	.8453	.8469	.8485	.8500	.8516	.8532	.8547
6.4	.8563	.8579	.8594	.8610	.8625	.8641	.8656	.8672	.8687	.8703
6.5	1.8718	.8733	.8749	.8764	.8779	.8795	.8810	.8825	.8840	.8856
6.6	.8871	.8886	.8901	.8916	.8931	.8946	.8961	.8976	.8991	.9006
6.7	.9021	.9036	.9051	.9066	.9081	.9095	.9110	.9125	.9140	.9155
6.8	.9169	.9184	.9199	.9213	.9228	.9242	.9257	.9272	.9286	.9301
6.9	.9315	.9330	.9344	.9359	.9373	.9387	.9402	.9416	.9430	.9445
7.0	1.9459	.9473	.9488	.9502	.9516	.9530	.9544	.9559	.9573	.9587
7.1	.9601	.9615	.9629	.9643	.9657	.9671	.9685	.9699	.9713	.9727
7.2	.9741	.9755	.9769	.9782	.9796	.9810	.9824	.9838	.9851	.9865
7.3	.9879	.9892	.9906	.9920	.9933	.9947	.9961	.9974	.9988	2.0001
7.4	2.0015	.0028	.0042	.0055	.0069	.0082	.0096	.0109	.0122	.0136
7.5	2.0149	.0162	.0176	.0189	.0202	.0215	.0229	.0242	.0255	.0268
7.6	.0281	.0295	.0308	.0321	.0334	.0347	.0360	.0373	.0386	.0399
7.7	.0412	.0425	.0438	.0451	.0464	.0477	.0490	.0503	.0516	.0528
7.8	.0541	.0554	.0567	.0580	.0592	.0605	.0618	.0631	.0643	.0656
7.9	.0669	.0681	.0694	.0707	.0719	.0732	.0744	.0757	.0769	.0782
8.0	2.0794	.0807	.0819	.0832	.0844	.0857	.0869	.0882	.0894	.0906
8.1	.0919	.0931	.0943	.0956	.0968	.0980	.0992	.1005	.1017	.1029
8.2	.1041	.1054	.1066	.1078	.1090	.1102	.1114	.1126	.1138	.1150
8.3	.1163	.1175	.1187	.1199	.1211	.1223	.1235	.1247	.1258	.1270
8.4	.1282	.1294	.1306	.1318	.1330	.1342	.1353	.1365	.1377	.1389
8.5	2.1401	.1412	.1424	.1436	.1448	.1459	.1471	.1483	.1494	.1506
8.6	.1518	.1529	.1541	.1552	.1564	.1576	.1587	.1599	.1610	.1622
8.7	.1633	.1645	.1656	.1668	.1679	.1691	.1702	.1713	.1725	.1736
8.8	.1748	.1759	.1770	.1782	.1793	.1804	.1815	.1827	.1838	.1849
8.9	.1861	.1872	.1883	.1894	.1905	.1917	.1928	.1939	.1950	.1961
9.0	2.1972	.1983	.1994	.2006	.2017	.2028	.2039	.2050	.2061	.2072
9.1	.2083	.2094	.2105	.2116	.2127	.2138	.2148	.2159	.2170	.2181
9.2	.2192	.2203	.2214	.2225	.2235	.2246	.2257	.2268	.2279	.2289
9.3	.2300	.2311	.2322	.2332	.2343	.2354	.2364	.2375	.2386	.2396
9.4	.2407	.2418	.2428	.2439	.2450	.2460	.2471	.2481	.2492	.2502
9.5	2.2513	.2523	.2534	.2544	.2555	.2565	.2576	.2586	.2597	.2607
9.6	.2618	.2628	.2638	.2649	.2659	.2670	.2680	.2690	.2701	.2711
9.7	.2721	.2732	.2742	.2752	.2762	.2773	.2783	.2793	.2803	.2814
9.8	.2824	.2834	.2844	.2854	.2865	.2875	.2885	.2895	.2905	.2915
9.9	.2925	.2935	.2946	.2956	.2966	.2976	.2986	.2996	.3006	.3016

Use the properties of logarithms and ln 10 ≈ 2.3026 to find logarithms of numbers less than 1 or greater than 10.

ANSWERS TO SELECTED EXERCISES

Exercise 1.1 (Page 5)

1. 5 **3.** 8 **5.** 12 **7.** 12 **9.** 5 **11.** 13 **13.** 10 **15.** $\sqrt{104} = 2\sqrt{26}$ **17.** 5
19. $d(AB) = 4\sqrt{2}$ and $d(BC) = 4\sqrt{2}$. Hence, the triangle is isosceles. **21.** Pick any point P on line CD. Let P have coordinates (x, y). Find the distances PA and PB, set them equal to each other, and simplify. The result is $3x + y = 24$.

Exercise 1.2 (Page 10)

1. Domain is $\{1, 2\}$; range is $\{2, 3\}$. **3.** Domain is $\{3, 5, 8, 10\}$; range is $\{4, 6, 9, 12\}$. **5.** Domain is the real number set; range is the real number set. **7.** Domain is the real number set; range is all real numbers greater than or equal to 1. **9.** Domain is the real number set; range is the set of nonnegative numbers. **11.** Domain is all real numbers greater than or equal to 3; range is the set of nonnegative numbers. **13.** a function **15.** not a function **17.** a function **19.** a function **21.** not a function **23.** a function

25. $\frac{3}{2}$ **27.** $\frac{7}{3}$ **29.** no value with 0 in denominator **31.** $\frac{2h-1}{h}$ **33.** 0

35. no real number value **37.** $\frac{\sqrt{a+h-1}}{2}$

39.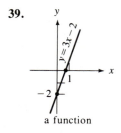
a function

41.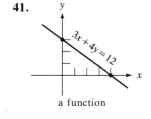
a function

43.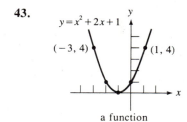
a function

45.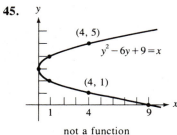
not a function

47.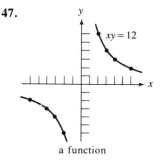
a function

49.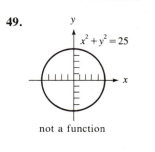
not a function

51.

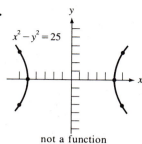

$x^2 - y^2 = 25$

not a function

53.

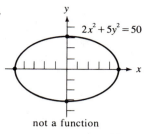

$2x^2 + 5y^2 = 50$

not a function

55. Let any point P on the circle have coordinates of (x, y). Find the distance r between point P and the origin: $r = \sqrt{(x - 0)^2 + (y - 0)^2} = \sqrt{x^2 + y^2}$. Square both sides to get $x^2 + y^2 = r^2$.

Exercise 1.3   (Page 15)

1. $\{(1, 2), (2, 3), (3, 4)\}$; a function   **3.** $\{(3, 2), (4, 2), (5, 2)\}$; a function   **5.** $\{(1, 5), (1, 4), (1, 3), (1, 2)\}$; not a function   **7.** $\{(1, 1), (8, 2), (27, 3), (64, 4)\}$; a function   **9.** $y = \dfrac{x + 2}{-2}$; a function

11. $y = 3 - x$; a function   **13.** $y = \pm\sqrt{x - 2}$; not a function   **15.** $y = \pm\sqrt{x + 1}$ and $y \geq 0$; a function

17.

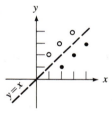

19.

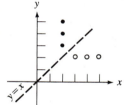

21.

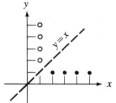

23.

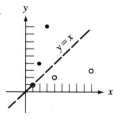

25.

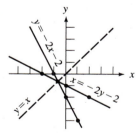

27.

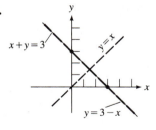

29.

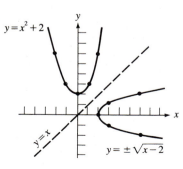

$y = x^2 + 2$

$y = \pm\sqrt{x - 2}$

31.

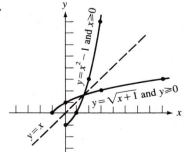

$y = x^2 - 1$ and $x \geq 0$

$y = \sqrt{x + 1}$ and $y \geq 0$

33. one-to-one
35. not one-to-one
37. one-to-one
39. one-to-one
41. one-to-one
43. one-to-one
45. not one-to-one
47. all values except 1
49. all values except 3

REVIEW EXERCISES (Page 16)

1. 7 **2.** 10 **3.** 17 **4.** 35 **5.** 5 **6.** 13 **7.** 5 **8.** 5 **9.** $5\sqrt{2}$ **10.** $5\sqrt{2}$
11. $\sqrt{170}$ **12.** $2\sqrt{17}$ **13.** $d(BC) = \sqrt{5}$ and $d(CA) = \sqrt{5}$. Hence, triangle ABC is isosceles.
14. Because $d(AB) + d(BC) = d(CA)$, points A, B, and C lie on a line. **15.** Domain = $\{2, 4, 5, 6, 7\}$;
range = $\{1, 2, 6, 7, 8\}$ **16.** Domain = $\{-2, 1, 3\}$; range = $\{-2, 2, 3, 4, 5\}$ **17.** Domain is the set of
real numbers; range is the set of real numbers. **18.** Domain is the set of real numbers; range is the set of all
real numbers greater than or equal to -10. **19.** Domain is the set of all real numbers greater than or equal to
10; range is the set of nonnegative real numbers. **20.** Domain is the set of all real numbers; range is the set of
nonnegative real numbers. **21.** not a function **22.** a function **23.** not a function **24.** a function
25. not a function **26.** not a function **27.** -4 **28.** 11 **29.** $h^2 + 2h - 4$ **30.** $2ah + 2h + h^2$

31.

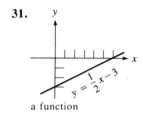

a function

32.

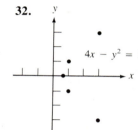

$4x - y^2 = 3$

33.

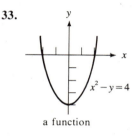

$x^2 - y = 4$

a function

34.

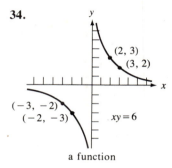

(2, 3)
(3, 2)
$(-3, -2)$
$(-2, -3)$
$xy = 6$

a function

35. $\{(1, 2), (1, 3), (1, 4)\}$; not a function

36. $\{(10, 1), (20, 2), (30, 3), (40, 4)\}$; a function **37.** $y = \dfrac{x - 12}{7}$; a function **38.** $y = 4x - 17$; a function

39. $y = \pm\sqrt{3x}$; not a function **40.** $yx = 36$; a function **41.** $y = \dfrac{x^2 + 1}{2}$, $x \geq 0$; a function

42. $y = \sqrt{x - 4}$; a function **43.**

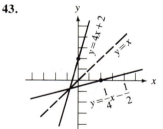

$y = 4x + 2$
$y = x$
$y = \frac{1}{4}x - \frac{1}{2}$

44.

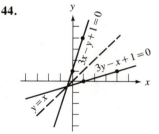

$3x - y + 1 = 0$
$3y - x + 1 = 0$
$y = x$

45.

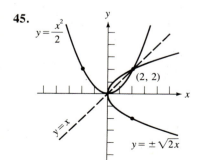

$y = \dfrac{x^2}{2}$
(2, 2)
$y = x$
$y = \pm\sqrt{2x}$

46.

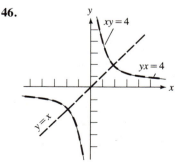

$xy = 4$
$yx = 4$
$y = x$

47. one-to-one
48. not one-to-one
49. not one-to-one
50. one-to-one
51. one-to-one
52. one-to-one

Exercise 2.1   (Page 26)

1. yes; positive   **3.** no; positive   **5.** no; negative   **7.** yes; negative   **9.** no; positive
11. yes; positive   **13.** QII   **15.** QIV   **17.** QII   **19.** QIII   **21.** QIII   **23.** QIV   **25.** QIII
27. yes   **29.** no   **31.** yes   **33.** yes

35.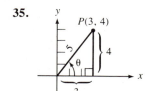
$$\begin{cases} \sin\theta = \frac{4}{5}, \cos\theta = \frac{3}{5}, \tan\theta = \frac{4}{3}, \\ \csc\theta = \frac{5}{4}, \sec\theta = \frac{5}{3}, \cot\theta = \frac{3}{4} \end{cases}$$

37.
$$\begin{cases} \sin\theta = \frac{-12}{13}, \cos\theta = \frac{-5}{13}, \tan\theta = \frac{12}{5}, \\ \csc\theta = \frac{13}{-12}, \sec\theta = \frac{13}{-5}, \cot\theta = \frac{5}{12} \end{cases}$$

39.
$$\begin{cases} \sin\theta = \frac{40}{41}, \cos\theta = \frac{-9}{41}, \tan\theta = \frac{40}{-9}, \\ \csc\theta = \frac{41}{40}, \sec\theta = \frac{41}{-9}, \cot\theta = \frac{-9}{40} \end{cases}$$

41.
$$\begin{cases} \sin\theta = \frac{\sqrt{2}}{2}, \cos\theta = \frac{\sqrt{2}}{2}, \tan\theta = 1, \\ \csc\theta = \sqrt{2}, \sec\theta = \sqrt{2}, \cot\theta = 1 \end{cases}$$

43.
$$\begin{cases} \sin\theta = \frac{4}{5}, \cos\theta = \frac{-3}{5}, \tan\theta = -\frac{4}{3} \\ \csc\theta = \frac{5}{4}, \sec\theta = -\frac{5}{3}, \cot\theta = \frac{-3}{4} \end{cases}$$

45.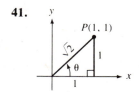
$$\begin{cases} \sin\theta = \frac{3}{5}, \cos\theta = \frac{-4}{5}, \tan\theta = -\frac{3}{4} \\ \csc\theta = \frac{5}{3}, \sec\theta = -\frac{5}{4}, \cot\theta = \frac{-4}{3} \end{cases}$$

47.

$$\begin{cases} \sin\theta = \dfrac{5\sqrt{34}}{34}, \cos\theta = \dfrac{3\sqrt{34}}{34}, \tan\theta = \dfrac{5}{3} \\ \csc\theta = \dfrac{\sqrt{34}}{5}, \sec\theta = \dfrac{\sqrt{34}}{3}, \cot\theta = \dfrac{3}{5} \end{cases}$$

49.

$$\begin{cases} \sin\theta = \dfrac{-5}{13}, \cos\theta = \dfrac{12}{13}, \tan\theta = \dfrac{-5}{12} \\ \csc\theta = -\dfrac{13}{5}, \sec\theta = \dfrac{13}{12}, \cot\theta = -\dfrac{12}{5} \end{cases}$$

51. $\cos\theta = \frac{4}{5}$, $\tan\theta = \frac{3}{4}$, $\csc\theta = \frac{5}{3}$, $\sec\theta = \frac{5}{4}$, $\cot\theta = \frac{4}{3}$ **53.** $\sin\theta = \frac{-12}{13}$, $\tan\theta = \frac{12}{5}$,
$\csc\theta = \frac{13}{-12}$, $\sec\theta = \frac{13}{-5}$ **55.** $\cos\theta = \frac{40}{41}$, $\tan\theta = \frac{-9}{40}$, $\csc\theta = \frac{41}{-9}$, $\sec\theta = \frac{41}{40}$, $\cot\theta = \frac{40}{-9}$
57. $\sin\theta = \frac{4}{5}$, $\cos\theta = \frac{-3}{5}$, $\tan\theta = \frac{-4}{3}$, $\cot\theta = \frac{-3}{4}$ **59.** $\sin\theta = \frac{-40}{41}$, $\csc\theta = \frac{41}{-40}$, $\sec\theta = \frac{41}{9}$,
$\cot\theta = \frac{9}{-40}$

Exercise 2.2 (Page 34)

1. $\tan\theta \cos\theta = \dfrac{y}{x} \cdot \dfrac{x}{r} = \dfrac{y}{r} = \sin\theta$ **3.** $\tan\theta = \dfrac{y}{x} = \dfrac{1}{\dfrac{x}{y}} = \dfrac{1}{\dfrac{x}{r} \cdot \dfrac{r}{y}} = \dfrac{1}{\cos\theta \csc\theta}$

5. $\sin^2\theta + \sin^2\theta\cot^2\theta = \dfrac{y^2}{r^2} + \dfrac{y^2}{r^2}\cdot\dfrac{x^2}{y^2} = \dfrac{y^2}{r^2} + \dfrac{x^2}{r^2} = \dfrac{y^2+x^2}{r^2} = \dfrac{r^2}{r^2} = 1$

7. $\cot^2\theta + \sin^2\theta = \dfrac{x^2}{y^2} + \dfrac{y^2}{r^2} = \dfrac{r^2-y^2}{y^2} + \dfrac{r^2-x^2}{r^2} = \dfrac{r^2}{y^2} - 1 + 1 - \dfrac{x^2}{r^2} = \dfrac{r^2}{y^2} - \dfrac{x^2}{r^2} = \csc^2\theta - \cos^2\theta$

9. $\tan\theta\cos\theta = \dfrac{\sin\theta}{\cos\theta}\cos\theta = \sin\theta$ **11.** $\tan\theta = \dfrac{\sin\theta}{\cos\theta} = \dfrac{1}{\cos\theta\csc\theta}$

13. $\sin^2\theta + \sin^2\theta\cot^2\theta = \sin^2\theta(1 + \cot^2\theta) = \sin^2\theta\csc^2\theta = 1$

15. $\cot^2\theta + \sin^2\theta = \csc^2\theta - 1 + 1 - \cos^2\theta = \csc^2\theta - \cos^2\theta$

17. $\cos\theta = \frac{3}{5}$, $\tan\theta = \frac{4}{3}$, $\csc\theta = \frac{5}{4}$, $\sec\theta = \frac{5}{3}$, $\cot\theta = \frac{3}{4}$

19. $\sin\theta = \frac{12}{13}$, $\tan\theta = \frac{-12}{5}$, $\csc\theta = \frac{13}{12}$, $\sec\theta = \frac{-13}{5}$, $\cot\theta = \frac{-5}{12}$

21. $\sin\theta = \frac{4}{5}$, $\cos\theta = \frac{-3}{5}$, $\csc\theta = \frac{5}{4}$, $\sec\theta = \frac{-5}{3}$, $\cot\theta = \frac{-3}{4}$

23. $\sin\theta = \frac{40}{41}$, $\cos\theta = \frac{9}{41}$, $\tan\theta = \frac{40}{9}$, $\csc\theta = \frac{41}{40}$, $\sec\theta = \frac{41}{9}$

25. $\sin\theta = \frac{-12}{13}$, $\cos\theta = \frac{-5}{13}$, $\tan\theta = \frac{12}{5}$, $\csc\theta = \frac{-13}{12}$, $\cot\theta = \frac{5}{12}$ **27.** odd **29.** even **31.** odd

33. neither **35.** neither **37.** $y = \csc x = 1/\sin x$. Because $\sin x$ is an odd function, so is $1/\sin x = y$.
39. yes **41.** no

43.

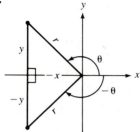

45.

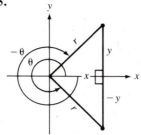

47. negative   **49.** positive

Exercise 2.3   (Page 42)

1.

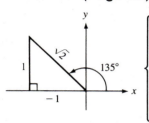

$$\sin 135° = \frac{1}{\sqrt{2}} = \frac{\sqrt{2}}{2}$$

$$\cos 135° = \frac{-1}{\sqrt{2}} = -\frac{\sqrt{2}}{2}$$

$$\tan 135° = \frac{1}{-1} = -1$$

3.

$$\sin 240° = -\frac{\sqrt{3}}{2}$$

$$\cos 240° = -\frac{1}{2}$$

$$\tan 240° = \sqrt{3}$$

5.

$$\sin 315° = \frac{-1}{\sqrt{2}} = -\frac{\sqrt{2}}{2}$$

$$\cos 315° = \frac{1}{\sqrt{2}} = \frac{\sqrt{2}}{2}$$

$$\tan 315° = \frac{-1}{1} = -1$$

7.

$$\csc 225° = -\sqrt{2}$$
$$\sec 225° = -\sqrt{2}$$
$$\cot 225° = 1$$

9.

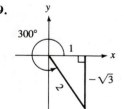

$$\csc 300° = \frac{2}{-\sqrt{3}} = -\frac{2\sqrt{3}}{3}$$

$$\sec 300° = 2$$

$$\cot 300° = \frac{1}{-\sqrt{3}} = -\frac{\sqrt{3}}{3}$$

11.

$$\csc 330° = -2$$

$$\sec 330° = \frac{2}{\sqrt{3}} = \frac{2\sqrt{3}}{3}$$

$$\cot 330° = -\sqrt{3}$$

13. $\sin 390° = \frac{1}{2}$, $\cos 390° = \frac{\sqrt{3}}{2}$, $\tan 390° = \frac{\sqrt{3}}{3}$ **15.** $\sin 510° = \frac{1}{2}$, $\cos 510° = -\frac{\sqrt{3}}{2}$, $\tan 510° =$

$-\frac{\sqrt{3}}{3}$ **17.** $\sin(-45°) = -\frac{\sqrt{2}}{2}$, $\cos(-45°) = \frac{\sqrt{2}}{2}$, $\tan(-45°) = -1$ **19.** 1   **21.** 0   **23.** 2

25. $-\frac{23}{4}$ **27.** $\frac{\sqrt{6}}{4} + 1$ **29.** $\frac{8\sqrt{3} - \sqrt{6}}{3}$ **31.** $-\sqrt{3}$ **33.** 1   **35.** $\theta = 30°$ **37.** $\theta = 300°$

39. $\theta = 210°$ **41.** $\theta = 315°$ **43.** $\theta = 135°$ **45.** $\theta = 90°, 270°$ **47.** $\theta = 330°$ **49.** $30°, 150°$
51. $30°, 210°$ **53.** $120°, 240°$ **55.** impossible   **57.** $210°$ **59.** $30°, 60°, 210°, 240°$

Exercise 2.4 (Page 47)

1. $\sin 50° \approx 0.77$, $\cos 50° \approx 0.64$, $\tan 50° \approx 1.2$ **3.** $\sin 235° \approx -0.82$, $\cos 235° \approx -0.57$, $\tan 235° \approx 1.4$
5. $\sin 12° \approx 0.21$, $\cos 12° \approx 0.98$, $\tan 12° \approx 0.21$ **7.** $\sin 115° \approx 0.91$, $\cos 115° \approx -0.42$, $\tan 115° \approx -2.1$
9. $\sin(-260°) \approx 0.98$, $\cos(-260°) \approx -0.17$, $\tan(-260°) \approx -5.7$ **11.** $\sin 17° \approx 0.2924$, $\cos 17°$
≈ 0.9563, $\tan 17° \approx 0.3057$ **13.** $\sin 62° \approx 0.8829$, $\cos 62° \approx 0.4695$, $\tan 62° \approx 1.881$ **15.** $\sin 119°$
≈ 0.8746, $\cos 119° \approx -0.4848$, $\tan 119° \approx -1.804$ **17.** $\sin(-233°) \approx 0.7986$, $\cos(-233°) \approx -0.6018$,
$\tan(-233°) \approx -1.327$ **19.** $\sin 1723° \approx -0.9744$, $\cos 1723° \approx 0.2250$, $\tan 1723° \approx -4.331$
21. $\sin 20° \approx 0.3420$, $\cos 70° \approx 0.3420$ **23.** $\cos 5° \approx 0.9962$, $\sin 85° \approx 0.9962$ **25.** $\sec 84°$
≈ 9.567, $\csc 6° \approx 9.567$ **27.** false   **29.** true   **31.** true   **33.** true   **35.** false   **37.** 0.3923
39. -0.6909 **41.** 0.9490   **43.** -1.3542 **45.** 1.4774   **47.** -0.1871 **49.** 14.0°
51. 110.0°   **53.** 207.0°   **55.** 280.0°   **57.** 49.0°   **59.** 220.0°

Exercise 2.5 (Page 56)

1. angle $B = 53°$, $a = 12$, $b = 16$ **3.** angle $B = 21.3°$, $a = 206$, $c = 221$ **5.** 15 ft   **7.** 721 ft
9. 4.0°   **11.** 29 ft   **13.** 319 mi   **15.** S 62.9° W   **17.** 5.8 mi   **19.** 685 mph   **21.** 60.7 ft
23. 9180 ft; 1.74 mi   **25.** 48.3°

Exercise 2.6 (Page 61)

1. 1450 ft   **3.** 556 ft   **5.** 631 ft   **7.** 143 ft   **9.** 123 ft   **11.** $\dfrac{r}{\tan\dfrac{\theta}{2}}$ **13.** $b = 2k\cos\alpha$
15. $H = k(\tan\alpha + \tan\beta)$ **17.** $D(\tan\phi - \tan\theta)$ **19.** $a = d\tan\beta$

Exercise 2.7 (Page 68)

1. 3 mph; 7 mph   **3.** 63°   **5.** 352 mph   **7.** N 11.6° W   **9.** 323 lb   **11.** 384 lb   **13.** 1160 lb
15. 13.7°   **17.** 224 lb; 48.1°   **19.** 13°   **21.** $(F_1)_y = 80$ lb; $(F_1)_x = 95$ lb   **23.** N 79.9° E
25. 58 lb

Exercise 2.8 (Page 73)

1. 45°   **3.** 1.31   **5.** 130 ohms   **7.** 46.4°   **9.** 120 ohms   **11.** $\dfrac{\sqrt{3}k}{2}$ **13.** 6% less

REVIEW EXERCISES (Page 76)

1. yes   **2.** no   **3.** no   **4.** yes   **5.** no   **6.** no   **7.** $\cos\theta = -\dfrac{\sqrt{51}}{10}$, $\tan\theta = \dfrac{7\sqrt{51}}{51}$, $\csc\theta =$
$-\dfrac{10}{7}$, $\sec\theta = -\dfrac{10\sqrt{51}}{51}$, $\cot\theta = \dfrac{\sqrt{51}}{7}$ **8.** $\sin\theta = -\dfrac{7\sqrt{130}}{130}$, $\cos\theta = -\dfrac{9\sqrt{130}}{130}$, $\csc\theta = -\dfrac{\sqrt{130}}{7}$, $\sec\theta$
$= -\dfrac{\sqrt{130}}{9}$, $\cot\theta = \dfrac{9}{7}$ **9.** $\sin\theta = \dfrac{\sqrt{51}}{10}$, $\tan\theta = -\dfrac{\sqrt{51}}{7}$, $\csc\theta = \dfrac{10\sqrt{51}}{51}$, $\sec\theta = -\dfrac{10}{7}$, $\cot\theta = -\dfrac{7\sqrt{51}}{51}$
10. $\sin\theta = \dfrac{-8\sqrt{145}}{145}$, $\cos\theta = \dfrac{9\sqrt{145}}{145}$, $\tan\theta = \dfrac{-8}{9}$, $\csc\theta = -\dfrac{\sqrt{145}}{8}$, $\sec\theta = \dfrac{\sqrt{145}}{9}$ **11.** $\dfrac{1}{\sec\theta} = \cos\theta$
$= \cos\theta \dfrac{\sin\theta}{\sin\theta} = \sin\theta\cot\theta$ **12.** $\cos\theta\csc\theta = \cos\theta\dfrac{1}{\sin\theta} = \dfrac{\cos\theta}{\sin\theta} = \cot\theta$ **13.** $\dfrac{\sqrt{6}}{4}$ **14.** $\dfrac{1}{2}$

15. $\frac{9}{16}$ **16.** 0   **17.** $\sin 930° = \frac{-1}{2}$, $\cos 930° = \frac{-\sqrt{3}}{2}$, $\tan 930° = \frac{\sqrt{3}}{3}$ **18.** $\sin 1380° = -\frac{\sqrt{3}}{2}$,

$\cos 1380° = \frac{1}{2}$, $\tan 1380° = -\sqrt{3}$ **19.** $\sin(-300°) = \frac{\sqrt{3}}{2}$, $\cos(-300°) = \frac{1}{2}$, $\tan(-300°) = \sqrt{3}$

20. $\sin(-585°) = \frac{\sqrt{2}}{2}$, $\cos(-585°) = \frac{-\sqrt{2}}{2}$, $\tan(-585°) = -1$ **21.** $\sin 15° \approx 0.26$, $\cos 15° \approx 0.97$,

$\tan 15° \approx 0.27$ **22.** $\sin 160° \approx 0.34$, $\cos 160° \approx -0.94$, $\tan 160° \approx -0.36$ **23.** $\sin 265° \approx -0.99$,

$\cos 265° \approx -0.09$, $\tan 265° \approx 11.4$ **24.** $\sin 340° \approx -0.34$, $\cos 340° \approx 0.94$, $\tan 340° \approx -0.36$

25. $\sin 15° \approx 0.2588$, $\cos 15° \approx 0.9659$, $\tan 15° \approx 0.2679$ **26.** $\sin 160° \approx 0.3420$, $\cos 160° \approx$

-0.9397, $\tan 160° \approx -0.3640$ **27.** $\sin 265° \approx -0.9962$, $\cos 265° \approx -0.0872$, $\tan 265° \approx 11.430$

28. $\sin 340° \approx -0.3420$, $\cos 340° \approx 0.9397$, $\tan 340° \approx -0.3640$ **29.** $\sin(-160°) \approx -0.3420$,

$\cos(-160°) \approx -0.9397$, $\tan(-160°) \approx 0.3640$ **30.** $\sin(-340°) \approx 0.3420$, $\cos(-340°) \approx 0.9397$,

$\tan(-340°) \approx 0.3640$ **31.** $119°$ **32.** $211°$ **33.** $317°$ **34.** $57.7°$ **35.** $100°$ **36.** $287°$

37. $17.3°$ **38.** 59.7 ft   **39.** 55 mi   **40.** 150 mi   **41.** 10 lb   **42.** $11.5°$

Exercise 3.1   (Page 85)

1. $\frac{1}{12}\pi$ **3.** $\frac{2}{3}\pi$ **5.** $\frac{7}{6}\pi$ **7.** $\frac{5}{3}\pi$ **9.** $\frac{13}{3}\pi$ **11.** $-\frac{26}{9}\pi$ **13.** $135°$ **15.** $450°$ **17.** $240°$

19. $1080°/\pi \approx 343.77°$ **21.** $-1800°/\pi \approx -572.96°$ **23.** $2250°/\pi \approx 716.20°$ **25.** $\sqrt{3}/2$ **27.** 1

29. $-\sqrt{2}/2$ **31.** $-\frac{1}{2}$ **33.** -2 **35.** $\sqrt{3}/3$ **37.** 39 cm   **39.** 76°   **41.** 2970 mi   **43.** $38.63°$ N

45. 104.72   **47.** 221,000 mi   **49.** 74 m   **51.** 1.15 mi

Exercise 3.2   (Page 89)

1. $2\pi\frac{\text{rads}}{\text{hr}}$ **3.** $\frac{\pi}{1800}\frac{\text{rads}}{\text{sec}}$ **5.** $\frac{4\pi}{59}\frac{\text{rads}}{\text{day}}$ **7.** $\frac{176}{5\pi} \approx 11.2\frac{\text{rev}}{\text{sec}}$ **9.** $\frac{1716}{\pi} \approx 546$ rpm   **11.** $\frac{75}{16}\pi$

≈ 14.7 ft/sec   **13.** $\frac{300}{\pi} \approx 95$ rpm   **15.** $\frac{50}{3}$ rpm   **17.** $\frac{225}{88}\pi \approx 8$ mph   **19.** approximately 933 mph

21. $44.9°$ N   **23.** Use the results of Exercise 22. $W = \frac{R_1}{R} \cdot W_1$. Hence, $W_2 = \frac{R}{R_2} \cdot W$

$= \frac{R}{R_2} \cdot \frac{R_1}{R} \cdot W_1 = \frac{R_1}{R_2} \cdot W_1$

Exercise 3.3   (Page 95)

1. $\frac{1}{2}$ **3.** $-\frac{\sqrt{3}}{2}$ **5.** 1   **7.** 2   **9.** $-\frac{\sqrt{3}}{2}$ **11.** -1 **13.** 0.9093   **15.** -0.1455 **17.** 0.8163

19. 3.5253   **21.** 0.6421   **23.** -0.1411 **25.** $(0, -1)$ **27.** $(-1, 0)$ **29.** $(-1, 0)$ **31.** $(0, 1)$

33. $\left(\frac{\sqrt{2}}{2}, \frac{\sqrt{2}}{2}\right)$ **35.** $\left(-\frac{\sqrt{2}}{2}, -\frac{\sqrt{2}}{2}\right)$ **37.** $\left(\frac{1}{2}, \frac{\sqrt{3}}{2}\right)$ **39.** $\left(\frac{-1}{2}, \frac{-\sqrt{3}}{2}\right)$ **41.** $\left(\frac{-1}{2}, \frac{-\sqrt{3}}{2}\right)$

43. $\left(\frac{1}{2}, \frac{\sqrt{3}}{2}\right)$ **45.** $\left(\frac{\sqrt{3}}{2}, \frac{-1}{2}\right)$ **47.** $\left(\frac{\sqrt{3}}{2}, \frac{-1}{2}\right)$ **49.** The formula for area of a sector of a circle is

$A = \frac{1}{2}r^2\theta$. In the unit circle, $r = 1$. Hence, $A = \frac{1}{2}(1)^2\theta = \frac{1}{2}\theta$.   **51.** $A = \frac{1}{2}bh$ with $b = 1$ and $h = \tan\theta$.

Hence, $A = \frac{1}{2}\tan\theta$.

Exercise 3.4   (Page 101)

1. $2; 2\pi$ **3.** $1; \dfrac{2\pi}{9}$ **5.** $1; 6\pi$ **7.** $1; 10\pi$ **9.** $3; 4\pi$ **11.** $\dfrac{1}{2}; 2$ **13.** $3; 1$ **15.** $\dfrac{1}{3}; \dfrac{2\pi^2}{3}$

17.
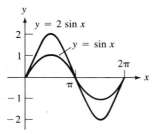

19.
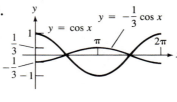
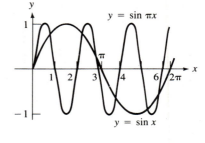

21.
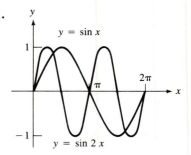

23.
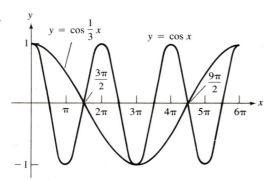

25.
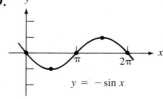

27.
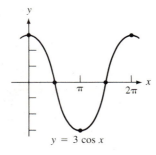

29.
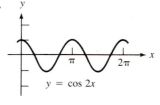

31.

33.
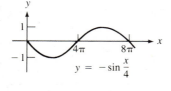

35.
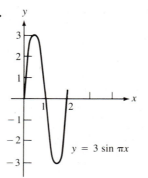

37.
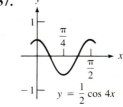

39.

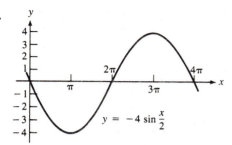

$y = -4 \sin \dfrac{x}{2}$

Exercise 3.5   (Page 106)

1. π **3.** 2π **5.** $\dfrac{\pi}{3}$ **7.** 2   **9.** 6π **11.** 3/2   **13.** 4   **15.** $2\pi^2$ **17.** $\dfrac{2}{3}\pi^2$

19.

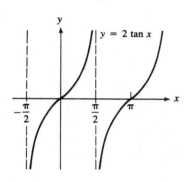

$y = 2 \tan x$

21.

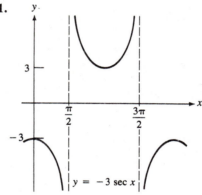

$y = -3 \sec x$

23.

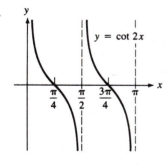

$y = \cot 2x$

25.

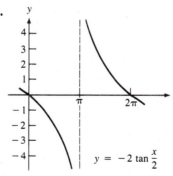

$y = -2 \tan \dfrac{x}{2}$

27.

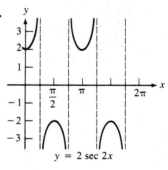

$y = 2 \sec 2x$

29.

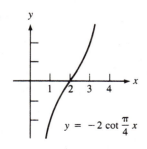

$y = -2 \cot \dfrac{\pi}{4} x$

31. The value of y can be as large as you want.
33. $\ldots, -\pi, 0, \pi, 2\pi, 3\pi, \ldots$

Exercise 3.6   (Page 112)

1. 2 units up; 2π **3.** 1 unit down; π **5.** 7 units up; $\dfrac{2\pi}{5}$ **7.** 3 units up; 2π **9.** 5 units down; π

11. 6 units up; 1   **13.** 2π; $\dfrac{\pi}{3}$ to the right   **15.** 2π; $\dfrac{\pi}{6}$ to the left   **17.** 1; no phase shift

19. π; π to the right   **21.** 2π; $\dfrac{\pi}{4}$ to the left   **23.** π; $\dfrac{\pi}{2}$ to the left   **25.** 2; $\dfrac{1}{2}$ to the left

27. 6π; 18π to the right   **29.** $\dfrac{\pi}{7}$; $\dfrac{3}{2}\pi$ to the right

31.

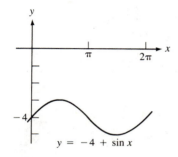

$y = -4 + \sin x$

33.

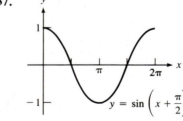

$y = 3 - \sec x$

35.

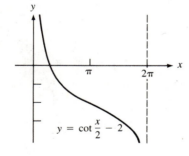

$y = \cot \dfrac{x}{2} - 2$

37.

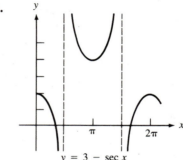

$y = \sin\left(x + \dfrac{\pi}{2}\right)$

39.

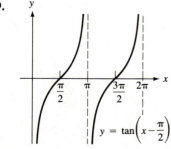

$y = \tan\left(x - \dfrac{\pi}{2}\right)$

41.

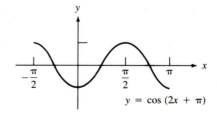

$y = \cos(2x + \pi)$

43.

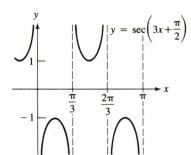

$$y = \sec\left(3x + \frac{\pi}{2}\right)$$

45.

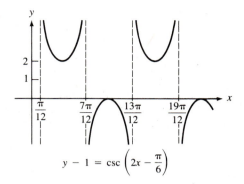

$$y - 1 = \csc\left(2x - \frac{\pi}{6}\right)$$

Exercise 3.7   (Page 115)

1.

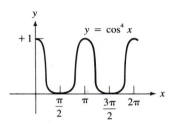

$y = \cos^4 x$

3.

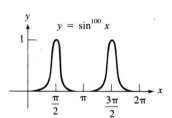

$y = -\sin^3 x$

5.

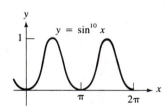

$y = \sin^{10} x$

7.

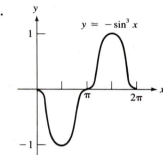

$y = \sin^{100} x$

9.

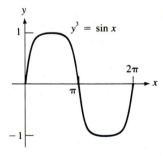

$y^3 = \sin x$

11.

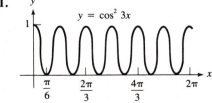

$y = \cos^2 3x$

13.

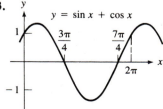

$y = \sin x + \cos x$

15.

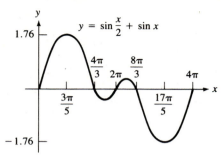

$$y = \sin\frac{x}{2} + \sin x$$

17.

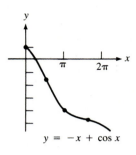

$$y = -x + \cos x$$

19.

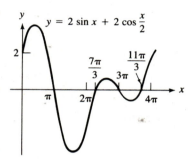

$$y = 2\sin x + 2\cos\frac{x}{2}$$

21.

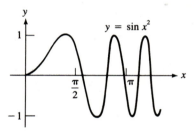

$$y = 2\sin x \cos x$$

23.

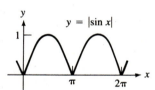

$$y = |\sin x|$$

25.

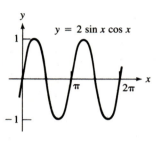

$$y = \sin x^2$$

27.

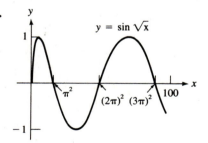

$$y = \sin\sqrt{x}$$

Exercise 3.8   (Page 119)

1. $V = 310 \sin 100\pi\, t$ **3.** 4π sec   **5.** 48 dynes per cm   **7.** period of 2 seconds per cycle; frequency of $\frac{1}{2}$ cycle per second   **9.** 286 hertz

11.

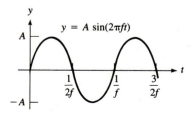

$$y = A\sin(2\pi f t)$$

REVIEW EXERCISES (Page 121)

1. $\frac{7}{12}\pi$ **2.** $\frac{65}{36}\pi$ **3.** $\frac{53}{30}\pi$ **4.** $\frac{-7}{12}\pi$ **5.** $570°$ **6.** $-150°$ **7.** $1260°$ **8.** $458.366°$ **9.** $\frac{1}{2}$

10. $\frac{\sqrt{3}}{2}$ **11.** $-\sqrt{3}$ **12.** 2 **13.** 2750 mi **14.** $41.5°$ **15.** 19 sq cm **16.** $\frac{\pi}{43,200}\frac{\text{rads}}{\text{sec}}$

17. $\frac{1815}{\pi}$ rpm **18.** about 12 mph **19.** $\left(\frac{-\sqrt{3}}{2}, \frac{-1}{2}\right)$ **20.** $\left(-\frac{\sqrt{2}}{2}, -\frac{\sqrt{2}}{2}\right)$ **21.** 0.0000

22. 0.7539 **23.** -2.1850 **24.** undefined **25.** 1.8508 **26.** -0.5774 **27.** -0.3983

28. -0.2003 **29.** $4; \frac{2\pi}{3}$ **30.** $\frac{1}{8}; \frac{\pi}{2}$ **31.** $\frac{1}{3}; 6\pi$ **32.** $0.875; 8\pi$ **33.** 2 units up **34.** $\frac{\pi}{6}$ to the left

35. 4 units up; $\frac{21}{2}$ to the left **36.** 1 unit down; $\frac{5}{2}\pi$ to the right

37.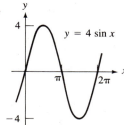

38.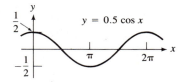

39.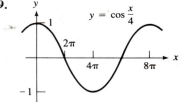

40.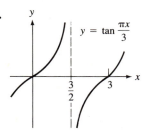

41.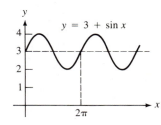

42.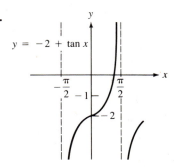

43.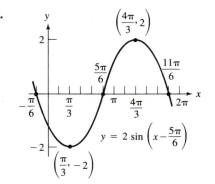

44.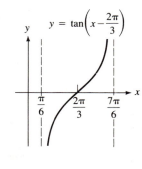

45.
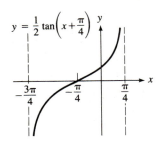
$$y = \frac{1}{2}\tan\left(x + \frac{\pi}{4}\right)$$

46.
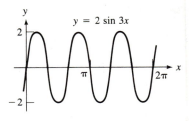
$$y = 2\sin 3x$$

47.
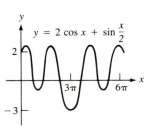
$$y = 2\sin^2 x$$

48.
$$y = \frac{1}{2}\cos^2 x$$

49.
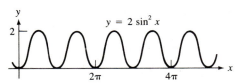
$$y = 2\cos x + \sin\frac{x}{2}$$

50.
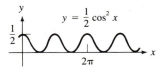
$(4\pi, 2\pi + 1)$
$\left(3\pi, \frac{3\pi}{2}\right)$
$\left(\pi, \frac{\pi}{2}\right)$
$y = \frac{x}{2} + \cos\frac{x}{2}$
$(2\pi, \pi - 1)$

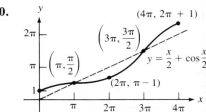

Exercise 4.1   (Page 128)

1. yes   **3.** no   **5.** no   **7.** yes   **9.** no   **11.** a meaningless statement   **13.** no   **15.** yes
17. no   **19.** no   **21.** yes   **23.** yes   **25.** no   **27.** no   **29.** no

Exercise 4.2   (Page 136)

1. $\sin 195° = \sin(45° + 150°) = \dfrac{-\sqrt{6} + \sqrt{2}}{4}$ **3.** $\tan 195° = \tan(225° - 30°) = \dfrac{3 - \sqrt{3}}{3 + \sqrt{3}}$

5. $\cos\dfrac{11\pi}{12} = \cos\left(\dfrac{\pi}{6} + \dfrac{3\pi}{4}\right) = \dfrac{-\sqrt{6} - \sqrt{2}}{4}$ **7.** $\cos\dfrac{19\pi}{12} = \cos\left(\dfrac{11\pi}{6} - \dfrac{\pi}{4}\right) = \dfrac{\sqrt{6} - \sqrt{2}}{4}$

9. $\sin 255° = \sin(210° + 45°) = \dfrac{-\sqrt{2} - \sqrt{6}}{4}$ **11.** $\tan 105° = \tan(60° + 45°) = \dfrac{\sqrt{3} + 1}{1 - \sqrt{3}}$

13. $\cos\dfrac{\pi}{12} = \cos\left(\dfrac{\pi}{3} - \dfrac{\pi}{4}\right) = \dfrac{\sqrt{2} + \sqrt{6}}{4}$ **15.** $\sin\dfrac{5\pi}{12} = \sin\left(\dfrac{2\pi}{3} - \dfrac{\pi}{4}\right) = \dfrac{\sqrt{6} + \sqrt{2}}{4}$

17. $\sin(60° + \theta) = \sin 60° \cos\theta + \cos 60° \sin\theta = \dfrac{\sqrt{3}}{2}\cos\theta + \dfrac{1}{2}\sin\theta$ **19.** $\tan(\pi + x)$

$= \dfrac{\tan\pi + \tan x}{1 - \tan\pi\tan x} = \dfrac{\tan x}{1} = \tan x$ **21.** $\cos(\pi - x) = \cos\pi\cos x + \sin\pi\sin x = -\cos x$

23. $\sin(10° + 30°) = \sin 40°$ **25.** $\tan(75° + 40°) = \tan 115°$ **27.** $\cos(120° - 40°) = \cos 80°$
29. $\sin(x + 2x) = \sin 3x$ **31.** $\sin(\alpha + \beta) = -\dfrac{56}{65}; \cos(\alpha - \beta) = \dfrac{63}{65}$ **33.** $\tan(\alpha + \beta) = \dfrac{140}{171};$
$\tan(\alpha - \beta) = -\dfrac{220}{21}$ **35.** $\sin\alpha = \dfrac{416}{425}; \cos\alpha = \dfrac{87}{425}$

Exercise 4.3   (Page 141)

1. $\sin 2\alpha$ **3.** $\sin 6\theta$ **5.** $\cos 2\beta$ **7.** $\cos \beta$ **9.** $2 \sin 2\theta$ **11.** $(\sin 4\theta)^2 = \sin^2 4\theta$ **13.** $\frac{1}{2}\cos 2\alpha$

15. $\cos 18\theta$ **17.** $\sin^2 10\theta$ **19.** $\tan 8C$ **21.** $\tan A$ **23.** $\cos 8x$ **25.** $-\cos 10x$ **27.** $\frac{\sqrt{3}}{2}$

29. undefined   **31.** -0.5 **33.** 0 **35.** $\frac{\sqrt{3}}{2}$ **37.** 0.5 **39.** $\sin 2\theta = \frac{120}{169}$, $\cos 2\theta$

$= -\frac{119}{169}$, $\tan 2\theta = -\frac{120}{119}$ **41.** $\sin 2\theta = \frac{120}{169}$, $\cos 2\theta = -\frac{119}{169}$, $\tan 2\theta = -\frac{120}{119}$ **43.** $\sin 2\theta = \frac{24}{25}$, $\cos 2\theta$

$= \frac{7}{25}$, $\tan 2\theta = \frac{24}{7}$ **45.** $\sin 2\theta = -\frac{336}{625}$, $\cos 2\theta = -\frac{527}{625}$, $\tan 2\theta = \frac{336}{527}$ **47.** $\sin 2\theta = \frac{720}{1681}$, $\cos 2\theta$

$= -\frac{1519}{1681}$, $\tan 2\theta = -\frac{720}{1519}$ **49.** $\sin 2\theta = -\frac{720}{1681}$, $\cos 2\theta = \frac{1519}{1681}$, $\tan 2\theta = -\frac{720}{1519}$

Exercise 4.4   (Page 150)

1. $\frac{\sqrt{2 + \sqrt{3}}}{2}$ **3.** $-2 - \sqrt{3}$ **5.** $\frac{\sqrt{2 - \sqrt{2}}}{2}$ **7.** $\frac{\sqrt{2 + \sqrt{3}}}{2}$ **9.** $\sqrt{3} - 2$ **11.** 1

13. $\sin \frac{\theta}{2} = \frac{\sqrt{10}}{10}$, $\cos \frac{\theta}{2} = \frac{3\sqrt{10}}{10}$, $\tan \frac{\theta}{2} = \frac{1}{3}$ **15.** $\sin \frac{\theta}{2} = \frac{2\sqrt{5}}{5}$, $\cos \frac{\theta}{2} = -\frac{\sqrt{5}}{5}$, $\tan \frac{\theta}{2} = -2$

17. $\sin \frac{\theta}{2} = \frac{3\sqrt{34}}{34}$, $\cos \frac{\theta}{2} = -\frac{5\sqrt{34}}{34}$, $\tan \frac{\theta}{2} = -\frac{3}{5}$ **19.** $\sin \frac{\theta}{2} = \frac{\sqrt{82}}{82}$, $\cos \frac{\theta}{2} = \frac{9\sqrt{82}}{82}$, $\tan \frac{\theta}{2} = \frac{1}{9}$

21. $\sin \frac{\theta}{2} = \frac{4\sqrt{17}}{17}$, $\cos \frac{\theta}{2} = \frac{\sqrt{17}}{17}$, $\tan \frac{\theta}{2} = 4$ **23.** $\sin \frac{\theta}{2} = \frac{\sqrt{6}}{6}$, $\cos \frac{\theta}{2} = \frac{-\sqrt{30}}{6}$, $\tan \frac{\theta}{2} = \frac{-\sqrt{5}}{5}$

25. $\cos 15°$ **27.** $\tan 100°$ **29.** $\tan 40°$ **31.** $\tan \pi = 0$ **33.** $\tan \frac{x}{4}$ **35.** $\tan 5A$

Exercise 4.5   (Page 156)

1. $\frac{1}{4}$ **3.** $\frac{1}{4}$ **5.** $\frac{\sqrt{2}}{4}$ **7.** $\frac{1}{4}$ **9.** $\frac{\sqrt{3} - 2}{4}$ **11.** $2 \sin 35° \cos 5°$ **13.** $2 \cos 165° \cos 55°$

15. $2 \cos 75° \sin 25°$ **17.** $-2 \sin 192\frac{1}{2}° \sin 87\frac{1}{2}°$ **19.** $2 \sin 3\theta \cos 2\theta$ **21.** $\frac{\sqrt{6}}{2}$ **23.** $-\frac{\sqrt{2}}{2}$

25. $\frac{-\sqrt{2}}{2}$ **27.** $\frac{\sqrt{6}}{2}$ **29.** 0 **31.** $\frac{1}{2}\left(1 + \frac{\sqrt{3}}{2}\right)$ **33.** $\frac{1}{4}$ **35.** $\frac{1}{2}\left[1 + \frac{\sqrt{3}}{2}\right]$ **37.** $\frac{1}{2}\left[-1 + \frac{\sqrt{3}}{2}\right]$

39. $\frac{1}{4}$

Exercise 4.6   (Page 160)

1. $10 \sin(x + 53.1°)$ **3.** $10 \sin(x - 53.1°)$ **5.** $\sqrt{5} \sin(x + 26.6°)$ **7.** $\sqrt{2} \sin(x + 45°)$

9. $\sqrt{26} \sin(x + 101.3°)$ or $-\sqrt{26} \sin(x - 78.7°)$ **11.** $2\sqrt{3} \sin(x - 60°)$

13. $\left(\frac{A}{\sqrt{A^2 + B^2}}\right)^2 + \left(\frac{B}{\sqrt{A^2 + B^2}}\right)^2 = \frac{A^2}{A^2 + B^2} + \frac{B^2}{A^2 + B^2} = \frac{A^2 + B^2}{A^2 + B^2} = 1$

15.

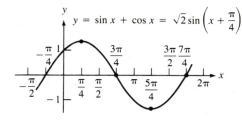

$$y = \sin x + \cos x = \sqrt{2}\sin\left(x + \frac{\pi}{4}\right)$$

17.

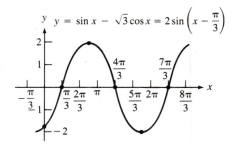

$$y = \sin x - \sqrt{3}\cos x = 2\sin\left(x - \frac{\pi}{3}\right)$$

19.

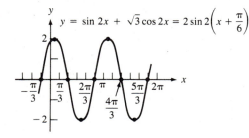

$$y = \sin 2x + \sqrt{3}\cos 2x = 2\sin 2\left(x + \frac{\pi}{6}\right)$$

Exercise 4.7   (Page 164)

1. $8.1°$ **3.** If L_1 is perpendicular to L_2, then the angle between the lines is $90°$. Because $\tan 90°$ is undefined, the denominator of $\dfrac{m_2 - m_1}{1 + m_2 m_1}$ must be zero. Hence, $1 + m_2 m_1 = 0$, or $m_1 = -\dfrac{1}{m_2}$.   **5.** 13   **7.** $\frac{5}{2}$ cycles per second apart.   **9.** $\dfrac{x^2}{a^2} + \dfrac{y^2}{b^2} = 1$ **11.** $xy = 12$

REVIEW EXERCISES   (Page 166)

1. $\dfrac{\sqrt{2} - \sqrt{6}}{4}$ **2.** $\dfrac{\sqrt{2} + \sqrt{6}}{4}$ **3.** $\sqrt{3} - 2$ **4.** $\dfrac{-\sqrt{6} - \sqrt{2}}{4}$ **5.** $\dfrac{\sqrt{2} - \sqrt{6}}{4}$

6. $\dfrac{-\sqrt{3} - 1}{1 - \sqrt{3}} = 2 + \sqrt{3}$ **7.** $-\dfrac{\sqrt{3}}{2}$ **8.** $-\dfrac{1}{2}$ **9.** $\sqrt{3}$ **10.** $\dfrac{\sqrt{2 - \sqrt{2}}}{2}$ **11.** $\dfrac{\sqrt{2 + \sqrt{2}}}{2}$

12. $\sqrt{\dfrac{2 - \sqrt{2}}{2 + \sqrt{2}}} = \sqrt{2} - 1$ **13.** $\sin 71°$ **14.** $\tan(-31°) = -\tan 31°$ **15.** $\cos 20°$ **16.** $\cos\frac{4}{11}\pi$

17. $\sin\dfrac{\pi}{7}$ **18.** $\sin 34°$ **19.** $\cos 34°$ **20.** $\tan 434° = \tan 74°$ **21.** $2\sin\dfrac{3\pi}{4} = \sqrt{2}$

22. $\cos\dfrac{2\pi}{3} = -\dfrac{1}{2}$ **23.** $\cos\dfrac{2\pi}{13}$ **24.** $2\sin^2\theta$ **25.** $\sin\theta\cos\alpha + \cos\theta\sin\alpha$

26. $\cos\theta\cos\alpha - \sin\theta\sin\alpha$ **27.** $\dfrac{\tan\theta + \tan\alpha}{1 - \tan\theta\tan\alpha}$ **28.** $\sin\theta\cos\alpha - \cos\theta\sin\alpha$

29. $\cos\theta\cos\alpha + \sin\theta\sin\alpha$ **30.** $\dfrac{\tan\theta - \tan\alpha}{1 + \tan\theta\tan\alpha}$ **31.** $2\sin\theta\cos\theta$

32. $\cos^2\theta - \sin^2\theta$, or $2\cos^2\theta - 1$, or $1 - 2\sin^2\theta$ **33.** $\dfrac{2\tan\theta}{1 - \tan^2\theta}$ **34.** $\pm\sqrt{\dfrac{1 - \cos\theta}{2}}$

35. $\pm\sqrt{\dfrac{1+\cos\theta}{2}}$ **36.** $\pm\sqrt{\dfrac{1-\cos\theta}{1+\cos\theta}}$,   or $\dfrac{1-\cos\theta}{\sin\theta}$,   or $\dfrac{\sin\theta}{1+\cos\theta}$

37. $\cos(60°+\theta)=\cos 60°\cos\theta-\sin 60°\sin\theta=\frac{1}{2}(\cos\theta-\sqrt{3}\sin\theta)$

38. $\sin\left(\dfrac{3\pi}{2}-\theta\right)=\sin\dfrac{3\pi}{2}\cos\theta-\cos\dfrac{3\pi}{2}\sin\theta=-\cos\theta$

39. $\tan(180°-\theta)=\dfrac{\tan 180°-\tan\theta}{1+\tan 180°\tan\theta}=-\tan\theta$

40. $\sin(120°+\theta)=\sin 120°\cos\theta+\cos 120°\sin\theta=\frac{1}{2}(\sqrt{3}\cos\theta-\sin\theta)$

41. $\cos(300°-\theta)=\cos 300°\cos\theta+\sin 300°\sin\theta=\frac{1}{2}(\cos\theta-\sqrt{3}\sin\theta)$

42. $\tan\left(\dfrac{\pi}{4}+\theta\right)=\dfrac{\tan\dfrac{\pi}{4}+\tan\theta}{1-\tan\dfrac{\pi}{4}\tan\theta}=\dfrac{1+\tan\theta}{1-\tan\theta}$ **43.** $\dfrac{\sin 2x}{2\cos x}=\dfrac{2\sin x\cos x}{2\cos x}=\sin x$

44. $\pm\sqrt{\sin^2\theta+\cos 2\theta}=\pm\sqrt{\sin^2\theta+\cos^2\theta-\sin^2\theta}=\pm\sqrt{\cos^2\theta}=\pm\cos\theta$

45. $\sin 2\theta=-\dfrac{120}{169}$, $\cos 2\theta=\dfrac{119}{169}$, $\tan 2\theta=-\dfrac{120}{119}$ **46.** $\sin 2\theta=\dfrac{120}{169}$, $\cos 2\theta=\dfrac{119}{169}$, $\tan 2\theta=\dfrac{120}{119}$

47. $\sin 2\theta=-\dfrac{840}{841}$, $\cos 2\theta=\dfrac{41}{841}$, $\tan 2\theta=-\dfrac{840}{41}$ **48.** $\sin 2\theta=\dfrac{840}{841}$, $\cos 2\theta=\dfrac{41}{841}$, $\tan 2\theta=\dfrac{840}{41}$

49. $\sin 2\theta=\dfrac{24}{25}$, $\cos 2\theta=\dfrac{-7}{25}$, $\tan 2\theta=\dfrac{-24}{7}$ **50.** $\sin 2\theta=-\dfrac{24}{25}$, $\cos 2\theta=-\dfrac{7}{25}$, $\tan 2\theta=\dfrac{24}{7}$

51. $\sin\dfrac{\theta}{2}=\dfrac{\sqrt{26}}{26}$, $\cos\dfrac{\theta}{2}=\dfrac{5\sqrt{26}}{26}$, $\tan\dfrac{\theta}{2}=\dfrac{1}{5}$ **52.** $\sin\dfrac{\theta}{2}=\dfrac{\sqrt{26}}{26}$, $\cos\dfrac{\theta}{2}=-\dfrac{5\sqrt{26}}{26}$, $\tan\dfrac{\theta}{2}=-\dfrac{1}{5}$

53. $\sin\dfrac{\theta}{2}=\dfrac{\sqrt{26}}{26}$, $\cos\dfrac{\theta}{2}=-\dfrac{5\sqrt{26}}{26}$, $\tan\dfrac{\theta}{2}=-\dfrac{1}{5}$ **54.** $\sin\dfrac{\theta}{2}=\dfrac{2\sqrt{5}}{5}$, $\cos\dfrac{\theta}{2}=\dfrac{\sqrt{5}}{5}$, $\tan\dfrac{\theta}{2}=2$

55. $\sin\dfrac{\theta}{2}=\dfrac{2\sqrt{5}}{5}$, $\cos\dfrac{\theta}{2}=\dfrac{\sqrt{5}}{5}$, $\tan\dfrac{\theta}{2}=2$ **56.** $\sin\dfrac{\theta}{2}=\dfrac{2\sqrt{5}}{5}$, $\cos\dfrac{\theta}{2}=-\dfrac{\sqrt{5}}{5}$, $\tan\dfrac{\theta}{2}=-2$ **57.** $-\dfrac{2+\sqrt{3}}{4}$

58. $\dfrac{1}{2}\left(-\dfrac{\sqrt{3}}{2}+1\right)$ **59.** $\dfrac{1-\sqrt{3}}{4}$ **60.** $\dfrac{1}{2}\left(-\dfrac{1}{2}-\dfrac{\sqrt{3}}{2}\right)$ **61.** $2\sin 6°\cos 1°$ **62.** $2\cos 226°\sin 86°$

63. $-2\sin\dfrac{2\pi}{5}\sin\dfrac{\pi}{5}$ **64.** $2\cos\dfrac{5\pi}{14}\cos\dfrac{\pi}{14}$ **65.** $-\dfrac{\sqrt{2}}{2}$ **66.** $\dfrac{\sqrt{6}}{2}$ **67.** $-\dfrac{\sqrt{2}}{2}$ **68.** $\dfrac{\sqrt{6}}{2}$

Exercise 5.1   (Page 175)

1. ..., $-300°$, $-240°$, $60°$, $120°$, $420°$, $480°$, ...   **3.** ..., $-180°$, $-90°$, $0°$, $90°$, $180°$, $270°$, $360°$, $450°$, ...   **5.** $120°$ **7.** $0°$, $90°$, $180°$, $270°$ **9.** $90°$, $270°$ **11.** $0°$, $180°$ **13.** $45°$, $135°$, $225°$, $315°$

15. $0°$, $120°$, $240°$ **17.** $210°$, $330°$ **19.** $90°$, $210°$, $270°$, $330°$ **21.** $45°$, $225°$ **23.** $0°$, $240°$

25. $0°$, $120°$, $240°$ **27.** $0°$, $180°$ **29.** $30°$, $150°$, $210°$, $330°$ **31.** $45°$ **33.** $0°$, $30°$, $150°$, $180°$

35. $90°$, $120°$, $240°$, $270°$ **37.** $0°$, $30°$, $90°$, $150°$, $180°$, $210°$, $270°$, $330°$ **39.** $0°$, $240°$

41. $60°$, $120°$, $240°$, $300°$ **43.** $0°$, $180°$ **45.** $45°$, $135°$, $315°$ **47.** $0°$, $120°$ **49.** $30°$

51. $0°$, $240°$ **53.** $\dfrac{\pi}{4},\dfrac{5\pi}{4}$ **55.** $0,\dfrac{\pi}{2},\pi,\dfrac{3\pi}{2}$ **57.** $0,\dfrac{\pi}{4},\dfrac{\pi}{2},\dfrac{3\pi}{4},\pi,\dfrac{5\pi}{4},\dfrac{3\pi}{2},\dfrac{7\pi}{4}$

59. $0,\dfrac{\pi}{3},\dfrac{\pi}{2},\dfrac{2\pi}{3},\pi,\dfrac{4\pi}{3},\dfrac{3\pi}{2},\dfrac{5\pi}{3}$

61. $\dfrac{\pi}{12},\dfrac{\pi}{6},\dfrac{\pi}{4},\dfrac{5\pi}{12},\dfrac{\pi}{2},\dfrac{7\pi}{12},\dfrac{3\pi}{4},\dfrac{5\pi}{6},\dfrac{11\pi}{12},\dfrac{13\pi}{12},\dfrac{7\pi}{6},\dfrac{5\pi}{4},\dfrac{17\pi}{12},\dfrac{3\pi}{2},\dfrac{19\pi}{12},\dfrac{7\pi}{4},\dfrac{11\pi}{6},\dfrac{23\pi}{12}$ **63.** $\dfrac{\pi}{4},\dfrac{3\pi}{4},\dfrac{5\pi}{4},\dfrac{7\pi}{4}$

65. $\dfrac{\pi}{6}, \dfrac{\pi}{2}, \dfrac{5\pi}{6}, \dfrac{7\pi}{6}, \dfrac{3\pi}{2}, \dfrac{11\pi}{6}$ **67.** $\dfrac{\pi}{2}, \dfrac{3\pi}{2}$ **69.** $\dfrac{7\pi}{6}, \dfrac{11\pi}{6}$ **71.** $\dfrac{\pi}{20}, \dfrac{3\pi}{20}, \dfrac{\pi}{4}, \dfrac{7\pi}{20}, \dfrac{9\pi}{20}, \dfrac{11\pi}{20}, \dfrac{13\pi}{20}, \dfrac{3\pi}{4}, \dfrac{17\pi}{20},$

$\dfrac{19\pi}{20}, \dfrac{21\pi}{20}, \dfrac{23\pi}{20}, \dfrac{5\pi}{4}, \dfrac{27\pi}{20}, \dfrac{29\pi}{20}, \dfrac{31\pi}{20}, \dfrac{33\pi}{20}, \dfrac{7\pi}{4}, \dfrac{37\pi}{20}, \dfrac{39\pi}{20}$ **73.** $0, \dfrac{\pi}{4}, \pi, \dfrac{5\pi}{4}$ **75.** $\dfrac{\pi}{6}, \dfrac{5\pi}{6}, \dfrac{7\pi}{6}, \dfrac{11\pi}{6}$

Exercise 5.2 (Page 179)

1. $30°, 150°$ **3.** $135°, 315°$ **5.** $135°, 225°$ **7.** $30°, 210°$ **9.** $0°, 180°$ **11.** no values

13. $\dfrac{2\pi}{3}, \dfrac{4\pi}{3}$ **15.** $\dfrac{\pi}{6}, \dfrac{5\pi}{6},$ **17.** $\dfrac{\pi}{3}, \dfrac{4\pi}{3}$ **19.** $\dfrac{2\pi}{3}, \dfrac{5\pi}{3}$ **21.** no values **23.** $\dfrac{\pi}{2}$ **25.** $1.000, 2.142$

27. $1.000, 4.141$ **29.** $2.000, 4.283$ **31.** $2.000, 5.142$ **33.** $0.142, 3.000$ **35.** $3.000, 6.142$

37. $55°, 125°$ **39.** $125°, 305°$ **41.** $55°, 305°$ **43.** $130°, 310°$ **45.** $50°, 310°$ **47.** $15°, 165°$

Exercise 5.3 (Page 187)

1. $\dfrac{\pi}{6}$ **3.** $\dfrac{\pi}{2}$ **5.** $\dfrac{\pi}{4}$ **7.** no value **9.** $\dfrac{3\pi}{4}$ **11.** $\dfrac{\pi}{3}$ **13.** $\sin \dfrac{\pi}{6} = \dfrac{1}{2}, \cos \dfrac{\pi}{6} = \dfrac{\sqrt{3}}{2}, \tan \dfrac{\pi}{6} = \dfrac{\sqrt{3}}{3}$

15. $\sin 0 = 0, \cos 0 = 1, \tan 0 = 0$ **17.** $\sin \dfrac{5\pi}{6} = \dfrac{1}{2}, \cos \dfrac{5\pi}{6} = -\dfrac{\sqrt{3}}{2}, \tan \dfrac{5\pi}{6} = -\dfrac{\sqrt{3}}{3}$

19. $\sin \dfrac{\pi}{2} = 1, \cos \dfrac{\pi}{2} = 0, \tan \dfrac{\pi}{2}$ is undefined **21.** $\sin \pi = 0, \cos \pi = -1, \tan \pi = 0$

23. $\sin \dfrac{\pi}{4} = \dfrac{\sqrt{2}}{2}, \cos \dfrac{\pi}{4} = \dfrac{\sqrt{2}}{2}, \tan \dfrac{\pi}{4} = 1$ **25.** $\dfrac{1}{2}$ **27.** 1 **29.** 1 **31.** $\dfrac{1}{2}$ **33.** $-\sqrt{3}$ **35.** $\dfrac{\sqrt{2}}{2}$

37. $\dfrac{3}{5}$ **39.** $\dfrac{12}{13}$ **41.** $\dfrac{-4}{3}$ **43.** $\dfrac{12}{5}$ **45.** $\dfrac{12}{13}$ **47.** $\dfrac{40}{41}$ **49.** $\sin\left(\dfrac{\pi}{6} + \dfrac{\pi}{3}\right) = 1$

51. $\cos\left(\dfrac{\pi}{6} - \dfrac{\pi}{3}\right) = \dfrac{\sqrt{3}}{2}$

53. $\sin 2\left(\dfrac{\pi}{4}\right) = 1$ **55.** $\tan 2\left(\dfrac{\pi}{4}\right)$ is undefined. **57.** $\sin \dfrac{1}{2}\left(\dfrac{\pi}{3}\right) = \dfrac{1}{2}$ **59.** $\tan \dfrac{1}{2}\left(\dfrac{\pi}{3}\right) = \dfrac{\sqrt{3}}{3}$

61. $\dfrac{x}{\sqrt{1 + x^2}}$ **63.** $\dfrac{x}{\sqrt{1 - x^2}}$ **65.** $\sqrt{1 - x^2}$ **67.** $2x\sqrt{1 - x^2}$ **69.** $\dfrac{2x}{1 - x^2}$ **71.** $1 - 2x^2$

73. $\sqrt{\dfrac{1 + x}{2}}$ **75.** $\sqrt{\dfrac{1 - x}{2}}$ **77.** $-\dfrac{\pi}{6}$ **79.** $\dfrac{\pi}{4}$

Exercise 5.4 (Page 194)

1. $\dfrac{\pi}{6}$ **3.** π **5.** $-\dfrac{\pi}{2}$ **7.** 3 **9.** $\dfrac{5}{3}$ **11.** $\dfrac{24}{25}$ **13.** $-\dfrac{33}{65}$ **15.** $\dfrac{1}{x}$ **17.** $\dfrac{\sqrt{1 + x^2}}{x}$ **19.** $\pm\sqrt{x^2 - 1}$

Exercise 5.5 (Page 201)

1. $48.8°$ **3.** $38.6°$ **5.** $54.7°$ **7.** $(2, 2)$ **9.** $(-1, \sqrt{3})$ **11.** $(2, 2)$ **13.** $\dfrac{u^2}{4} - \dfrac{v^2}{4} = 1; 45°$

15. $\dfrac{5u^2}{4} - \dfrac{v^2}{4} = 1; 45°$

REVIEW EXERCISES   (Page 203)

1. 30°, 330°   **2.** 150°, 330°   **3.** 45°, 225°   **4.** 90°, 270°   **5.** no solutions in the given interval
6. 300°   **7.** 0°, 180°, 270°   **8.** 0°, 180°   **9.** 0°, 90°   **10.** 180°, 210°, 330°   **11.** 0°, 45°, 225°
12. 0°, 60°, 180°, 300°   **13.** 0°, 30°, 150°, 180°   **14.** 90°, 210°, 330°   **15.** 60°, 180°, 300°
16. 0°, 120°, 240°   **17.** 120°, 240°   **18.** 45°, 225°   **19.** 45°, 135°, 225°, 315°   **20.** an identity

21. no solutions   **22.** 60°, 180°, 300°   **23.** 45°   **24.** 330°   **25.** $\dfrac{\pi}{3}, \dfrac{2\pi}{3}$ **26.** $\dfrac{\pi}{6}, \dfrac{7\pi}{6}$ **27.** $\dfrac{2\pi}{3}, \dfrac{4\pi}{3}$

28. $\dfrac{4\pi}{3}, \dfrac{5\pi}{3}$ **29.** $\dfrac{2\pi}{3}, \dfrac{5\pi}{3}$ **30.** no values   **31.** no values   **32.** $\dfrac{\pi}{6}, \dfrac{7\pi}{6}$ **33.** $\dfrac{\pi}{2}, \dfrac{3\pi}{2}$ **34.** $\dfrac{\pi}{6}, \dfrac{5\pi}{6}$

35. no values   **36.** $\dfrac{2\pi}{3}, \dfrac{4\pi}{3}$ **37.** 0.629, 2.513   **38.** no values   **39.** 1.557, 4.699   **40.** 0.013, 3.155

41. no values   **42.** no values   **43.** 0.013, 3.155   **44.** 0.003, 3.145   **45.** no values

46. 1.369, 4.914   **47.** 1.047, 5.236   **48.** no values   **49.** $-\dfrac{\pi}{6}$ **50.** $\dfrac{2\pi}{3}$ **51.** $\dfrac{\pi}{6}$ **52.** $-\dfrac{\pi}{6}$

53. $\dfrac{\pi}{2}$ **54.** $-\dfrac{\pi}{2}$ **55.** $-\dfrac{\pi}{4}$ **56.** $\dfrac{\pi}{3}$ **57.** 0   **58.** no values   **59.** 0   **60.** $\dfrac{\pi}{2}$ **61.** $\dfrac{4}{5}$ **62.** $\dfrac{1}{3}$

63. $\dfrac{\sqrt{3}}{2}$ **64.** $\dfrac{\sqrt{3}}{2}$ **65.** 17   **66.** $-\sqrt{3}$ **67.** 1   **68.** $\dfrac{\sqrt{2}}{2}$ **69.** $\dfrac{\sqrt{2}}{2}$ **70.** $\dfrac{2\sqrt{3}}{3}$ **71.** 2

72. $\sqrt{2}$ **73.** 1   **74.** $\dfrac{1}{2}$ **75.** $\dfrac{\sqrt{3}}{2}$ **76.** 0   **77.** $\dfrac{\sqrt{3}}{2}$ **78.** $\dfrac{1}{2}$ **79.** $\sqrt{3}$ **80.** $-\dfrac{\sqrt{3}}{3}$ **81.** $\dfrac{\sqrt{3}}{2}$

82. $\dfrac{\sqrt{2}}{2}$ **83.** 0   **84.** $\dfrac{\sqrt{3}}{3}$ **85.** $\sqrt{1-u^2}$ **86.** $\sqrt{1-u^2}$ **87.** $\dfrac{u}{\sqrt{1-u^2}}$ **88.** $\dfrac{u}{\sqrt{1+u^2}}$

89. $\sqrt{1-u^2}$ **90.** $\dfrac{\sqrt{1-u^2}}{u}$ **91.** $2u\sqrt{1-u^2}$ **92.** $2u^2 - 1$ **93.** $\dfrac{2u}{1-u^2}$ **94.** $2u\sqrt{1-u^2}$

95. $1 - 2u^2$ **96.** $\pm\sqrt{\dfrac{1+u}{2}}$ **97.** $\dfrac{\pi}{3}$ **98.** $\dfrac{\pi}{4}$ **99.** $\dfrac{4}{5}$ **100.** $\dfrac{12}{13}$

Exercise 6.1   (Page 212)

1. 61.0 cm   **3.** 1410 km   **5.** 65.9 cm   **7.** 54°   **9.** 90°   **11.** 37.85°   **13.** 210 lb
15. 5°, if 210 lbs is used for the resultant force.   **17.** 1090 lb   **19.** 36.7 nautical miles   **21.** 131 m
23. 69.3°, 64.4°, 46.3°   **27.** 86°   **29.** 85.2°   **31.** 90.0°   **33.** 1 hr

Exercise 6.2   (Page 219)

1. 67 km   **3.** 256 m   **5.** 2.55 m   **7.** 305 m   **9.** 49.2 cm   **11.** 1.0 mi   **13.** 218   **15.** 3.97
17. 180 yd   **19.** 420 ft   **21.** 7.1°   **23.** 2.5 nautical miles   **25.** approximately 2:50 P.M.

Exercise 6.3   (Page 224)

1. 31.5°   **3.** 61.6°   **5.** No triangle.   **7.** 136°   **9.** 156.19° or 4.09°   **11.** No triangle.
13. 12 or 2.1   **15.** No triangle.   **17.** 2900 ft or 780 ft   **19.** $h = 957$ ft

Exercise 6.4   (Page 229)

1. A triangle exists.   **3.** A triangle exists.   **5.** A triangle exists.   **7.** No triangle.   **9.** No triangle.
11. 74.1   **13.** 86   **15.** 44.39   **17.** 10.3   **19.** 19   **21.** 850

Exercise 6.5   (Page 235)

1. 190 sq units   **3.** 301 sq units   **5.** 6.5 sq units   **7.** 6 sq units   **9.** 42 sq units   **11.** 960 sq units
13. 0.001 sq units   **15.** 195,000 sq km   **17.** 31,400 sq ft   **19.** 72 sq m   **21.** $\sqrt{2295}$ sq cm

23. 126,000 sq ft   **25.** $\dfrac{s(s-a)}{bc} = \dfrac{(a+b+c)(b+c-a)}{4bc} = \dfrac{1}{2}\left(\dfrac{(a+b+c)(b+c-a)}{2bc}\right) = \frac{1}{2}(1 + \cos A)$

$= \cos^2 \dfrac{A}{2}$ **27.** Area of $\triangle\, ABC = \frac{1}{2}(AC)\frac{1}{2}(DB) \sin \alpha = \frac{1}{4}(AC)(DB) \sin \alpha$

$\qquad\qquad$ Area of $\triangle\, ACD = \frac{1}{2}(AC)\frac{1}{2}(DB) \sin \alpha = \frac{1}{4}(AC)(DB) \sin \alpha$

$\qquad\qquad\qquad\qquad\qquad\qquad\qquad\qquad$ $\frac{1}{2}(AC)(DB) \sin \alpha$

29. $A = \frac{1}{4} b^2 \cot \dfrac{\alpha}{2} = \dfrac{b^2 \cos^2 (\alpha/2)}{2 \sin \alpha}$ **31.** $\frac{1}{4} b^2 \sqrt{15}$

Exercise 6.6   (Page 243)

1. $\langle 7, -5 \rangle$ **3.** $\langle 6, -9 \rangle$ **5.** $\langle 9, -8 \rangle$ **7.** $\sqrt{13}$ **9.** $\sqrt{5}$ **11.** $\sqrt{13} + \sqrt{2}$ **13.** $8\mathbf{i} + 8\mathbf{j}$
15. $5\sqrt{3}\,\mathbf{i} + 5\mathbf{j}$ **17.** $18.6\mathbf{i} + 14.1\mathbf{j}$ **19.** 9   **21.** 7   **23.** 0   **25.** 45°   **27.** 150°   **29.** 36.9°
31. perpendicular   **33.** not perpendicular   **35.** perpendicular   **37.** $\dfrac{63}{13}$ **39.** 0

REVIEW EXERCISES   (Page 246)

5. 7.6   **6.** 32   **7.** 0.6   **8.** 11.1   **9.** 25°   **10.** 37°, 143°   **11.** 18.6°, 161.4°
12. 51.31, 128.69°   **13.** 65.7   **14.** 70.8   **15.** 14°, 82°   **16.** 42.4°   **17.** a triangle exists
18. a triangle exists   **19.** 640 sq units   **20.** 1200 sq units   **21.** 39 sq units   **22.** 110 sq units
23. 1.4 sq units   **24.** 12,000 sq units   **25.** 2400 sq units   **26.** 67.3   **27.** 558 mi
28. about 27.2°, 48.8°, and 104.0°   **29.** 180 ft   **30.** 611 ft   **31.** 280 sq m   **32.** 24 sq units
33. $\langle 0, 29 \rangle$ **34.** $3\sqrt{58} - \sqrt{29}$ **35.** $\langle 25, 10 \rangle$ **36.** $\sqrt{145}$ **37.** 90°   **38.** 0°   **39.** 75°
40. 120°

Exercise 7.1   (Page 254)

1. i **3.** -1 **5.** $-i$ **7.** i **9.** 1   **11.** i **13.** $x = -\frac{1}{2}; y = -\frac{1}{2}$ **15.** $x = 0; y = 0$
17. $x = \frac{2}{3}; y = -\frac{2}{9}$ **19.** $5 - 6i$ **21.** $-2 - 10i$ **23.** $4 + 10i$ **25.** $6 - 17i$ **27.** $52 + 56i$
29. $-6 + 17i$ **31.** $-5 + 12i$ **33.** $2 - 11i$ **35.** $\frac{2}{5} - \frac{1}{5}i$ **37.** $-\frac{3}{4} + 0i$ **39.** $\frac{1}{25} + \frac{7}{25}i$

41. $\frac{1}{2} + \frac{1}{2}i$ **43.** $\dfrac{-2 - 2\sqrt{5}}{17} + \dfrac{16 - \sqrt{5}}{34}i$ **45.** $\dfrac{6 + \sqrt{3}}{10} + \dfrac{3\sqrt{3} - 2}{10}i$
47. $(1 - i)(1 - i) = 1 - i - i - 1 = -2i$ **49.** $(1 + 2i)(1 + 2i) = 1 + 2i + 2i - 4 = -3 + 4i$
53. $-1 + i, -1 - i$ **55.** $-2 + i, -2 - i$ **57.** $\frac{1}{3} + \frac{1}{3}i, \frac{1}{3} - \frac{1}{3}i$ **59.** $34.87 + 32.69i$
61. approximately $-6.92 + 9.26i$

Exercise 7.2   (Page 258)

1.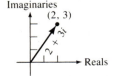

3.

5.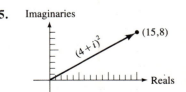

7.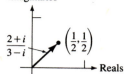

9.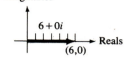

11.

13. $\sqrt{13}$ **15.** $7\sqrt{2}$ **17.** 6   **19.** 5   **21.** $\dfrac{3\sqrt{5}}{5}$ **23.** 1

Exercise 7.3   (Page 262)

1. $6(\cos 0° + i \sin 0°)$ **3.** $3(\cos 270° + i \sin 270°)$ **5.** $\sqrt{2}\,(\cos 225° + i \sin 225°)$
7. $6(\cos 60° + i \sin 60°)$ **9.** $2(\cos 240° + i \sin 240°)$ **11.** $2(\cos 210° + i \sin 210°)$ **13.** $\sqrt{3} + i$

15. $0 + 7i$ **17.** $1 - \sqrt{3}i$ **19.** $-\dfrac{1}{2} + 0i$ **21.** $\dfrac{3\sqrt{2}}{2} + \dfrac{3\sqrt{2}}{2}i$ **23.** $\dfrac{11\sqrt{3}}{2} - \dfrac{11}{2}i$

25. $8(\cos 90° + i \sin 90°)$ **27.** $\cos 300° + i \sin 300°$ **29.** $6(\cos 2\pi + i \sin 2\pi)$

31. $6\left(\cos \dfrac{\pi}{2} + i \sin \dfrac{\pi}{2}\right)$ **33.** 30 cis 116°   **35.** 36 cis $\dfrac{13\pi}{12}$ **37.** $6(\cos 30° + i \sin 30°)$

39. $\dfrac{3}{2}\left(\cos \dfrac{\pi}{2} + i \sin \dfrac{\pi}{2}\right)$ **41.** $\dfrac{12}{5}$ cis 130°   **43.** $\dfrac{1}{2}$ cis $\dfrac{\pi}{2}$ **45.** cis 40°   **47.** 4 cis $\dfrac{5\pi}{6}$

Exercise 7.4   (Page 268)

1. $27(\cos 90° + i \sin 90°)$ **3.** $\cos 180° + i \sin 180°$ **5.** 3125 cis 10°   **7.** $81(\cos \pi + i \sin \pi)$
9. $256(\cos 12 + i \sin 12)$ **11.** $\dfrac{1}{27}$ cis $\dfrac{3\pi}{2}$ **13.** $1 + \sqrt{3}\,i$ **15.** $\dfrac{1}{2} + \dfrac{\sqrt{3}}{2}\,i$ **17.** $1 + \sqrt{3}\,i$

19. $1 + \sqrt{3}\,i,\ -2,\ 1 - \sqrt{3}\,i$

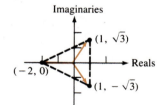

21. $\dfrac{\sqrt{2}}{2} + \dfrac{\sqrt{2}}{2}\,i,\ -\dfrac{\sqrt{2}}{2} - \dfrac{\sqrt{2}}{2}\,i;$

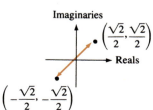

23. $\cos 54° + i \sin 54°$, $\cos 126° + i \sin 126°$, $\cos 198° + i \sin 198°$, $\cos 270° + i \sin 270°$, $\cos 342° + i \sin 342°$;

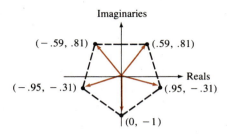

25. $\sqrt{3} + i$, $-1 + i\sqrt{3}$, $-\sqrt{3} - i$, $1 - i\sqrt{3}$;

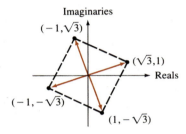

Exercise 7.5 (Page 275)

1. $(\sqrt{3}, 1)$ **3.** $\left(\dfrac{7}{2}, -\dfrac{7\sqrt{3}}{2}\right)$ **5.** $\left(\dfrac{-3}{2}, \dfrac{-3\sqrt{3}}{2}\right)$ **7.** $(0, 2)$ **9.** $(-\sqrt{3}, -1)$

11. $\left(\dfrac{-5\sqrt{2}}{2}, \dfrac{-5\sqrt{2}}{2}\right)$ **13.** $(\sqrt{3}, -1)$ **15.** $(0, 10)$ **17.** $(0, 0)$ **19.** $(-3\sqrt{3}, 3)$ **21.** $(\sqrt{2}, 45°)$

23. $(4, 330°)$ **25.** $(2, 210°)$ **27.** $(2, 150°)$ **29.** $(0, \theta°)$ **31.** $(5, 180°)$ **33.** $(3\sqrt{2}, 315°)$
35. $(14, 60°)$ **37.** $r \cos \theta = 3$ **39.** $r(3 \cos \theta + 2 \sin \theta) = 3$ **41.** $r = 9 \cos \theta$

43. $r^2 = 4 \cos^2 \theta \sin^2 \theta$ **45.** $\theta = \dfrac{\pi}{2}$ or $r = \sec \theta$ **47.** $r^2 \cos^2 \theta - 2r \sin \theta = 1$

49. $x^2 + y^2 = 9$ **51.** $x = 5$ **53.** $\sqrt{x^2 + y^2} + y = 1$ **55.** $(x^2 + y^2)^2 = 2xy$ **57.** $y = 0$
59. $2\sqrt{x^2 + y^2} - x = 2$

61.

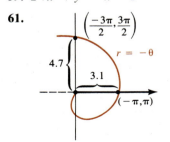

63.

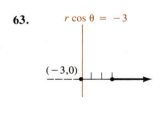

65.

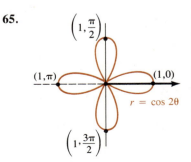

67.

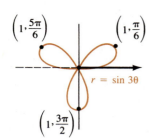

$\left(1, \frac{5\pi}{6}\right)$ $\left(1, \frac{\pi}{6}\right)$

$r = \sin 3\theta$

$\left(1, \frac{3\pi}{2}\right)$

69.

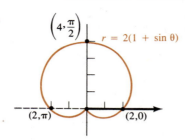

$\left(4, \frac{\pi}{2}\right)$ $r = 2(1 + \sin \theta)$

$(2, \pi)$ $(2, 0)$

71.

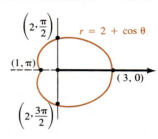

$\left(2, \frac{\pi}{2}\right)$ $r = 2 + \cos \theta$

$(1, \pi)$ $(3, 0)$

$\left(2, \frac{3\pi}{2}\right)$

73.

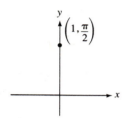

y

$\left(1, \frac{\pi}{2}\right)$

x

Exercise 7.6   (Page 282)

1.

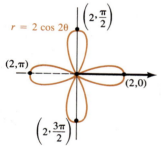

$r = 2 \cos 2\theta$ $\left(2, \frac{\pi}{2}\right)$

$(2, \pi)$

$(2, 0)$

$\left(2, \frac{3\pi}{2}\right)$

3.

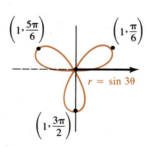

$\left(1, \frac{5\pi}{6}\right)$ $\left(1, \frac{\pi}{6}\right)$

$r = \sin 3\theta$

$\left(1, \frac{3\pi}{2}\right)$

5.

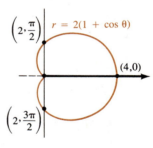

$\left(2, \frac{\pi}{2}\right)$ $r = 2(1 + \cos \theta)$

$(4, 0)$

$\left(2, \frac{3\pi}{2}\right)$

7.

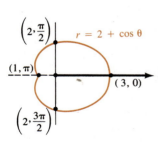

$\left(2, \frac{\pi}{2}\right)$ $r = 2 + \cos \theta$

$(1, \pi)$ $(3, 0)$

$\left(2, \frac{3\pi}{2}\right)$

9.

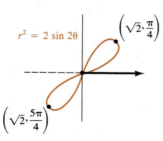

$r^2 = 2 \sin 2\theta$ $\left(\sqrt{2}, \frac{\pi}{4}\right)$

$\left(\sqrt{2}, \frac{5\pi}{4}\right)$

11.

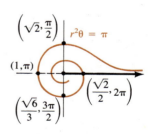

$\left(\sqrt{2}, \frac{\pi}{2}\right)$ $r^2\theta = \pi$

$(1, \pi)$

$\left(\frac{\sqrt{6}}{3}, \frac{3\pi}{2}\right)$ $\left(\frac{\sqrt{2}}{2}, 2\pi\right)$

13.

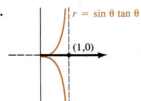

$r = \sin \theta \tan \theta$

$(1, 0)$

15.

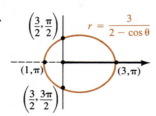

$\left(\frac{3}{2}, \frac{\pi}{2}\right)$ $r = \dfrac{3}{2 - \cos \theta}$

$(1, \pi)$ $(3, \pi)$

$\left(\frac{3}{2}, \frac{3\pi}{2}\right)$

17.

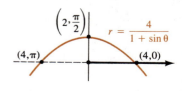

$\left(2, \frac{\pi}{2}\right)$ $r = \dfrac{4}{1 + \sin \theta}$

$(4, \pi)$ $(4, 0)$

Exercise 7.7   (Page 285)

1. 42.9; 24.0°   **3.** 8 units

REVIEW EXERCISES   (Page 287)

1. $-i$ **2.** $-i$ **3.** $-i$ **4.** $\dfrac{1}{i^{1812}} = 1$ **5.** $\left(\frac{4}{3}, \frac{2}{3}\right)$ **6.** $(-11, -18)$ **7.** $-4 + i$ **8.** $-5 + i$

9. $-5 - 5i$ **10.** $21 + i$ **11.** $0 - \frac{1}{5}i$ **12.** $0 + \frac{13}{6}i$ **13.** $\frac{8}{17} - \frac{2}{17}i$ **14.** $-\frac{3}{2} - \frac{1}{2}i$ **15.** $0 + i$

16. $\frac{2}{5} - \frac{1}{5}i$ **17.** $\dfrac{2 - 3\sqrt{2}}{3} + \dfrac{3 + 2\sqrt{2}}{3}i$ **18.** $\dfrac{3 + \sqrt{3}}{4} + \dfrac{3\sqrt{3} - 1}{4}i$ **19.**   **20.**   **21.**   **22.**

19. Imaginaries

20.

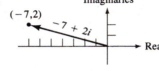

21. Imaginaries

22. Imaginaries

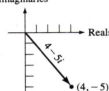

23. $\sqrt{73}$ **24.** $10\sqrt{2}$ **25.** $\dfrac{3\sqrt{10}}{10}$ **26.** 1   **27.** $2\sqrt{2}(\cos 135° + i \sin 135°)$

28. $5\sqrt{2}(\cos 315° + i \sin 315°)$ **29.** $6(\cos 60° + i \sin 60°)$ **30.** $4(\cos 0° + i \sin 0°)$ **31.** $\dfrac{3}{2} + \dfrac{3\sqrt{3}}{2}i$

32. $\sqrt{3} - i$ **33.** $-\dfrac{3}{2} - \dfrac{3\sqrt{3}}{2}i$ **34.** $-\dfrac{7\sqrt{3}}{2} + \dfrac{7}{2}i$ **35.** cis 110°   **36.** 6 cis 570°

37. $6\left(\cos\dfrac{\pi}{4} + i\sin\dfrac{\pi}{4}\right)$ **38.** $21(\cos \pi + i \sin \pi)$ **39.** 2 cis 50°   **40.** $\dfrac{2}{3}(\cos 10° + i \sin 10°)$

41. $\cos 20° + i \sin 20°$ **42.** $\sqrt[4]{14}\left(\cos\dfrac{45°}{4} + i\sin\dfrac{45°}{4}\right)$ **43.** 5, $\dfrac{-5 + 5\sqrt{3}i}{2}, \dfrac{-5 - 5\sqrt{3}i}{2}$

44. 3, 3i, -3, -3i   **45.** $\left(\dfrac{5}{2}, \dfrac{5\sqrt{3}}{2}\right)$ **46.** $(-\sqrt{3}, -1)$ **47.** $\left(\dfrac{\sqrt{3}}{2}, \dfrac{1}{2}\right)$ **48.** $(-5\sqrt{2}, 5\sqrt{2})$

49. (2, 135°)   **50.** (2, 150°)   **51.** (1, 0°)   **52.** (2, 300°)   **53.** $r^2 \cos \theta \sin \theta = 1$
54. $r(\cos \theta + 2 \sin \theta) = 2$ **55.** $r \cos^2 \theta = 3 \sin \theta$ **56.** $r^2 = 4 \cos \theta \sin \theta$
57. $(x^2 + y^2)^2 = 9x^2 - 9y^2$ **58.** $x^2 + y^2 = 5y$ **59.** $4\sqrt{x^2 + y^2} + y = 1$ **60.** $\sqrt{x^2 + y^2} - x = 2$

61.

$$r = \dfrac{6}{1 + \sin \theta}$$

62.

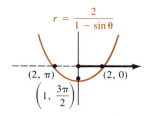

63.

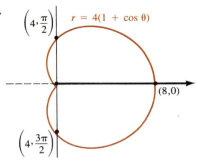

64.

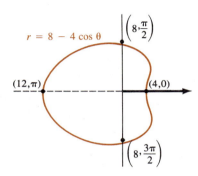

Exercise I.1   (Page A-6)

1. angle *ABC* or *GBE*   **3.** angle *A*, *GAC*, *CAG*, or *CAB*   **5.** angle *BDE* or *EDB*   **7.** acute   **9.** obtuse
11. right   **13.** none   **15.** straight   **17.** 130°   **19.** 96°   **21.** 63°   **23.** 60°   **25.** 123°   **27.** yes
29. no   **31.** yes   **33.** 50 m   **35.** $\sqrt{5}$ cm   **37.** $\sqrt{a^2 + b^2}$ ft   **39.** $BC = 2$ cm; $AB = 2\sqrt{2}$ cm
41. $AC = BC = 5\sqrt{2}$ in.   **43.** $BC = 3\sqrt{3}$ km; $AB = 3\sqrt{6}$ km   **45.** $BC = 3\sqrt{3}$ cm; $AB = 6$ cm
47. $AC = \dfrac{5\sqrt{3}}{3}$ mi; $AB = \dfrac{10\sqrt{3}}{3}$ mi   **49.** $AC = 5$ cm; $BC = 5\sqrt{3}$ cm   **51.** $BC = 6$ cm; $AB = 4\sqrt{3}$ cm
53. $AC = 2$ ft; $AB = 4$ ft   **55.** $AC = \sqrt{3}$ cm; $BC = 3$ cm   **57.** $4|x|$ **59.** $25x^4$ **61.** x **63.** $|x|$

Exercise I.2   (Page A-9)

1. $\dfrac{60}{7}$ cm   **3.** $\dfrac{168}{25}$ cm   **5.** $37\frac{1}{2}$ ft   **7.** 60 ft   **9.** 43 ft $+ 5\frac{1}{2}$ ft $= 48\frac{1}{2}$ ft   **11.** 528 ft   **13.** 40 ft
15. 40°   **17.** Since angle *A* + angle 2 = 90° and angle 1 + angle 2 = 90°, angle *A* = angle 1. Use same
argument to show angle *B* = angle 2.   **19.** Triangles are similar because two angles of triangle *ACD* equal two
angles of triangle *ACB*.   **21.** 4.5

Exercise I.3   (Page A-13)

1. 12π in.   **3.** 9 units   **5.** $\dfrac{81}{2}$ units   **7.** $\sqrt{\dfrac{63}{\pi}} = \dfrac{\sqrt{63\pi}}{\pi}$ units   **9.** 2 units   **11.** 12 units

13. $10\sqrt{2}\,\pi$ cm   **15.** $\dfrac{2\sqrt{6\pi}}{\pi}$ sq in.   **17.** about 94 plants   **19.** about $177 million
21. about 747 revolutions   **23.** about $2094   **25.** 24 m   **27.** 54.76 cm   **29.** 3979 mi

Exercise I.4   (Page A-15)

1. $84.607°$ **3.** $25.802°$ **5.** $5.422°$ **7.** $23°8'24''$ **9.** $73°52'12''$ **11.** $123°17'24''$

Exercise II.1   (Page A-23)

1. 3   **3.** -3 **5.** 2   **7.** 2   **9.** 7   **11.** 4   **13.** $-\frac{3}{2}$ **15.** $\frac{2}{3}$ **17.** 5   **19.** $\frac{3}{2}$ **21.** $\frac{1}{9}$
23. 8

25. 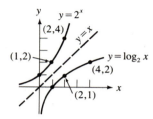   **27.**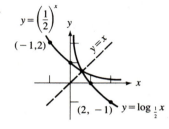

29. 5   **31.** $\frac{1}{50}$ **33.** no value   **35.** no value   **37.** 3   **39.** no value   **41.** T
43. F   **45.** F   **47.** F   **49.** T   **51.** F   **53.** F   **55.** T   **57.** F   **59.** F   **61.** F   **63.** F
65. T   **67.** 3.036   **69.** 1.142   **71.** 3.120×10^5 **73.** 0.2665   **75.** 0.3087

Exercise II.2   (Page A-30)

1. $\dfrac{\log 5}{\log 4}$ **3.** $\dfrac{1}{\log 2} + 1$ **5.** -2 **7.** $\sqrt{2}, -\sqrt{2}$ **9.** 0   **11.** $\dfrac{-\log 3}{\log 3 - \log 4}$ **13.** 9   **15.** 8
17. 2   **19.** 3   **21.** 7   **23.** 4   **25.** $10, -10$ **27.** 50   **29.** 20   **31.** 10   **33.** 10   **35.** 2
37. 10^{10} **39.** 1.771   **41.** -1 **43.** 1/6   **45.** 0.3028   **47.** 0.6045   **49.** 0.8782   **51.** 1.638
53. 0.631

Exercise II.3   (Page A-35)

1. 4.77   **3.** 0.710 volts   **5.** 24.77 db   **7.** \$5,368,709.12   **9.** It will be 2.828 times larger.
11. $A_n = D(1 + r)^n + D(1 + r)^{n-1} + \cdots + D(1 + r)$ **13.** 6.213 years

Exercise II.4   (Page A-38)

1. $\dfrac{\ln 2}{r}$ years   **3.** $L = L_0 + k \ln 2$ **5.** 6 places

Exercise III.1   (Page A-41)

1. 0.3963   **3.** 0.1542   **5.** 1.116   **7.** 26.39°   **9.** 9.03°   **11.** 13.18°

INDEX

4.5 Product-to-Sum and Sum-to-Product Formulas

$$\sin A \cos B = \tfrac{1}{2}[\sin(A + B) + \sin(A - B)]$$

$$\cos A \sin B = \tfrac{1}{2}[\sin(A + B) - \sin(A - B)]$$

$$\sin A \sin B = \tfrac{1}{2}[\cos(A - B) - \cos(A + B)]$$

$$\cos A \cos B = \tfrac{1}{2}[\cos(A + B) + \cos(A - B)]$$

$$\sin A + \sin B = 2 \sin\frac{A + B}{2}\cos\frac{A - B}{2}$$

$$\sin A - \sin B = 2 \cos\frac{A + B}{2}\sin\frac{A - B}{2}$$

$$\cos A + \cos B = 2 \cos\frac{A + B}{2}\cos\frac{A - B}{2}$$

$$\cos A - \cos B = -2 \sin\frac{A + B}{2}\sin\frac{A - B}{2}$$

4.6 Sums of Form $A \sin x + B \cos x$

$$A \sin x + B \cos x = k \sin(x + \phi)$$

$$\text{where } k = \sqrt{A^2 + B^2}$$

$$\text{and } \sin\phi = \frac{B}{\sqrt{A^2 + B^2}}$$

$$\text{and } \cos\phi = \frac{A}{\sqrt{A^2 + B^2}}$$

5.3, 5.4 The Graphs of the Inverse Trigonometric Functions

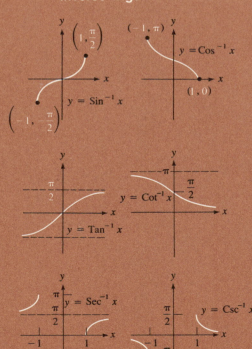

6.1 The Law of Cosines

$$a^2 = b^2 + c^2 - 2bc \cos A$$

$$b^2 = a^2 + c^2 - 2ac \cos B$$

$$c^2 = a^2 + b^2 - 2ab \cos C$$

6.2 The Law of Sines

$$\frac{a}{\sin A} = \frac{b}{\sin B} = \frac{c}{\sin C}$$

6.4 Mollweide's Equations

$$\frac{a + b}{c} = \frac{\cos\frac{1}{2}(A - B)}{\sin\frac{1}{2}C}$$

$$\frac{a - b}{c} = \frac{\sin\frac{1}{2}(A - B)}{\cos\frac{1}{2}C}$$

6.4 The Law of Tangents

$$\frac{a - b}{a + b} = \frac{\tan\frac{1}{2}(A - B)}{\tan\frac{1}{2}(A + B)}$$

6.5 Areas of Triangles

$$A = \tfrac{1}{2}bh$$

$$\text{Area} = k = \tfrac{1}{2}ab\sin C$$
$$= \tfrac{1}{2}ac\sin B \left.\vphantom{\begin{matrix}a\\b\\c\end{matrix}}\right\} \text{SAS}$$
$$= \tfrac{1}{2}bc\sin A$$

$$k = \frac{c^2 \sin A \sin B}{2 \sin C}$$
$$= \frac{b^2 \sin C \sin A}{2 \sin B} \left.\vphantom{\begin{matrix}a\\b\\c\end{matrix}}\right\} \begin{matrix}\text{AAS}\\\text{and}\\\text{ASA}\end{matrix}$$
$$= \frac{a^2 \sin B \sin C}{2 \sin A}$$

$$k = \sqrt{s(s - a)(s - b)(s - c)} \left.\vphantom{\begin{matrix}a\\b\end{matrix}}\right\} \text{SSS}$$
$$\text{where } s = \tfrac{1}{2}(a + b + c)$$

6.6 Vectors

If $\mathbf{V} = \langle a, b \rangle$ and $\mathbf{W} = \langle c, d \rangle$, then

$$|\mathbf{V}| = \sqrt{a^2 + b^2}$$

$$\mathbf{V} + \mathbf{W} = \langle a + c, b + d \rangle$$

$$k\mathbf{V} = k \langle a, b \rangle = \langle ka, kb \rangle$$

$$\mathbf{V} \cdot \mathbf{W} = |\mathbf{V}||\mathbf{W}|\cos\theta$$